Water Governance—Concepts, Methods, and Practice

Series editors

Claudia Pahl-Wostl, Osnabrück, Germany
Joyeeta Gupta, Amsterdam, The Netherlands

About this Series

This book series aims at providing a platform for developing an integrated perspective on major advances in the field of water governance. Contributions will build bridges across established fields of expertise, across disciplinary perspectives, across levels from global to local and across geographical regions. Topics to be covered include conceptual and methodological advances capturing the complexity of water governance systems, institutional settings, actor constellations, diagnostic approaches, and comparative studies of governance systems. The book series encompasses monographs, textbooks and edited, coordinated volumes addressing one theme from different perspectives. The book series will address mainly a wider scientific audience, but will also provide valuable knowledge to interested practitioners. The sustainable management of fresh water resources and the ecosystem services that they provide is one of the key challenges of the 21st century. Many water related problems can be attributed to governance failure at multiple levels of governance rather than to the resource base itself. At the same time our knowledge on water governance systems and conditions for success of water governance reform is still quite limited. The notion of water governance aims at capturing the complexity of processes that determine the delivery of water related services for societal needs and that provide the context within which water management operates. Water governance is a fast growing field of scholarly expertise which has largely developed over the past decade. The number of publications in peer reviewed journals has increased from less than 20 in the year 2000 to nearly 400 in the year 2013. The increasing popularity of the term in science and policy has not lead to conceptual convergence but rather to an increasing vagueness and competing interpretations of how the concept should be understood, studied, and analyzed; which disciplines are involved; and what methodological approaches are most suitable for the study and analysis of water governance. The time seems to be ripe for comprehensive synthesis and integration.

More information about this series at http://www.springer.com/series/13400

Claudia Pahl-Wostl

Water Governance in the Face of Global Change

From Understanding to Transformation

Springer

Claudia Pahl-Wostl
Institute for Environmental Systems
Research
University of Osnabrück
Osnabrück
Germany

ISSN 2365-4961 ISSN 2365-297X (electronic)
Water Governance—Concepts, Methods, and Practice
ISBN 978-3-319-21854-0 ISBN 978-3-319-21855-7 (eBook)
DOI 10.1007/978-3-319-21855-7

Library of Congress Control Number: 2015947412

Springer Cham Heidelberg New York Dordrecht London

Printed on acid-free paper

Springer International Publishing AG Switzerland is part of Springer Science+Business Media
(www.springer.com)

To Wolfgang, for your patience when I worked even more than usual and for your love, and to the underprivileged in the Global South who deserve to live in a better world

Preface

Writing a book is always a major undertaking. And it seems to become increasingly difficult given the plethora of information and events that compete for one's attention and reduce the amount of concentrated time for one activity. Why then did I decide to write this book? After having written numerous journal articles, I decided that the point has come to take stock and weave all these individual contributions together into a more coherent whole. This book summarizes more than a decade of my conceptual and empirical work on water governance and management.

"Water Governance in the Face of Global Change"—the title of the book is timely. When I started my work on water governance and management, water crises were already a topical theme in academic, policy, and practitioner circles. Since then, the situation has deteriorated rather than improved, and water governance, in its current form, cannot cope with the challenges ahead.

From understanding to transforming—the subtitle of the book conveys several messages. I believe that an improved scientific understanding of the complex dynamics of water governance systems will also strengthen the guidance for urgently needed reforms in water governance. And water governance requires a sweeping transformation rather than small, incremental changes.

This book makes a contribution to the development of a theory of water governance that is built on a systemic and complete understanding of water governance. I am convinced that only a systemic and broad approach can do justice to the complexity of the phenomena we are facing in this domain. Strong theoretical foundations must be established in close exchange with the phenomena under consideration. How can a theory of the dynamics of societal processes of change be developed without closely observing and even engaging in such processes? My own research program and theoretical reasoning have been inspired mainly by what I have observed as problems and unexpected challenges rather than pursuing one school of thought.

My initial work focused on the role of processes of social learning and participatory, adaptive approaches in water management. In a field dominated by technocratic approaches and instrumental thinking, it was not a trivial undertaking to communicate the importance of the human dimension and participatory management. Innovative approaches in management were constrained by a whole range of factors that were characteristic of and stabilized the prevailing way of doing business. I realized that "adaptive management requires integrated system design to build and sustain enabling structural conditions." One could also rephrase this and argue that the introduction of innovative management approaches requires major structural transformations. What are the structural conditions and how are they interrelated? Being a system scientist, I noted that the entire design of and logic underlying the delivery of a societal function such as water management is strongly influenced by the reigning paradigm. In water management, this has traditionally been a command and control approach that has dominated regulatory frameworks, the design of large-scale technical infrastructure, or management practices. Hence in my work, the importance of paradigms—or expressed in another way, cultural–cognitive institutions—has always played a key role.

The emphasis of my research activities shifted from social learning in actor groups to structural change and societal learning, to governance of transformation. To what extent can transformative change be governed? This is an as-yet open question which I attempt to at least partly answer in this book. My own reasoning builds on an evolutionary understanding of structural change which combines purposeful design with instances of self-organization and emergence.

In trying to draw together the scholarly work that is relevant for developing the overall argument of this book, I cover a lot of ground. At the same time, this coverage cannot be complete. I have included what I consider important theoretical and empirical contributions that influenced my work. I summarize major streams of scientific discourse in a field and give credit to eminent scholars that collectively shaped an important line of reasoning. Being a system scientist, I try to integrate different perspectives to achieve a holistic understanding of governance systems and their dynamics. In adopting such a broad understanding, seemingly incompatible theories may start to look complementary rather than contradictory. A problem orientation supports integration and openness.

For whom did I write this book? My main target audience is the interdisciplinary and diverse community of scholars working on water governance and management issues. In particular, I would like to reach young scholars who seek inspiration for their own work. I can only encourage those early-career researchers not to follow trodden paths but to be creative and to escape narrow disciplinary thinking. This may also imply not pursuing what looks in the short term to be the most promising path for a successful academic career. Incentive systems in science do not necessarily encourage unconventional and interdisciplinary thinking. Admittedly, the situation has improved over the past decade, and many excellent journals with a strong reputation are now available for publishing interdisciplinary research. Yet securing tenured positions in academia remains difficult for those with an interdisciplinary background. But I am optimistic that this situation will change soon as we urgently need such people both in research and education.

In my career, I have been fortunate in having the opportunity for exchange with quite diverse communities, which I have always experienced as enriching and inspiring for my own thinking. My efforts in community-building work are reflected in my role as editor of three books and twelve special issues in peer-reviewed journals. During these undertakings, I have always been exposed to both natural and social scientists from numerous disciplines. The water community dominated by natural scientists and engineers has a focus on a mechanistic understanding, on well-defined problems and on instrumental approaches to problem solving. With regard to the scale of the phenomena under investigation, the global change community constitutes the other end of the spectrum. As part of the scientific steering committee and subsequently co-chair of the Global Water System Project, I have had substantial interactions with this community. Most of the human dimensions scholars working on global change questions come from an interdisciplinary background. What unites this community is the desire to develop an improved understanding of and responses to the challenges posed by global change to sustainable development. Working both in developed and developing countries has proven to be extremely important by permitting reflection on the potential for and limitations of the transferability of insights from one place to another. And it convinced me of the importance to developing frameworks that facilitate comparisons across cases.

My work on frameworks profited from exchanges with the SES (social-ecological systems) Club, an interdisciplinary informal group of scholars. Over the past decade, the SES Club has worked on developing and applying a framework for social-ecological systems to overcome fragmentation and to facilitate comparative analyses of case study research. Our most prominent member was the late Elinor Ostrom with whom I had a lot of productive exchanges. We did not always agree on the underlying theoretical arguments, but discussions with Lin were always inspiring and even when we disagreed she remained constructive and never dismissed alternative ways of thinking.

Writing this book would not have been possible without the contributions of the many enthusiastic members of my research team at the Osnabrück University. The Institute of Environmental Systems Research provides the freedom to conduct interdisciplinary and unconventional Ph.D. research. Conducting numerous empirical analyses, introducing new conceptual and methodological ideas, and challenging my thinking, my research team has always provided a very enriching environment for my scientific work. I would like to thank all of my colleagues for their hard work and numerous inspiring discussions.

A sabbatical last year provided the distance from everyday business that was required to start writing the book. I would like to thank STIAS, the Stellenbosch Institute for Advanced Studies in South Africa, for the invitation to use their facilities as a base and for their generous support. STIAS is a stimulating and tranquil place to focus on writing. Discussions with other fellows from entirely different fields during the lunch breaks and wine receptions often provided unexpected inspiration. The spectacular landscape offered many opportunities for exploration for a nature lover and bike enthusiast like me. At the same time, I was exposed to the harsh realities of a developing country and the challenges that water governance and management is facing in such a context.

The book profited from the numerous comments provided by my colleagues during the various stages of the writing process. I would like, in particular, to thank Janos Bogardi, Stefanie Engel, Louis Lebel, Andrea Lenschow, Oran Young, and Andreas Thiel for reading and commenting on draft chapters.

I would also like to thank Caroline van Bers for her meticulous review of the book, for trying to understand what I wanted to say, for making suggestions for linguistic improvement, and for pointing out vague statements and inconsistencies. The book profited a lot from this thorough check!

This book is the inaugural volume of an entire series on "Water Governance—Concepts, Methods and Practice." I trust that this series will contribute to the strengthening of the reputation of water governance scholarship in science and in practice.

Contents

Chapter 1
The Challenge of Water Governance

Water is the source of life on Earth. Is has been a source of inspiration for artists. People have always sought land for settlements and leisure activities along rivers and coastal areas. We are dependent on water for a multitude of purposes the most important being drinking water, farming, transport, manufacturing, and recreation and cultural. The downside of our dependence on water and of the importance we place on it is that competing water uses are the source of many conflicts. Conflicts arise not only between different human uses but also between water for nature and water for human activities. Over a decade of global water research has provided clear evidence of the global dimension of the water challenge and the role of humans as a major force influencing the global water cycle. Mounting evidence suggests that major trends such as increasing water appropriation by humans and nutrient pollution are, for the most part, irreversible over the next half century and will be intensified by water problems of pandemic proportions (Vörösmarty et al. 2013; Pahl-Wostl et al. 2013).

Global change poses unprecedented challenges to scientific and policy communities. But neither science nor policy has yet been able to provide effective responses. Until relatively recently, research has emphasized the identification of problems more than the development of solutions. In their summary of a global consultative process on the priorities for Earth System Science, Reid et al. (2010) pointed out that the most pressing research questions were quite different from those that initially shaped global change programmes, and that social science is increasingly necessary as the balance of attention shifts from quantifying the impacts of human activities on the environment to identifying the alternative pathways that societal change may take. Among the five "Grand Challenges" identified, responding (determining what institutional, economic and behavioural changes can enable effective steps towards global sustainability) and innovating (encouraging innovation in developing technological, policy and social response to achieve global sustainability) are those areas in which knowledge gaps are the most prevalent. Such problems are particularly pronounced in the sustainable management of global water resources (Pahl-Wostl et al. 2013).

At the global scale freshwater resources are not yet scarce. However, their uneven distribution at different scales (among world regions, countries, societal groups) provides multiple sources of tension. Technological progress has allowed

C. Pahl-Wostl, *Water Governance in the Face of Global Change*,
Water Governance—Concepts, Methods, and Practice,
DOI 10.1007/978-3-319-21855-7_1

the cultivation of deserts and floodplains. However, pushing human activities towards or even beyond the limits of environmental systems has resulted in increased vulnerability to environmental extremes, unsustainable land use patterns and degradation of ecosystems. Biodiversity in aquatic ecosystems is in decline—even faster than in terrestrial ecosystems (Dudgeon et al. 2006). Climate change and the concomitant increase of extreme weather events with massive consequences for human populations, economic assets and critical physical infrastructures have exposed weaknesses in current water governance and management systems. This has raised awareness of uncertainties, the complexity of the systems to be managed, and the need for profound changes in policy and management paradigms, as well as governance systems (Pahl-Wostl 2007b; Opperman et al. 2009). Experts studying human-environment interactions have stressed the need for a radical paradigm shift to replace the prevailing mechanistic and technocratic strategies that have proven to be inadequate for responding to recent challenges because they neglect complexity and the human dimension (Holling and Meffe 1996; Gleick 2003; Pahl-Wostl 2007a, b).

Historically, water resources management focused on technical solutions to well-defined problems, an approach that gained urgency with the increasing concentration of urban populations and the intensification of industrial and agricultural productivity in the 19th and 20th centuries. Health and hygiene problems within cities and the seemingly insatiable demand for more water has driven major efforts in urban water management to improve water quality and ensure reliable supplies. Rivers were controlled to protect cities and dryland agriculture from flooding. Technological fixes such as increasingly sophisticated wastewater treatment plants proved to be highly effective and efficient in solving many pressing problems in terms of both water quality and quantity in the short run. However, interventions which worked in the past are, in many cases, proving to be inappropriate for addressing the challenges of the present and the future.

In a review of the literature on paradigm shifts in water management we identified several recurring topics in the calls for a fundamental paradigmatic change in water management (Pahl-Wostl et al. 2006, 2011a):

- management of problem sources not effects,
- increased integration of issues and sectors,
- explicit inclusion of environment in management goals,
- decentralized and more flexible management approaches,
- participatory management and collaborative decision making,
- more attention to management of human behaviour through "soft" measures,
- open and shared information sources,
- incorporation of iterative learning cycles.

There is a clear emphasis on the need to pay more attention to the human dimension of water management which had been largely ignored in the past. The strategy of simplifying complex issues to make them manageable and to reduce problems with multifactorial causation to single causes in order to make them amenable to technical solutions provided short-term success but has proven to be unsustainable in the long-term. A dominant driver of demands for change has thus been the need to develop a better understanding of water resources and their management as complex systems. Increasing awareness of the complexity of environmental problems has encouraged the development of new management approaches. Such approaches take into account the inter-connectedness of human-technology-environment systems which are complex, non-predictable and characterized by unexpected responses to intervention (Pahl-Wostl 2002; Prato 2003). What needs to be managed are 'Complex Adaptive Systems' (CASs) which can be characterized as hierarchies of components interacting within and across scales. CASs have emergent properties that cannot be predicted based on knowing the components alone (Lansing 2003). Rather than trying to change the structure of CASs to control them by external intervention, innovative management approaches make use of the self-organizing properties of the systems to be managed. This increased awareness of complexity together with the fundamental change in our understanding of what management implies is not limited to the field of natural resources (Pahl-Wostl 2007b). It leads to a somewhat different research agenda for improving our understanding of the function of management and how it can best be practised. The now-fading water management paradigm was characterised by a command and control approach. Typically, control is exerted centrally, adhering to rigid and detailed plans for the fulfilment of established goals. Command and control infers that management interventions can be optimised and their impact calculated. This is facilitated by the disaggregation of the system to be controlled into separable elements. Uncertainties are either ignored or assessed quantitatively and dealt with by the establishment of technical norms such as security margins. Such measures are effective for a roughly stable system with regularly recurring phenomena such as seasonal water shortages. They fall short, however, in adequately dealing with the types of non-linear change and unprecedented (judged by human time scales) system behaviour which characterise many river basins, such as extreme droughts (e.g. California 2014). More adaptive management approaches do not aim for short-term optimization of single objectives such as profit from agriculture, but for long-term resilience of the system as a whole (Pahl-Wostl 2007b, 2011a, b). They nurture the capacity to steer the system to a certain degree without destroying the ability of the system to respond flexibly to unexpected developments and surprises. An adaptive management approach may set a general direction for the achievement of certain goals, but allows greater freedom in the interventions deemed appropriate to achieve those goals. Such an approach is demanding with respect to coordinating the actions of actors who contribute to policy development and implementation, but it is even more demanding of governance. Despite the introduction of adaptive management in the field of ecology several decades ago (Holling 1978) its broader adoption by the water management community happened only in recent years when climate change and associated uncertainties made it imperative to reflect more on the practice of water management (Pahl-Wostl 2007b).

Over the past few decades IWRM (Integrated Water Resources Management) moved to centre stage as the approach to achieve sustainable water resources management and overcome the deficiencies of technocratic management approaches that have largely ignored the human dimension. IWRM promotes: (1) an integrated approach across sectors and different uses and users; (2) a balance of the three pillars of sustainability—economic, social and environmental concerns; (3) participatory approaches and the involvement of women. IWRM thus clearly recognises the importance of so-called 'soft' strategies and the need for governance reform (GWP Technical Committee 2004). In 2004 we tried to convince major proponents in the field of IWRM that this approach needs to be combined with an adaptive management approach. This happened in the context of the NeWater[1] project that focused on integrated and adaptive water management. Our suggestions were met with scepticism and refuted with the argument that a "new" paradigm might confuse practitioners. The situation changed entirely with the sharp increase in the awareness of climate change adaptation in the water sector in the mid to late 2000s (Sadoff and Muller 2009). I argue though that problems encountered with the implementation of IWRM can at least partly be attributed to the fact that IWRM did not (yet) overcome a command and control approach and that an adaptive approach is a prerequisite for any kind of integrated and systemic water management.

Despite the fact that IWRM has been strongly promoted as a new path-breaking paradigm, progress in its operationalization has been slow and incremental, and has not yet led to major transformations in the governance and management of freshwater. A report commissioned by UN-Water arrives at more positive conclusions (UNEP 2012). According to the findings of a global survey, the majority of countries have adopted integrated resource management principles in their laws and policies. However, implementation on the ground which would translate principles into management practice and ultimately into an improved state of water resources and sustainable use of water services is still slow. Many obstacles may be impeding implementation. Adopting IWRM principles in laws and policies on paper does not overcome the lack of adherence to good governance principles in practice (Pahl-Wostl et al. 2012). Large-scale infrastructural development may have become an end in itself, rather than a means to an end, fuelling rent-seeking by powerful elites and symbolising state power in what may be called hydraulic bureaucracies (Molle et al. 2009). Furthermore, goals of an integrated approach are less tangible and more difficult to communicate and to operationalize than focusing on single pressing problems in isolation.

In recent years the concept of water security has come to the fore to express the goals of sustainable water resources management. The Global Water Partnership argued in their framework for action to achieve water security: "*Water security, at any level from the household to the global, means that every person has access to enough safe water at affordable cost to lead a clean, healthy and productive life, while ensuring that the natural environment is protected and enhanced*" (Global

[1]NeWater (www.newater.uni-osnabrueck.de) was funded by the 6th European Framework Program. NeWater developed new methods for adaptive and integrated water management and focused in particular on the transition from current regimes of water management in a river basin to more integrated, adaptive approaches with strong stakeholder participation. The project had case studies in Europe, Africa and Central Asia and involved forty partner organizations.

Water Partnership (GWP) 2000, p. 12). The increasing focus on sustainable pathways towards increased water security led Grey and Sadoff (2007, p. 545) to define water security as "… *the availability of an acceptable quantity and quality of water for health, livelihoods, ecosystems and production, coupled with an acceptable level of water-related risks to people, environments and economies*". This all-encompassing definition embraces a risk-based perspective and addresses the role of water as both a source of services and a threat. It also makes evident that trying to enhance water security entails trade-offs and a perception-based assessment of risks. What constitutes acceptable risk is interpreted in different ways by different groups. Without respect for good governance principles, the introduction of the notion of water security will not lead to a more equitable and sustainable management of the resource.

All of these issues make it evident that sustainable water resources management and enhancing water security is first of all a governance challenge which cannot be separated from politics. Governance and management are often used synonymously. As a consequence the meaning of governance and that of management become quite fuzzy. From a theoretical perspective it is more appropriate, albeit not easy to operationalize, to make a distinction between the two concepts. As defined by Pahl-Wostl (2009, p. 355): "*'Resources management' refers to the activities of analysing and monitoring, developing and implementing measures to keep the state of a resource within desirable bounds. The notion of 'resource governance' takes into account the different actors and networks that help formulate and implement environmental policy and/or policy instruments.*" Governance sets the rules under which management operates. A modern interpretation of governance embraces the full complexity of regulatory processes and their interaction as reflected in the definition of water governance by the UN (United Nations 2002, p. 47): "*The governance of water in particular can be said to be made up of the range of political, social, economic and administrative systems that are in place, which directly or indirectly affect the use, development and management of water resources and the delivery of water services at different levels of society*".

The notion of government as the single decision-making authority, where state authorities exert sovereign control over the people and groups making up civil society, has been replaced by the notion of multi-level governance where many actors in different institutional settings contribute to policy development and implementation (Mayntz 1998). Claims about the legitimacy of intervention and change no longer reside exclusively in the realms of authority and privileged knowledge. Legitimacy now depends on shared visions of both the problem and equitable solutions to it and desirable and undesirable outcomes. Governance systems are not mainly a product of rational design but are characterised by self-organisation, emergence and diverse leadership in complex networks with a multitude of interests and power relations. What does "managing change" towards a more desirable state mean in such diffuse, complex and multi-level networks? How do all these processes act in concert and under which conditions do they lead to sustainable management of environmental resources? These remain open questions. Given the worrying trends observed globally serious doubts prevail that the changes in water management paradigms that have taken place over the past few decades have, in fact, initiated a path-breaking transformation towards significant improvement.

The water community has witnessed the waxing and waning of a whole suite of principles guiding water governance reform—hierarchical centralization, coordinated river basin planning and management, devolution and decentralization, markets and privatization (Ingram 2011). The discourse on the need for a paradigm shift has promoted more participatory integrated approaches and more recently adaptive management, and water security. Are these the latest fads which will be replaced by other trendy concepts without really leading to substantive change on the ground? I argue that there is an evolution in the nature of the concepts promoted which increasingly embraces more of the complexity of the real-world situation. The paradigm shift is a response to insights about the changing reality of water management and dissatisfaction with prevailing strategies and water governance arrangements. However, I agree with arguments that stress the need to focus more on politics and power relations than on new paradigms (Ingram 2011; Molle et al. 2009).

These then are the fundamental questions that need to be posed: Is the discourse on paradigm shifts, in principle, flawed? Does it mainly fuel symbolic politics and detract from the problems encountered in the implementation of new governance and management approaches? Under which conditions can the discourse on the need for change be translated to the urgently needed transformation in water governance and management systems?

To answer these questions major knowledge gaps in the conceptual and empirical foundations of water governance and in particular of governance of transformative change need to be closed. Through this book I endeavour to make a significant contribution to closing the gaps in the scientific understanding of water governance to support sustainability transformations and to translate such understanding into actionable knowledge for transformative change.

The chapters in this book follow a logical order to develop the overall argument. Chapters may be read individually but readers should be aware that the chapters build upon each other. Concepts and terms may be used that were introduced in earlier chapters. To facilitate navigation and locating related information I provide many chapter and section cross-references in the book.

To do justice to the development of a broad and systemic understanding of water governance, the book covers a lot of ground. Doing so is always a tightrope walk between avoiding too much detail and being too superficial. Many of the research themes addressed would deserve their own book. I tried to avoid falling in either trap by being selective in the theories reviewed. Nevertheless I stand by my claim that the book considers all relevant scientific discourses with respect to water governance and its transformation towards sustainability.

The guiding principles of water policy experienced significant changes over the past few decades. Chapter 2 discusses and illustrates the progression from a command-and-control approach to market-based policies and then community-based approaches. These transitions have not been accompanied by critical analysis and reflections on the lessons learned from these experiences. Despite progress in the scholarship of environmental governance I identify clear tasks for science in order to close the knowledge gaps in water governance.

Chapter 2 sets the scene for the introduction of the key elements of the conceptual framework for analysing water governance and transformative change in Chap. 3. It was not an easy task to select appropriate categories for these central elements. My choice was guided by what I considered necessary for a comprehensive representation of governance systems and their dynamics: institutions, actors, governance modes, multi-level interactions and processes of social and societal learning. All of these aspects are dealt with in more detail in subsequent chapters. Furthermore, the Management and Transition Framework (MTF) is introduced as a methodological approach to operationalize theoretical concepts underpinning water governance and make them amenable to empirical analysis. The MTF is further developed and extended in subsequent chapters.

Understanding the interplay between structure and agency and between institutions and actors, is still a key challenge for governance research. Chapter 4 describes the role of institutions and actors in processes of change and their role with respect to the adaptive and transformative capacity of governance systems. It elaborates on important theories for understanding institutions and their role in governance systems, as well as the main theories of human behaviour and their implications for understanding social learning and societal change. An important concept is cultural cognitive institutions, such as paradigms. Paradigms strongly influence meaning, understanding and perception of reality and of problem situations, how boundaries are delineated, and how the space for identifying problems and developing solutions is determined.

Chapter 5 introduces the notion of governance modes. Bureaucratic hierarchies, markets and network governance are introduced as Weberian ideal types of governance modes reflecting different logic reasoning for organizing governance processes. The chapter discusses an approach for developing an improved understanding of how these governance modes can be combined in hybrid modes, thus maximizing synergies and reducing potential conflicts between these governance modes. The set of challenges associated with multi-level governance and spatial scale are addressed in Chap. 6. The chapter discusses attempts to confine water governance to preferred spatial scales. However, multi-level governance is a necessity rather than an option. Polycentric governance combining the decentralization of power with effective coordination of decision centres is identified as a promising guiding principle for the structural design of governance systems. However, the development of governance systems including both instances of purposeful design and processes of self-organization is clearly taken into account.

Up to this point I mainly concentrate on the conceptualization of governance systems and their dynamics since I consider that the significant knowledge gaps are to be found in this domain. However, the relationship between humans and the natural environment is the key to sustainable resource governance and management. Chapter 7 introduces ecosystem services and environmental hazards as boundary concepts supporting an integrated and interdisciplinary approach to understanding and analysing social-ecological systems and their governance and management. Water security is introduced as a risk-based concept integrating the two perspectives. Based on this conceptualization I argue for and highlight the challenges associated with adaptive governance and management for the increased resilience of social-ecological systems.

Having developed the foundations, Chap. 8 presents a theoretical framework for describing and explaining the dynamics of governance systems and transformative change towards sustainability. It builds on and integrates concepts and insights developed in preceding chapters. It is largely a framework of analysis but also entails a normative dimension by identifying characteristics that are considered here to be essential for dealing with complex governance challenges. Theories without empirical foundations are futile. Chapter 9 therefore introduces a methodological framework within which the empirical foundations for understanding the dynamics of governance systems and transformative change are built. The chapter promotes a comparative case-study approach and methodological pluralism. Chapter 10 illustrates how these approaches have been put into practice to test theoretical propositions of adaptive water governance and management, social learning and transformative change. Chapter 10 synthesizes the more significant results from a number of empirical studies that were conducted under the umbrella of the research programme on water governance and management that I developed.

Chapters 11 and 12 reflect a more future-oriented perspective and outlook. Experimentation is the key to developing and testing new concepts. Chapter 11 elaborates on the potential of virtual and real world experimentation to broaden the scope of analyses, to foster creativity and innovation and to explore the terrain beyond our experience. The chapter discusses the role of and experience with (simulation) models for exploratory analyses but also in supporting communication and social learning. Virtual and real world laboratories can be instrumental in developing the knowledge and capacity for transformative change. To realize this potential requires a stronger emphasis on inter- and transdisciplinary research with a new role for science-in-action research and real world laboratories.

Chapter 12 identifies and discusses several global discourses and developments that if combined in a synergistic way, could be central to providing the impetus for the urgently-needed sustainability transformation in water governance and management: the "water-energy-food nexus", "water security", "bioeconomy and green infrastructure" and "sustainable development goals".

References

Dudgeon, D., Arthington, A., Gessner, M., Kawabata, Z.-I., Knowler, D. J., Lévêque, C., et al. (2006). Freshwater biodiversity: importance, threats, status and conservation challenges. *Biological Reviews, 81*, 163–182.

Gleick, P. (2003). Global freshwater resources: Soft-path solutions for the 21st century. *Science, 302*, 1524–1528.

Global Water Partnership (GWP). (2000). *Towards water security: A framework for action*. Stockholm: Global Water Partnership.

Grey, D., & Sadoff, C. W. (2007). Sink or Swim? Water security for growth and development. *Water Policy, 9*, 545–571.

GWP Technical Committee. (2004). *Catalyzing Change: A handbook for developing integrated water resources management (IWRM) and water efficiency strategies*. Stockholm, Sweden: Global Water Partnership (GWP).

Holling, C., & Meffe, G. (1996). Command and control and the pathology of natural resource management. *Conservation Biology, 10*, 328–337.

Holling, C. S. (Ed.). (1978). *Adaptive environmental assessment and management*. New York: Wiley.

Ingram, H. (2011). Beyond universal remedies for good water governance. In A. Garrido, & H. Ingram (Eds.), *Water for Food in a changing World* (pp. 241–261, Contributions from the Rosenberg International Forum on Water Policy). Miltorn Park: Routledge.

Lansing, J. S. (2003). Complex adaptive systems. *Annual Review of Anthropology, 32*, 183–204.

Mayntz, R. (1998). New Challenges to Governance Theory. *Jean Monet Chair Papers* European University Institute.

Molle, F., Mollinga, P. P., & Wester, P. (2009). Hydraulic bureaucracies and the hydraulic mission: Flows of water. *Flows of Power. Water Alternatives, 2*(3), 328–349.

Opperman, J. J., Galloway, G. E., Fargione, J., Mount, J. F., Richter, B. D., & Secchi, S. (2009). Sustainable floodplain through large-scale reconnection to rivers. *Science, 326*, 1487–1488.

Pahl-Wostl, C. (2002). Towards sustainability in the water sector—The importance of human actors and processes of social learning. *Aquatic Sciences, 64*, 394–411.

Pahl-Wostl, C. (2007a). The implications of complexity for integrated resources management. *Environmental Modelling & Software, 22*, 561–569.

Pahl-Wostl, C. (2007b). Transition towards adaptive management of water facing climate and global change. *Water Resources Management, 21*(1), 49–62.

Pahl-Wostl, C. (2009). A conceptual framework for analysing adaptive capacity and multi-level learning processes in resource governance regimes. *Global Environmental Change, 19*, 354–365.

Pahl-Wostl, C., Isendahl, N., Möllenkamp, S., Brugnach, M., Jeffrey, P., Medema, W., et al. (2006). *Paradigms in water management*. Osnabrueck: European Commission.

Pahl-Wostl, C., Jeffrey, P., Isendahl, N., & Brugnach, M. (2011a). Maturing the new water management paradigm: Progressing from aspiration to practice. *Water Resources Management, 25*(3), 837–856. doi:10.1007/s11269-010-9729-2

Pahl-Wostl, C., Jeffrey, P., & Sendzimir, J. (2011b). Adaptive and integrated management of water resources. In R. Q. Quentin & K. Hussey (Eds.), *Water resources planning and management* (pp. 292–310). Cambridge, UK: Cambridge University Press.

Pahl-Wostl, C., Lebel, L., Knieper, C., & Nikitina, E. (2012). From applying panaceas to mastering complexity: Toward adaptive water governance in river basins. *Environmental Science & Policy, 23*, 24–34.

Pahl-Wostl, C., Vörösmarty, C., Bhaduri, A., Bogardi, J., Rockström, J., & Alcamo, J. (2013). Towards a sustainable water future: Shaping the next decade of global water research. *Current Opinion in Environmental Sustainability, 5*(6), 708–714. doi:10.1016/j.cosust.2013.10.012

Prato, A. (2003). Adaptive management of large rivers with special reference to the Missouri River. *Journal of the American Water Resources Association, 39*(4), 935–946.

Reid, W. V., Chen, D., & Goldfarb, L. (2010). Earth system science for global sustainability: Grand challenges. *Science, 330*, 916–917.

Sadoff, C., & Muller, M. (2009). Water management, water security and climate change adaptation: Early impacts and essential responses. In G. W. P. T. C. (TEC) (Ed.), *TEC Background Papers* (Vol. 14): Global Water Partnership.

UNEP. (2012). *The UN-water status report on the application of integrated approaches to water resources management*. Kenia, Nairobi: United Nations Environment Programme.

United Nations. (2002). Water: A shared responsibility. The United Nations World Water Development Report 2. Paris: UNESCO Publ.

Vörösmarty, C. J., Pahl-Wostl, C., Bunn, S., & Lawford, R. (2013). Global water, the anthropocene and the transformation of science. *Current Opinion in Environmental Sustainability, 5*(6), 539–550.

Chapter 2
Water Policy—From Panaceas Towards Embracing Complexity

The waxing and waning of paradigms discussed in the previous chapter has also been reflected in developments in water policy. This chapter summarizes major global trends in water policy over the past half-century with reference to scale, dominant rationality and logical reasoning, and the role of different societal groups in shaping and implementing water policy. The developments reflect the overall shift in our understanding of the role of government as the central actor in water policy to one that is embedded in a more comprehensive notion of water governance (Ingram 2011; Pahl-Wostl et al. 2006). This is exemplified by European water policy. Its evolution reflects the general trends of shifting from command and control as the guiding principle towards more market-based and, in recent years, more participatory approaches. Furthermore we observe a gradual shift from the promotion of simplistic panaceas for water governance reform towards more context-sensitive approaches. The chapter closes with some reflections on the state of scientific understanding of environmental governance and the ability of the scientific community to address the challenge of developing context-sensitive advice for water governance reform.

2.1 Major Trends in Water Policy Over Last Few Decades

In the 60s and 70s water policy was characterized worldwide by the strong role played by central government and central regulation, in essence a hierarchical top-down command and control approach. The late 80s and 90s saw a shift towards the principles of subsidiarity, decentralization and privatization, and the market became a key instrument for water management. The trend was particularly pronounced in urban water supply and sanitation. The late 90s and 2000s saw an upsurge in participatory approaches. Central roles were delegated to community groups and water user associations, a shift that was especially noticeable in irrigation management. Developments in water management over the last few decades have seen changes in the role of three major social agents: government, market/economy (production, consumption), and civil society/community (individual citizens and organized groups outside of government and market, i.e. public voice).

C. Pahl-Wostl, *Water Governance in the Face of Global Change*,
Water Governance—Concepts, Methods, and Practice,
DOI 10.1007/978-3-319-21855-7_2

In the 60s and 70s, economic activities were responsible for creating water management problems. Government was in the role of problem solver and service provider acting in a hierarchical governance mode (cf. Chap. 5) and pursuing a command and control approach. Water governance and management were in the hands of bureaucrats and technicians. Civil society was not actively participating—unless fundamental failures gave rise to public protest. In the subsequent phase government was ascribed the role of instigator of problems. Lack of efficiency, effectiveness and rent-seeking by powerful elites were diagnosed as reasons for the failure to deliver adequate water services and to address increasing water problems. The market economy was seen as the problem solver. Civil society had a role as an arena for mobilizing protest and voicing lack of satisfaction but was still not a major player in shaping policy. In a third phase, the 1990s and early 2000s, direct community involvement was supposed to make up for the failure of governments and markets. Civil society was assigned a leadership role in making progress towards more equitable, sustainable and effective resource governance.

However, roles have become increasingly blurred. Government, the economy and civil society all play a role of contributing to the problems associated with resource management, albeit to varying degrees and for different reasons. And, at the same time, in various kinds of collaborative partnerships they are all part of the problem-solving process. This blurring of roles and the emergence of diverse hybrid forms of governance are typical of a more all-encompassing understanding of societal steering both from a descriptive and a normative perspective.

Governance activities and responsibilities are increasingly distributed across spatial levels. The introduction of the river basin principle added to this complexity. The basin principle implies that the functionally-specific governance institutions are given jurisdiction over the hydrologically-defined spatial scale of the river basin in order to address the spatial 'misfit' between resource management issues and governance scales (Young 2002a). However, as pointed out by Moss (2003) problems of spatial misfit have often been solved at the expense of the interplay between institutions. On one hand, introducing another layer of bureaucracy is always associated with frictions. On the other, water is now governed at a different spatial scale than other sectors such as agriculture. This is not necessarily beneficial to the goal of increased integration of issues and cross-sectoral collaboration and poses considerable challenges to vertical and horizontal coordination.

The trends identified are global and manifest themselves in similar ways in numerous water policy frameworks in developed and developing countries. They are exemplified by the development of European water policy over the past decades.

2.2 Evolution of the European Union's Water Policy

The European Union (EU) is a unique political construct in the political world. It comprises 28 member states that give up part of their national sovereignty by being placed under binding EU laws. Member states retain considerable autonomy though and are represented in the major decision-making bodies at the EU level. The development of EU policy thus largely reflects developments in national policies and the priorities of a large number of European countries.

EU environmental policy in general and water policy in particular developed in the 1970s. Initially it was based on a clear command and control. The first phase of EU policy emphasized the prescription of binding water quality norms mainly to protect water for human uses (e.g., Surface Water Directive 1976; Bathing Water Directive 1976; Shellfish Water directive 1979; Drinking Water Quality Directive 1980) (Aubin and Varone 2004). These directives also prescribed methods of analysis and monitoring. They left little freedom to member states to tailor policy implementation to their national conditions. Furthermore, the number of specific directives illustrates the piecemeal and fragmented approach of the first phase of EU water policy that dealt with problems one by one in isolation.

In the late 80s the focus on immission standards was replaced by an emission-based policy[1] (Aubin and Varone 2004). Subsequent directives prescribed instruments to achieve water quality norms in order to improve the unsatisfactory progress in the implementation of directives already in place. The key problem to be tackled was eutrophication of freshwater bodies due to excessive nutrient loadings. Correspondingly, the two main directives implemented during that period focused on the major sources of nutrient inputs—domestic wastewater and agriculture. The Urban Waste Water Treatment Directive (Council Directive 91/271/EEC—1991) had domestic wastewater as a clear target, whereas the Nitrates Directive (Council **Directive 91**/676/EEC—1991) targeted diffuse pollution from agriculture. The instruments chosen revealed a slow drift away from a command and control approach by also allowing voluntary instruments such as the code of good agricultural practice. However, implementation proved to be difficult casting doubt on the effectiveness of decentralized and voluntary measures. Changes in the Common Agricultural Policy leading to a reduction in agricultural subsidies further undermined the willingness of farmers to comply with voluntary standards. EU member states felt the financial burden of implementation, in particular regarding the Urban Waste Water Directive. Some countries, in particular the UK, responded with privatization hoping to attract private capital into the urban water sector. Furthermore privatization was seen as a remedy to cure the inefficiencies and ineffectiveness of governmental policies. Implementation was lagging behind expectations leading to a number of court cases and triggering a rethinking which resulted in significant reforms in water policy.

The European Water Framework Directive (WFD) which came into force in 2000 (European Parliament 2000) reflected a clear change towards a more

[1]Immission-based policies refer to the upper limit of a concentration of a pollutant in the environment. Water quality standards may, for example, prescribe upper thresholds for the concentration of a pollutant in the aquatic environment. Emission-based standards refer to the amount of a pollutant that can be released into the environment. Water quality standards may, for example, prescribe concentrations of pollutants in the effluents of wastewater treatment plants can discharge to the environment.

comprehensive understanding of multi-level governance embracing a range of instruments and leaving more freedom to member states in policy implementation. This policy initiative promotes an integrated management approach with the goal of achieving "good status" for all European waters (surface waters and groundwater). The WFD introduced the basin principle by prescribing water management at a river-basin scale and has put an end to the increasing fragmentation of water policy in terms of both objectives and means. The WFD promotes sectoral integration and encourages trans-boundary cooperation in international river basins. River basin management plans are to be revised every 15 years, supporting an adaptive approach to developing and implementing measures. The WFD is also the first major European directive to formally prescribe the involvement of stakeholders and the public at large. In fact, consultation of organized stakeholder groups was openly invited by the Commission during development of the directive. Arguably, the process favoured well-organized and resourced interest groups. At the least, open consultation made the omnipresent government lobbying a more transparent process.

Despite its innovative character, implementation of the WFD has also encountered obstacles. A major loophole has resulted in delays in the implementation process and stems from the fact that the WFD allows exemptions to the achievement of a good state for water bodies classified as heavily-modified. Classification of water bodies is based on a concept of water quality that includes hydro-morphological, chemical and ecological indicators (Mostert 2003). The approach measures the multi-criteria quality status of a surface water body on a five-point scale, and requires member states to report on improvement in quality towards at least a "good" state through a programme of monitoring and restorative measures. However, quality targets are negotiable, as exemptions can be sought for 'heavily-modified water bodies' if costs for improvement would be excessive. As initial experience with the classifications of water bodies by member states has shown, exemptions abound (European Environmental Bureau 2010). A mechanism upon which to base such decisions using an explicit analysis of trade-offs is still lacking.

Furthermore, a good state is particularly compromised by hydro-morphological and ecological indicators. The WFD classification revealed major ecological deficits in water quality. For example, while 88 % of the surface water bodies in Germany have reached good chemical status only 10 % of these water bodies have good ecological status. As many as 34 % are classified as poor and 23 % have bad ecological status (BMU 2010). Improvements of the chemical status could largely be achieved by technical and often end-of-pipe measures even when high investments (e.g. wastewater treatment) were needed. Improving ecological status requires a profound shift towards more holistic landscape management integrating across sectors and among issues that influence aquatic ecosystems. Such a shift encounters considerable barriers since it requires significant transformations in institutional settings, actor networks and power constellations (Pahl-Wostl 2006, 2007).

This example of European water policy illustrates that despite undeniable progress and evolution towards more sophisticated policies, water policy reformers cannot pride themselves on having achieved comprehensive institutional transformation and substantial breakthroughs in halting the deterioration of aquatic ecosystems. This experience casts doubt on the prospects for implementing an effective policy framework to bring about a fundamental change. I argue that one major obstacle is the fact that the water policy community does not excel in learning from experience and has largely ignored the need to develop capacity for structured learning.

For long time water policy has been characterized by a waxing and waning of simplistic panaceas without much reflection on the conditions for success. Idealized design principles based on institutional and technological panaceas have been applied to water issues without long-term monitoring of their performance and effectiveness, and without revision and critical reflection on the practices that would have ensured the appropriate responses to failures at a much earlier stage (Gleick 2003; Meinzen-Dick 2007; Ingram 2011).

2.3 Neither Privatization nor Community Governance Can Meet the Water Governance Challenge

Regarding the various widely-praised water governance principles of hierarchical centralization, coordinated river basin planning and management, devolution and decentralization, markets and privatization (Ingram 2011), the push towards privatization and liberalization has been particularly controversial. In the 1990s, decentralization became the guiding principle of water policy reform. In particular the World Bank was instrumental in supporting and enforcing such trends (World Bank 1993). According to the principle of *subsidiarity*, the authority and responsibility for decision making and operations were transferred from national government to lower-level governmental organizations, community organizations and/or the private sector. Neoliberal thinking led to the connecting of such decentralization with deregulation and privatization (e.g. Achterhuis et al. (2010)). Market-based approaches were supposed to overcome the perceived lack of efficiency and effectiveness of governmental command and control policies and the failure of governments to deliver water services.

Decentralization of water governance to increase effectiveness and efficiency of water management was, for example, a centrepiece of water governance reform in many Latin American countries (Wilder and Romero Lankao 2006; OECD 2012). However, the huge costs of infrastructure exceeded governmental financial budgets even in developed countries. The anticipated costs for infrastructure for wastewater treatment to meet the standards set by the European Urban Wastewater Directive were, for example, a major driver of privatization in many European countries (Aubin and Varone 2004). Engaging the private sector was linked to the expectation of attracting external sources for financing infrastructure development.

Privatization did not meet the expectations that had been placed in it. Experience has been quite varied with some striking failures—notably in developing countries (Bakker 2010). As discussed by the various contributions in Boelens et al. (2010), decentralization—if guided by neoliberal thinking only—may have detrimental consequences and lead to distortions in power structures. Furthermore, such reform does not solve a systemic governance problem (Brown and Cloke 2004, 2005; Soliman and Cable 2011)—high levels of corruption and the dominance of informal institutions with goals that are often in conflict with sustainable resource management. In the absence of effective regulation and in the presence of rent-seeking elites in government, particularly in developing countries, privatization leads, in most cases, to dissatisfaction among both consumers and private enterprise. Furthermore, water infrastructure does not lend itself easily to private ownership and management. This has become particularly evident in the urban context. Due to the high costs of investment in building and maintaining urban water infrastructure with long-time scales for amortization it is difficult to make profits from water delivery services at a price that is affordable for all societal groups. The price of water is mainly determined by sunk costs of infrastructure rather than the amount of water provided to customers. Since water possesses the characteristics of a natural monopoly and has little competition governmental regulation is required. Otherwise companies may maximize profits by exploiting and not maintaining available infrastructure and by delivering services only to those privileged societal groups who can afford it.

Such developments characterized the privatization of drinking water supply in Cochabamba, Bolivia, a striking example of the failure of privatization (Shultz 2009). With the strong encouragement of the World Bank, the Bolivian government granted a concession to an international company to supply drinking water and wastewater treatment services to the city of Cochbamba in 1999. Shortly afterwards a law was passed to regulate the water supply and sanitation sector with an emphasis on promoting privatization. Many local communities regarded this law as a threat to their access to water resources. A massive increase in water tariffs enacted by the new private water supplier triggered massive protests in the whole country. As a result, the contract with the private water supplier had to be retracted. Cases such as Cochabamba mobilized those groups that had from the beginning opposed privatization in the water sector. Critical voices were as undifferentiated in their opposition as proponents had been in their advocating of the principle of privatization. Critical voices had always argued against the market system for the delivery of natural resources since they were not designed to guarantee fairness and adherence to just criteria for access to basic needs such as water, a common good essential for life. However, in many countries governments have not proven to be much better in allocating this resource. Hence another solution has had to be identified and pleas have been made for more direct community involvement in the distribution of urban water (Bakker 2009). Such pleas reflect general trends in a stronger reliance on participatory approaches in water governance and environmental governance in general (Lemos and Agrawal 2006). There is a real danger though that this "commons approach" is mistakenly seen as a panacea for all problems.

In a comprehensive review Bakker (2009) analysed various forms of the role of "community" that have been advocated as alternatives to private sector management of urban water supply. She makes the distinction between community ownership and community governance. Ownership and self-management by community groups is facilitated by the increasing popularity of low-cost, small-scale infrastructure. Large-scale cooperatives that own centralized water supply infrastructure are rare though. Community-based governance gives communities a central role through the establishment of customer service boards or community watershed boards and similar management structures. Bakker's analysis of the water supply sector demonstrates that the often-held assumption of changing behavioural patterns by introducing community-based management, thus solving all governance problems, is highly mistaken. She comes to the overall conclusion that "*'ownership' (i.e. public versus private) is less important than institutions (rules, norms, and laws) and governance (decision-making processes); it follows that the imposition of 'public' or 'community' management is not a sufficient condition for better water services.*" (Bakker 2010, p. 245). Again, these findings are a clear indication that moving to another panacea—in this case, community governance that delivers what governments and markets failed to do—cannot provide a universal solution to problems originating from complex and intertwined governance systems.

2.4 Environmental Governance—Shifting Away from Panaceas and Towards the Mastering of Complexity

As a response to the urgent need for effective water governance reform the OECD launched the OECD water governance initiative in 2013.[2] This initiative has established an international multi-stakeholder network from public, private and not-for profit sectors whose members gather regularly to share on-going reforms, projects, lessons and good practices in support of improved water governance. The OECD has also launched a series of comparative studies on water governance and the preparation of in-depth individual country reports (OECD 2011, 2012).

What can science offer to assist such developments and the urgent need to develop an improved knowledge base? Science has been slow in addressing the challenges posed by developments in environmental governance, in general, and water governance in particular. On one hand, water governance has not been a well-respected topic for scholary work in the social sciences and has thus been established by a number of resolute scientists as its own field of expertise only in

[2]http://www.oecd.org/env/watergovernanceprogramme.htm.

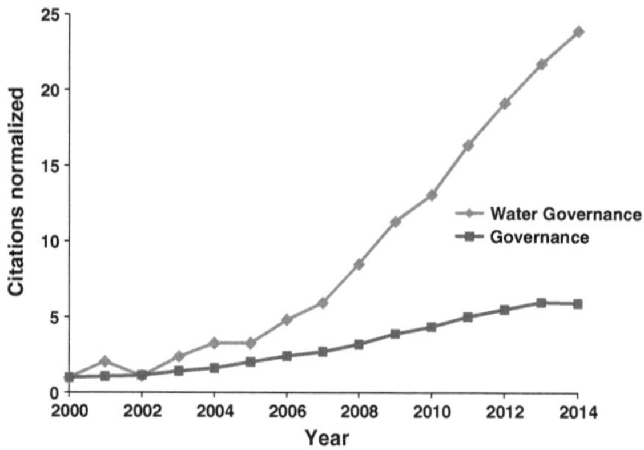

Fig. 2.1 Number of publications in peer reviewed journals containing the search terms "Water Governance" and Governance respectively in either title, abstract or keywords. In order to present the results on the same scale the number of publications in a given year was normalized to the number of publications in the year 2000 which was 18 for Water Governance and 1081 for Governance (SCOPUS Data Base 25.05.2015). The change in the number of publications referring to governance is shown as a benchmark for the development of scholary work on governance in general

recent years (cf. Fig. 2.1). On the other hand adequate concepts and methods to deal with the complexity of governance systems are missing in general.

The most relevant conceptual frameworks in the social sciences are weak in their ability to analyse the complex, context-dependent dynamics of governance systems. Most governance analyses focus on static descriptions and embrace only some of the important processes (e.g. the focus on institutions) from a disciplinary perspective. Looking back on the achievements of a decade of research under the umbrella of the IDGEC (Institutional Dimensions of Global Environmental Change) research program Young (2008) noted that "*Knowledge regarding the nature of change in the institutional dimensions of socio-ecological systems remains relatively underdeveloped*" (ibid, p. 140).

The work of Oran Young and the IDGEC program in general had a strong influence in shaping the scholary field of environmental governance. Young's work has been central to the development of international environmental governance and regime theory. As early as 2002, he promoted the importance of **institutional diagnostics** taking into account the need for institutions to fit the nature of the problem to the biophysical and societal settings in which they are assumed to operate (Young 2002b). In a contribution to a special issue summarizing the main achievements of his work over the past decades (Mitchell 2013), Young summarized the major insights that he could derive from his work and the main challenges that he foresees for future work on environmental governance (Young 2013). He noted that governance without government is quite common at all levels. Spontaneity and self-organizing properties are important characteristics of institutional dynamics. He argues in favour of a more integrative and comprehensive approach to studying environmental governance to overcome the still prevailing fragmentation of different water governance approaches. Research on the impacts of governance regimes on the behaviour of actors can be largely classified into two approaches: alleviating collective action problems based on utilitarian rational choice or influencing actor behaviour through the development of social practices (Young 2001). These two approaches are largely distinct and are sometimes even seen as being mutually exclusive. But more integration would be beneficial for a more in-depth and grounded understanding of the impacts of institutions on human behaviour. Young acknowledges that regimes are influenced by the dominant world view—the paradigm. The influence of paradigms became quite evident from the historical account of the evolution of water policy showing a succession of changes with respect to the role of government, of markets and so forth (cf. Sects. 2.1 and 2.2). One of the major contributions of Young's work was the research on fit and interplay. The success of regimes hinges on their fitting into the major biophysical and socioeconomic setting in which they operate. Young is clearly dismissing institutional panaceas and advocates a diagnostic approach. He highlights in particular the importance at the international level of the ability of governance regimes to deal with complexity and uncertainty and to adapt to rapid change and unexpected developments. He identified four key challenges for environmental governance: "*(1) How can we deepen our understanding of the complex causality involved in the operation of environmental governance systems? (2) How can we integrate the collective-action and the social-practice models of environmental governance? (3) How can we address needs for governance arising in the Anthropocene? (4) How can we improve our ability to design effective environmental and resource regimes?*" (Young 2013, p. 100).

Another pioneer and highly influential intellectual leader in the field of governance of social-ecological systems (SES) was the late Elinor Ostrom. In contrast to Young, she focused largely on the local level. Elinor Ostrom laid the foundations of scholarship on the governance of common pool resources. Vincent and Elinor Ostrom introduced common pool resources as a fourth type of good alongside public, private and club goods (Ostrom and Ostrom 1977). Common pool resources are characterized by subtractability of uses and thus competition. At the same time it is difficult to exclude potential users. This makes them different from private goods with private ownership and use rights. Water-related resource use possesses the typical properties of common pool resources—e.g., groundwater use or fisheries.

Despite being a political scientist by training, Elinor Ostrom received the Noble prize for Economics in 2009. Her prize-winning lecture "Beyond Markets and States: Polycentric Governance of Complex Economic Systems" conveys the essential pillars of her work (Ostrom 2010). Elinor Ostrom was less a theoretician than a sharp analytical observer. In numerous well-designed studies she provided evidence for the ability of local communities to self-organize and develop effective rules which contradicted conventional wisdom and Hardin's influential paper on the tragedy of the commons (Ostrom 1990). Her work paved the way for the increased recognition of community-based governance. In line with economic thinking, she embraced a rational choice model of human behaviour. But in contrast to mainstream, neo-classical economic approaches she addressed complexity by, among other things, identifying seven different types of rules in use in local settings (Ostrom 2005). Furthermore she demonstrated the importance of trust and reputation for cooperation and collective governance processes. From numerous studies of local user communities she distilled design principles for effective collective choice arrangements (Ostrom 1990, 2005). However, she was always strong in arguing against panaceas and recognized the need for rules to be tailored to the setting in which they operate (Ostrom 2007). One condition for ensuring the effectiveness of rules proved to be that communities need to develop the rules themselves.

Despite their different theoretical standpoints and levels of analysis both Ostrom and Young embrace complexity and acknowledge the importance of self-organizing processes in governance systems. Both have worked on governance systems where government is often absent. There exists no government at the international level with a global jurisdiction. Government is also often absent or ineffective at the local level. Both make strong pleas against panaceas and simplification and argue in favour of a generic but contextual diagnostic approach. Such approaches should take into account the complexity of social-ecological systems in a systematic fashion and support context-sensitive analysis and a transferability of insights among similar classes of problems and contexts. Such an analysis requires a systemic and interdisciplinary approach in the social sciences and across the social-natural science interface. In her later work, Ostrom made an attempt to move in this direction and suggested organising variables of interest in the study of SES in a nested, multi-tier framework (Ostrom 2007, 2009).

Another stream of interdisciplinary research has focused on an improved understanding of the requirements for adaptive resource governance, since the ability to respond to uncertain developments and surprise together with learning are considered as essential for governing social-ecological systems (Dietz et al. 2003; Folke et al. 2005; Pahl-Wostl 2009). Folke et al. (2005) point out that adaptive governance systems often self-organise as social networks with teams and actor groups that draw on various knowledge systems and experiences for the development of a common understanding and policies. Empirical evidence has shown that the formation of informal networks plays an important role (Olsson et al. 2006; Nooteboom 2006). Ostrom (2001) highlighted the importance of polycentricity for adaptive governance. Polycentric systems combine decentralization of power with effective coordination among the multiple centres of decision-making. They are assumed to enhance innovation, learning, adaptation, trustworthiness, level of cooperation among participants, and the achievement of more effective, equitable, and sustainable outcomes at multiple scales (Ostrom 2010; Pahl-Wostl and Knieper 2014). Pahl-Wostl (2009) developed a conceptual framework to analyse the adaptive capacity of resource governance systems and highlighted the importance of multi-level interactions, polycentric system architectures and the interplay between formal and informal networks. Armitage et al. (2008) deplored the fact that work on adaptive governance of SES did not sufficiently take scholary work in the more traditional social science disciplines into account. To remedy this situation they pointed out the links to political ecology by addressing the importance of power, scale and levels of organisation, the positioning of social actors and social constructions of nature, which might explain certain barriers to change and learning.

Despite such promising conceptual developments and an increasing number of case studies to exemplify them, empirical evidence is fragmented, and the different conceptual and methodological approaches for studying resource governance in SES are barely comparable. The field of water governance lacks both a systematic empirical base and theoretical understanding of governance systems. To date scarcely any large-scale comparative studies acknowledging the complexity of water governance and management systems exist. Notable exceptions are the study by Saleth and Dinar (2004) using an institutional economics approach to conduct an analysis of the performance of national water policy reform and the study by Pahl-Wostl et al. (2012) who conducted the first comprehensive comparative analysis of the performance of complex water governance and management systems in national river basins.

2.5 The Challenges Ahead

There is an urgent need to take stock of experiences with water policy reform in order to support learning and build capacity for transformative change. Science is not yet up to the challenge of playing a major role in this.

A major bottleneck for using the governance concept in scientific theorizing and analysis and in water policy reform seems to lie in the lack of sound conceptual foundations for an integrative approach that embraces the various dimensions of governance systems. Furthermore, a lack of analytical rigour and comparability in empirical analyses prevents the development of a sound and cumulative knowledge base.

A diagnostic approach seems to point to a middle way between simplistic governance panaceas applicable to all circumstances and the uniqueness of specific governance settings determined by societal and environmental context without transferability of lessons learned from one case to another. A diagnostic approach identifies links between characteristics of governance systems and the degree to which they fulfil their societal function taking into account the influence of context and path-dependence on these relationships. Diagnosis should also identify and suggest pathways to improvement. The results of this diagnosis should not provide blueprints for the properties of an ideal governance system, but strategies for implementing change which take into account historical context, and biophysical and societal characteristics.

It is a key challenge for science to move away from the quite static approaches still prevalent in governance research to an approach which focuses on an understanding of the dynamics of governance systems and the governance of transformation.

Key questions that need to be addressed include: To what extent can governance regimes be purposefully designed and steered in a particular direction? To what extent can one refer to intentionality in a governance system? How can science best capture the dynamic relationship between structure and agency? How can science support the fundamental transformations required for making significant progress towards sustainable water governance and management? These questions will be addressed in subsequent chapters of this book.

References

Achterhuis, H., Boelens, R. A., & Zwarteveen, M. Z. (2010). Water property relations and modern policy regimes: Neoliberal utopia and the disempowerment of collective action. In D. R. A. Boelens, D. Getches, & A. Guevara-Gil (Eds.), *Out of the mainstream. Water rights, politics and identity* (pp. 27–55). London, UK: Earthscan.

Armitage, D., Berkes, F., & Doubleday, N. (2008). Governance and the commons in a multi-level world. *International Journal of the Commons, 2*(1), 7–32.

Aubin, D., & Varone, F. (2004). The evolution of European water policy: Towards integrated resource management at EU level. In I. Kissling-Näf, & S. Kuks (Eds.), *The evolution of national water regimes in Europe. Transitions in water rights and water policies* (pp. 49–86). Dordrecht-Boston-London: Kluwer Academic Publishers.

Bakker, K. (2009). The ambiguity of Community: Debating alternatives to private-sector provision of urban water supply. *Water Alternatives, 1*(2), 236–252.

Bakker, K. (2010). *Privatizing water: Governance failure and the world's urban water crisis.* Ithaca, NY: Cornell University Press.

BMU. (2010). *Water framework directive: The way towards healthy waters. Results of the German river basin management plans.* Berlin: Federal Ministry for the Environment, Nature Conservation and Nuclear Safety, BMU.

Boelens, R. A., Getches, D., & Guevara-Gil, A. (Eds.). (2010). *Out of the mainstream. Water rights, politics and identity.* London, UK: Earthscan.

Brown, E., & Cloke, J. (2004). Neoliberal reform, governance and corruption in the South: Assessing the International Anti-Corruption Crusade. *Antipode, 36*, 272–294.

Brown, E., & Cloke, J. (2005). Neoliberal reform, governance and corruption in Central America: Exploring the Nicaraguan case. *Political Geography, 24*(5), 601–630.

Dietz, T., Ostrom, E., & Stern, P. C. (2003). The struggle to govern the commons. *Science, 302*, 1907–1912.

European Environmental Bureau. (2010). *10 years of the water framework directive: A toothless tiger? A snapshot assessment of EU environmental ambitions.* Brussels: European Environmental Bureau, Federation of Environmental Citizens Organisations.

European Parliament and of the Council. (2000). Directive 2000/60/EC of the European Parliament and of the Council of 23 October 2000 establishing a framework for Community action in the field of water policy. In E. Parliament (Ed.).

Folke, C., Hahn, T., Olsson, P., & Norberg, J. (2005). Adaptive governance of social-ecological systems. *Annual Review of Environment and Resources, 30*, 441–473.

Gleick, P. (2003). Global freshwater resources: Soft-path solutions for the 21st century. *Science, 302*, 1524–1528.

Ingram, H. (2011). Beyond universal remedies for good water governance. In A. Garrido, & H. Ingram (Eds.), *Water for Food in a changing World* (pp. 241–261, Contributions from the Rosenberg International Forum on Water Policy). Miltorn Park: Routledge.

Lemos, M. C., & Agrawal, A. (2006). Environmental governance. *Annual Review of Environment and Resources, 31*(1), 297–325. doi:10.1146/annurev.energy.31.042605.135621

Meinzen-Dick, R. (2007). Beyond panaceas in water institutions. *Proceedings of the National Academy Sciences USA, 104*, 15200–15205.

Mitchell, R. B. (2013). Oran Young and international institutions. *International Environmental Agreements: Politics, Law and Economics, 13*(1), 1–14. doi:10.1007/s10784-012-9200-3

Moss, T. (2003). Solving problems of 'Fit' at the expense of problems of 'Interplay'? The spatial reorganisation of water management following the EU water framework directive. In H. Breit, A. Engels, T. Moss, & M. Troja (Eds.), *How institutions change. Perspectives on social learning in global and local environmental contexts* (pp. 85–121). Opladen: Leske+Budrich.

Mostert, E. (2003). The European water framework directive and water management research. *Physics and Chemistry of the Earth, 28*, 523–527.

Nooteboom, S. (2006). *Adaptive networks—The governance for sustainable development*. Delft, The Netherlands: Eburon Academic Publishers.

OECD. (2011). Water governance in OECD countries: A multi-level approach, OECD studies on water. In OECD (Ed.), *OECD Studies on Water*. Paris.

OECD. (2012). Water Governance in Latin America and the Caribbean: A multi-level approach. In OECD (Ed.), *OECD Studies on Water*. Paris.

Olsson, P., Gunderson, L. H., Carpenter, S. R., Ryan, P., Lebel, L., Folke, C., et al. (2006). Shooting the rapids: Navigating transitions to adaptive governance of social-ecological systems. *Ecology and Society, 11*(1), 18.

Ostrom, E. (1990). *Governing the commons. The evolution of institutions for collective action* (Political economy of institutions and decisions). Cambridge: Cambridge University Press.

Ostrom, E. (2001). Vulnerability and polycentric governance systems. *IHDP (International Human Dimensions Programme on Global Environmental Change) Newsletter UPDATE, 3*(1), 3–4.

Ostrom, E. (2005). *Understanding institutional diversity*. Princeton: Princeton University Press.

Ostrom, E. (2007). A diagnostic approach for going beyond Panaceas. *Proceedings of the National Academy Sciences USA, 104*, 15181–15187.

Ostrom, E. (2009). A general framework for analyzing sustainability of social-ecological systems. *Science, 325*(5939), 419–422. doi:10.1126/science.1172133

Ostrom, E. (2010). Beyond markets and states: Polycentric governance of complex economic systems. *American Economic Review, 100*, 641–672. doi:10.1257/aer.100.3.641

Ostrom, V., & Ostrom, E. (1977). Public goods and public choices. In E. S. Savas (Ed.), *Alternatives for delivering public services: Toward improved performance* (pp. 7–49). Boulder, CO: Westview Press.

Pahl-Wostl, C. (2006). The importance of social learning in restoring the multifunctionality of rivers and floodplains. *Ecology and Society, 11*(1), 10. http://www.ecologyandsociety.org/vol11/iss11/art10/

Pahl-Wostl, C. (2007). Transition towards adaptive management of water facing climate and global change. *Water Resources Management, 21*(1), 49–62.

Pahl-Wostl, C. (2009). A conceptual framework for analysing adaptive capacity and multi-level learning processes in resource governance regimes. *Global Environmental Change, 19*, 354–365.

Pahl-Wostl, C., & Knieper, C. (2014). The capacity of water governance to deal with the climate change adaptation challenge: Using fuzzy set qualitative comparative analysis to distinguish between polycentric, fragmented and centralized regimes. *Global Environmental Change, 29*, 139–154.

Pahl-Wostl, C., Isendahl, N., Möllenkamp, S., Brugnach, M., Jeffrey, P., Medem, W., et al. (2006). *Paradigms in water management*. Osnabrueck: European Commission.

Pahl-Wostl, C., Lebel, L., Knieper, C., & Nikitina, E. (2012). From applying panaceas to mastering complexity: toward adaptive water governance in river basins. *Environmental Science & Policy, 23*, 24–34.

Saleth, R. M., & Dinar, A. (2004). The institutional economics of water: A cross country analysis of institutions and performance. Cheltenham: Elgar Publishing.

Soliman, H., & Cable, S. (2011). Sinking under the weight of corruption: Neoliberal reform, political accountability and justice. *Current Sociology, 59*(6), 735–753.

Shultz, J. (2009). The Cochabamba water revolt and its aftermath. In J. Shultz (Ed.), *Dignity and defiance—Stories from Bolivia's challenge to globalization* (pp. 9–42). Oakland, CA: University of California Press.

Wilder, M., & Romero Lankao, P. (2006). Paradoxes of decentralization: Water reform and social implications in Mexico. *World Development, 34*, 1977–1995.

World Bank. (1993). *Water resources management: A world bank policy paper.* Washington, D.C., USA: The World Bank.

Young, O. R. (2001). The behavioral effects of environmental regimes: Collective-action vs. Social-practice models. *International Environmental Agreements, 1*(1), 9–29. doi:10.1023/A:1010181007938

Young, O. R. (2002a). *The institutional dimensions of environmental change: Fit, interplay, and scale.* Cambridge: MIT Press.

Young, O. R. (2002b). Usable knowledge: Design principles and institutional diagnostics. In O. R. Young (Ed.), *The institutional dimensions of environmental change: Fit, interplay and scale* (pp. 165–190). Cambridge, MA: MIT Press.

Young, O. R. (2008). Building regimes for socioecological systems: Institutional diagnostics. In O. R. Young, L. A. King, & H. Schroeder (Eds.), *Institutions and environmental change: Principal findings, applications, and research frontiers* (pp. 115–144). Cambridge, MA: MIT Press.

Young, O. R. (2013). Sugaring off: Enduring insights from long-term research on environmental governance. *International Environmental Agreements: Politics, Law and Economics, 13*(1), 87–105. doi:10.1007/s10784-012-9204-z

Chapter 3
Conceptual and Analytical Framework

This chapter lays out the major pillars of the conceptual foundations required for improving our understanding of the complex dynamics of water governance systems. After clarifying terminology and providing some reflections on the nature of causality in social systems I sketch out what I consider to be the major building blocks for conceptualizing governance systems and processes of adaptive and transformative change. With this background, the *management and transition framework* which provides a structured and standardized approach to representing governance systems and processes is introduced.

3.1 Clarification of Guiding Assumptions and Terminology

In the scientific literature one finds a wide range of approaches for conceptualizing governance. Treib et al. (2007) classified governance according to its function being primarily politics, polity or policy. Within the *politics* dimensions governance emphasizes the means of making policy, the translation of different preferences into effective policy choices and the transformation of different interests into unitary action (Kohler-Koch 1999). Emphasis is placed on actor networks, power constellations and the role of private and public actors and their relationship in the context of policy making. Other governance forms more closely related to the *polity* dimension focus on institutions and conceive of governance as a system of rules that shape the actions of actors (e.g. Rosenau 1995; Ostrom 2005). Emphasis is on the various types of rules, their interdependence and the overall logic (e.g. hierarchy or market) which guides them. In political science the concept of governance has mainly been associated with this institutionalist approach and an emphasis on regulatory structures in contrast to the traditional guiding theories involving an actor and government-centred focus (Mayntz 2004a, b). Finally, governance may also be defined as modes of political steering and thus refer primarily to the *policy* dimension (e.g. Héritier 2002). The emphasis here is on the governance instruments employed such as hierarchical regulation, market-based instruments or voluntary agreements.

C. Pahl-Wostl, *Water Governance in the Face of Global Change*,
Water Governance—Concepts, Methods, and Practice,
DOI 10.1007/978-3-319-21855-7_3

From an analytical perspective it may be useful to make a distinction between these dimensions of governance. However, such a distinction may not do justice to the complexity of real-world governance systems where these dimensions are strongly entwined. It may be virtually impossible to determine the dependent variable and the independent variable. In particular the politics and polity dimensions are difficult to separate. Increasingly scholars are arguing in favour of developing an all-encompassing concept of governance that reflects the complexity of societal dynamics and that bridges social science disciplines (e.g. Kooiman 2000; Benz 2004; Schuppert 2006). Furthermore the prevailing emphasis on regulatory structures has led to a structural and static perspective on governance which impedes improved understanding of the dynamics of governance processes and transformative change of governance systems. To overcome such limitations broader conceptual foundations are required (Plummer et al. 2013; Pahl-Wostl 2009). The approach presented in this book to conceptualize water governance follows similar lines of reasoning.

An comprehensive approach to governance is reflected in the now widely used definition on water governance originally suggested by the Global Water Partnership: "*Water governance refers to the range of political, social, economic and administrative systems that are in place to regulate development and management of water resources and provisions of water services at different levels of society*" (Rogers and Hall 2003, p. 88). This definition is, however, determined more by practical considerations than by analytical rigour. It was developed to do justice to the increasing complexity of real-world policy processes. In his review of decades of work on environmental governance Young (2013, p. 88) makes the distinction between governance as "*a social function centred on steering human groups toward desired outcomes and away from undesirable outcomes*" and the governance system as "*an ensemble of elements performing the function of governance in a given setting. Institutional arrangements form the core of such a system, but the ensemble normally includes cognitive, cultural, and technological elements as well*". The definitions used in this book are based upon the work of Young in defining governance as a social function and build on the approach chosen by Pahl-Wostl (2007b).

3.1.1 Some Definitions

Water governance is the social function that regulates development and management of water resources and provisions of water services at different levels of society and guiding the resource towards a desirable state and away from an undesirable state.

A **water governance system** is the interconnected ensemble of political, social, economic and administrative elements that performs the function of water governance. These elements embrace institutions as well as actors and their interactions.

A **water governance regime** is the interdependent set of institutions (formal laws, societal norms or professional practices) that is the main structural component feature of a governance system.

By making an explicit use of the term "regulate" rather than "steer" in the definition of water governance I maintain a distinction between water governance and water management.

Water management refers to the activities of analysing and monitoring water resources, as well as developing and implementing measures to keep the state of a water resource within desirable bounds.

The notion of 'water governance' embraces the full complexity of regulatory processes and their interaction that set the context in which water management operates (Pahl-Wostl 2009). Applying this distinction in practice is not trivial and often management is treated as part of governance. However, such a broad categorization makes water governance analytically quite intractable. Governance is already exceedingly complex given the multitude of actors and processes at different levels. Including management would add another largely hidden layer of complexity. Despite advanced regulatory frameworks, water management may for example perform poorly due to the lack of appropriate skills among water managers or the financial resources to invest in certain technologies (Pahl-Wostl et al. 2012). It would then be the task of governance bodies to make efforts for improving capacities for the implementation of effective management.

In contrast to Pahl-Wostl (2009) I make a distinction here between adaptive and transformative capacity.

Adaptive capacity is defined as the ability of a governance system to alter processes and to adapt structural elements as a response to current or anticipated changes in the social or natural environment.

Transformative capacity is defined as the ability of a governance system to first adapt and, if required, transform structural elements as a response to current or anticipated changes in the social or natural environment.

Even when it is analytically difficult to develop operational measures that distinguish between adaptive and transformative capacity it makes sense from a conceptual point of view to be more explicit in this distinction.

3.2 Finding General Patterns

3.2.1 Causality and Dynamics in Complex Governance Systems

Governance systems and the human-technology-environment systems in which they are embedded are most appropriately described as complex adaptive systems

(CASs) characterized by self-organization, adaptation, heterogeneity across scales and distributed control (Kauffman 1995; Pahl-Wostl 1995; Casti 1997; Pahl-Wostl 2007a). In such systems it is difficult, often meaningless and sometimes entirely wrong to attribute a single cause to an observed phenomenon. An illustrative example that most people know from own experience are traffic jams on highways. Given a specific density of cars and a particular type of behaviour among drivers (speeding up and slowing down) traffic jams will be observed in regular but unpredictable intervals (Wilensky 1999). This is an emergent phenomenon of the constellation of the interacting parts of the system. It is futile to search for the accident that caused the traffic jam or any other single cause. In CASs system function is largely an emergent product of processes of self-organization, of feedback between emergent macro-level control variables and micro-level interactions. Societal norms are for example emergent macro-level control variables. Societal norms are not the product of purposeful design but emerge from repeated social interactions and constrain the behaviour of individual actors. Other important characteristics of CASs are path-dependence and threshold effects (Scheffer 2009). A system may accumulate impacts of a stressor without showing noticeable signs of change at the macro-level. Then suddenly another single event may bring the system to a tipping point and it might collapse and move to another undesirable state (Pahl-Wostl 1995). Such so-called regime shifts were first identified in ecological systems. One of the best known examples is the eutrophication of lakes due to excessive nutrient loadings. Once in an undesirable state a system may take very long to recover, if the change is reversible at all. In recent years a wider range of regime shifts have also been identified in social-ecological systems (Leadley et al. 2014). The role of governance should be to avoid such undesirable regime shifts, which is no trivial governance task.

At the same time, we have some expectations that social systems are anticipatory and purposeful even when intentionality at the level of a social system as a whole is quite hard to capture. Governance is defined as a social function with a certain purpose. Assuming that history and context have some influence leads to the conclusion that the function of governance can be fulfilled in more ways than one. The argument for a diagnostic approach (cf. Sect. 2.3) though is based on the assumption that the number of ways is not infinite. This implies that one can identify similarities across groups of cases falling in comparable categories. Hence, analysis has to take into account multifactorial causation—constellations of several factors are causes for a certain phenomenon—and equifinality—there may be more than one constellation of factors that can be linked to a certain phenomenon/outcome.

A pioneer promoting configurational rather than correlational analysis was (Ragin 1987). He not only introduced a method—Qualitative Comparative Analysis—but derived this methodological development from on epistemological reasoning, on the need for different ways of dealing with causality in social systems. He argued against simple correlational analysis and in favour of using set theoretic approaches, and combining formal analysis with qualitative reasoning based on a deep understanding of case studies. The overall approach advocated by Ragin resonates with the "causal reconstruction" approach advocated by Mayntz (2004a, b). According to Mayntz (ibid, p. 2) "*causal reconstruction does not look for statistical relationships among variables, but seeks to* explain *a given social phenomenon - a given event, structure, or development - by identifying the processes through which it is generated. Causal reconstruction may lead to a historical narrative, but in its theoretically more ambitious version, causal reconstruction aims at generalizations*". Generalisations can be made more plausible if similar patterns are found in different cases. The search for pattern should be informed by the current level of theoretical and empirical understanding of which factors are important for producing an outcome.

Such reasoning is consistent with the basic assumptions underlying a diagnostic approach which is a structured search for constellations of factors that lead to a certain outcome (Young 2002; Ostrom 2007). Such a search tries to identify processes of self-organization that lead to an (unintended) emergent outcome as well as outcomes resulting from purposeful design and, in particular, outcomes that result from the interaction between self-organization and purposefulness. Here it might be useful to recall the different kinds of causalities introduced by Aristotle: material (material base of factors), formal (nature of the arrangement of factors), efficient (executing agent) and final (purpose). Final causes are unique to anticipatory systems and are omnipresent in human systems. An important aspect of anticipatory purposeful systems is feedback between the final and the other causes, and between purpose and the means to achieve it. Such feedback is generally called learning, and poses another challenge for any kind of causal explanation.

3.2.2 Requirements for a Diagnostic Approach

According to the Merriam-Webster Dictionary, a diagnosis is defined as: "*The investigation or analysis of the cause or nature of a condition, situation, problem and a statement or conclusion from such an analysis. A diagnosis attempts to infer underlying causes for a problem from systematic analysis based on experience with similar cases and/or general mechanistic understanding on what might be the causes for a certain problem constellation*".

One may argue the relevance of a mechanistic understanding for explaining phenomena in complex social-ecological systems. Obviously, a diagnosis in a SES differs from diagnosing a mechanical failure of a car. A diagnosis in medicine seems to offer a more appropriate analogy. Modern medicine has followed a mechanistic approach and tried to reduce causality for disease patterns to simple cause-effect relationships. Such an approach has largely failed in diagnosing complex diseases such as rheumatism where configurational, context-sensitive analysis seems to be the method of choice. A diagnostic approach in water governance needs to take into account the complexity of social-ecological systems (SESs) in a systematic fashion in order to support context-sensitive analysis and a transferability of insights between similar classes of problems and contexts. A diagnostic approach needs to assess why a certain governance system can or cannot fulfil the social function of water governance given certain contextual

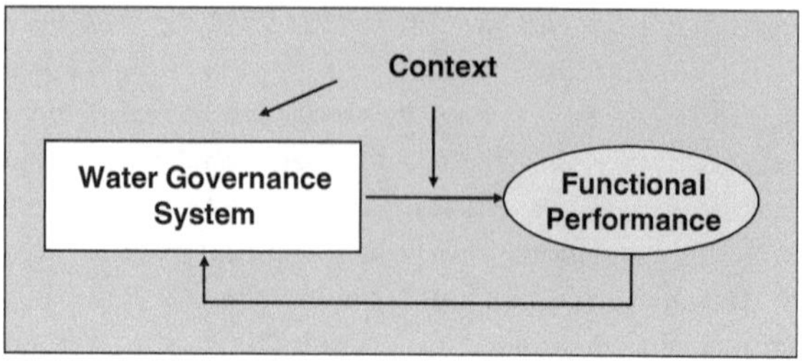

Fig. 3.1 Major elements of a diagnostic approach

conditions. This is schematically outlined in Fig. 3.1. To operationalize such an approach, one needs to derive conceptualizations of the relevant characteristics of the water governance system and of measures for its performance. Furthermore, the diagram includes feedback from functional performance to the water governance system. Such feedback is required for any kind of learning and adaptive governance approach.

Water governance has a certain purpose—it is a means to an end and not an end in itself. This does not necessarily imply that goals to be achieved are clearly articulated and operationalized. Monitoring the functional performance of water governance systems is not very common, quite in contrast to operational water management where measureable objectives such as water quality targets are, or at least should be, defined and monitored. But a failure of management can often be attributed to a failure of governance. Therefore, the performance of governance systems should be assessed based on some normative criteria. These may be defined either internally (by those involved in water governance) or externally (by widely-accepted international standards). Being 'internally defined' implies that goals for water policies are explicitly stated. These might also include specific goals for different groups. Normative criteria for performance would then refer to the achievement of the stated goals. Being 'externally defined' implies a general agreement and acceptance of normative standards for good water governance for the delivery of water services to society and for the sustainability of developing and managing water resources, and for water security for the economy, society and the environment. Without adhering to good governance principles,[1] it is unlikely that water governance takes the needs of the less powerful or the environment into account.

A diagnostic approach is guided on the one hand by the assumption that panaceas are futile. On the other, it is also assumed that policy outcomes are not entirely contingent on context and history which would make any generalizations largely impossible. The societal context may provide either an enabling or hindering environment. The biophysical context strongly influences and partly determines the nature of the resource governance challenge (e.g., physical water scarcity).

[1]Good governance is participatory, consensus oriented, accountable, transparent, responsive, equitable and inclusive, effective and efficient and follows the rule of law (UNESCAP 2009).

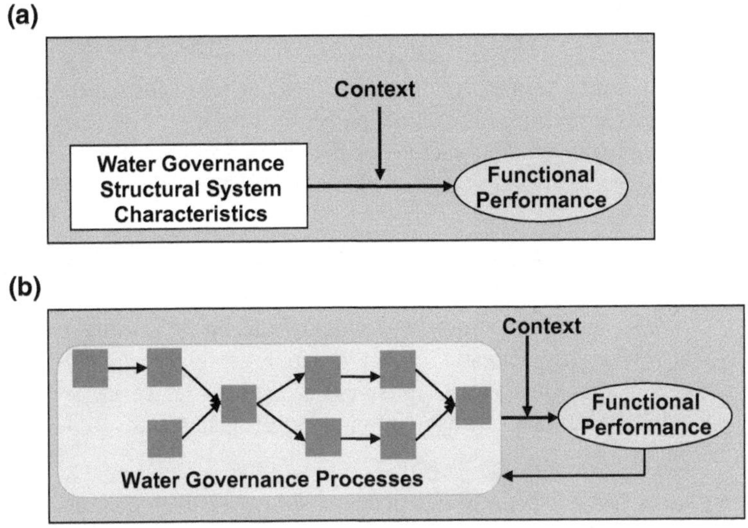

Fig. 3.2 Two possible representations of a diagnostic approach: (A) configuration-based and (B) process-based analysis

Two possible approaches for analysis are represented in Fig. 3.2:

(A) Configuration analysis (e.g., regulatory framework, coordination structure) and linking of structural system characteristics to functional performance.

(B) Process analysis (e.g., implementation of an innovative water policy framework) over time and linking of process characteristics of the water governance system to functional performance.

(A) Configuration analysis is a static approach in the sense that it does not resolve the dynamics of the configurations analysed. However, it cannot be a snapshot in time. Analysing the relationship between constellations of factors (e.g. vertical and horizontal coordination) and measures of functional performance (e.g. adaptive capacity) must take into account that a specific constellation must remain relatively stable over a sufficiently long period of time to become effective. An educated guess for the duration of such a period would be about a decade. But this should be assessed and justified on a case by case base.

(B) A process analysis aims at characterizing a multi-level policy/governance process over time. Such analysis might for example identify which actors participate in which role at different levels and in different phases of the process. How to decide the overall period of analysis, the degree of resolution and what should be included requires a systematic and well-justified approach.

Implementing such a diagnostic approach requires a sound conceptualization of both structural characteristics and processes of governance systems.

3.3 Conceptualizing Governance Systems

The governance regime formed by the interdependent set of institutions has been already introduced as the main structural characteristic of a governance system. To analyse the complexity of the structure of a governance system in a systematic fashion, I distinguish below between four classes of structural elements (Pahl-Wostl 2009):

- Institutions and the relationship and relative importance of formal and informal institutions;
- Actor networks and power structures with emphasis on the role and interactions of state and none-state actors;
- Governance modes—bureaucratic hierarchies, markets, networks;
- Multi-level interactions across administrative boundaries and vertical integration.

These classes are not entirely independent. However, they capture important characteristics that influence functional performance. In the following, these dimensions and their relevance for and expected influence on water governance are explained briefly. An in depth discussion follows in subsequent chapters.

3.3.1 *Institutions—Formal and Informal*

'Institution' is used here according to the convention in institutional analyses, within the social sciences, to denote rules governing the behaviour of actors (North 1990; Scott 2008). Institutions do not refer to organizations or physical structures. Formal and informal refer to the nature of processes of development, codification, communication and enforcement. Formal institutions are linked to the official channels of governmental bureaucracies. They are codified in regulatory frameworks or any kind of legally binding documents. Correspondingly, they can be enforced by legal procedures. Informal institutions refer to socially shared rules such as social or cultural norms. In most cases they are not codified or written down. Compliance is not enforced via legally sanctioned channels.

The relative strength of the relationship between formal and informal institutions has a major influence on a governance regime. Helmke and Levitsky (2004) derived a 2-dimensional typology for the relationship between formal and informal institutions based on the compatibility of goals and the effectiveness of formal institutions. Formal institutions are classified as either effective or ineffective. The goals of informal and formal institutions are classified as either compatible or conflicting with each other. If formal institutions are effective and the goals of formal and informal institutions are compatible, the two types of rule systems complement each other. The implementation of formal provisions for river restoration may, for example, be supported by voluntary community-based activities for nature protection. Creating such synergies is the goal of many participatory processes in policy implementation. As a consequence, the efficiency and effectiveness of governance processes are increased. If formal institutions are ineffective, and formal and informal institutions pursue conflicting goals, the two kinds of rule systems compete with each other. Formal rules regulating the allocation of scarce water resources may, for example, be replaced by allocation practices based on informal power structures and clan networks. As a consequence, the efficiency and effectiveness of governance processes are decreased. Such a governance system would most likely be characterized by a high degree of corruption, non-transparent decision-making processes and the dominance of established power structures (Pahl-Wostl and Knieper 2014). This is the case in many developing countries where strong environmental regulation exists on paper but is not at all implemented in practice. It is important to take these distinctions into account to fully understand the nature of potential governance failures, drivers and barriers for change and the role of informality in these processes.

3.3.2 Role of Actor Groups—State, Non-state Actors

A major characteristic of governance systems is the diverse roles of non-state actors. The past decades have seen a weakening of the influence and power of the nation state (Holton 1998). This may be attributed to increasing globalization, to the internationalization of economic activities and to a strengthening of civil society in general. Participatory approaches have become a major pillar in environmental resources management (Berkes and Folke 2002; Pahl-Wostl et al. 2007a). Such issue-specific participation in policy development and implementation complements representative democracy by reflecting the need for new modes of governance and knowledge generation to deal with increasing uncertainty and complexity (Berkes et al. 2003; Pahl-Wostl et al. 2007c). The participation of interested parties can reduce uncertainties in the policy implementation process by reducing the likelihood of unexpected resistance (Newig et al. 2005). As well, including a broader set of stakeholders gives access to different kinds of knowledge which may be vital for a full assessment of a resource governance problem and for developing innovative solutions.

Compared to the traditional form of governmental authority and control, the roles of actors become blurred in more complex and intertwined governance systems. Actors are involved in designing the institutions that (are supposed to) govern their behaviour. Ostrom (1990) showed convincingly that user communities of a common pool resource have the capacity for self-organization and self-governance and that there are many different viable combinations between the public and private. Involving actors in the design of formal institutions is expected to increase compliance and effectiveness. But this may come at the expense of decreased efficiency since participatory processes are resource consuming.

3.3.3 Governance Modes—Bureaucratic Hierarchies, Markets, Networks

Governance modes refer to the various forms through which governance can be realized. One attempt at classification is the distinction among bureaucratic hierarchies, networks and markets as the main governance modes (e.g. Thompson et al. 1991). They may be understood as ideal types in the Weberian sense since, in reality, an individual mode will rarely occur in isolation. These modes differ markedly along the dimensions of the degree of formality of institutions and the role of state versus non-state actors. In bureaucratic hierarchies, regulatory processes are mainly based on formal institutions and governmental actors play the dominant role. Markets are based on a combination of formal and informal institutions, and non-state actors dominate. Networks are largely governed by informal institutions and both state and non-state actors may participate. The informality and high flexibility in membership makes networks interesting with respect to processes of learning and change.

Given the complex nature of governance systems, change can be expected to be a combination of purposeful collective action and emergent phenomena resulting from self-organising processes and the interactions among a range of actors. Attention to the importance of networks in this respect has increased considerably over recent years (Kooiman 2003; Pahl-Wostl 2009). In particular, informal networks may be very flexible in terms of membership, role and the power of actors and connections. They may support learning by providing access to new kinds of knowledge and by supporting multiple interpretations. However, informal networks may also be closed to outsiders. Membership may not be representative and their legitimacy when dealing with an issue of public interest may be disputed. It is also not clear who is to be made accountable for failed governance notions in increasingly diffuse and complex governance networks where roles of governing and becoming governed become increasingly blurred. Hence, none of these governance modes should be seen as a panacea. Each has its strengths and weaknesses. It is of major interest to develop an improved understanding of hybrid forms of governance where these modes are combined (cf. Chap. 5).

3.3.4 Multi-level Interactions

The dispersion of authority away from the central state has drawn increasing attention to the multi-level nature of governance, both from an analytical and normative perspective. Multi-level water governance systems are particularly complex. To reduce problems of fit between administrative and biophysical boundaries, new formal institutions have been introduced in most countries of the world following hydrological principles. As Moss (2003) has highlighted, problems of fit have often been solved at the expense of problems of interplay. Problems of vertical and horizontal interplay between newly established institutions at basin scale and those organized at traditional administrative boundaries (e.g., spatial planning, agriculture) prove to be a barrier for implementing integrated management approaches and may lead to overly complex structures (Borowski et al. 2008). They are also an impediment to the adaptation to climate change which requires effective vertical coordination (Pahl-Wostl et al. 2012; Krysanova et al. 2010). The introduction of the hydrological principle is only one of a variety of barriers to the effective vertical coordination of governance levels. Innovative solutions to overcoming the potential barriers for vertical coordination are in high demand (cf. Chap. 6).

3.3.5 Processes of Social and Societal Learning

To capture the dynamics of governance systems, I adapted and further developed the triple-loop learning model to explain social learning and societal change from an evolutionary perspective (Pahl-Wostl 2009). Essential elements of this model are represented in Fig. 3.3.

The concept of triple-loop learning (Flood and Romm 1996) can be seen as a refinement of the popular concept of double-loop learning developed by Argyris and Schön (1978). The triple-loop learning concept is intended to support refinement of the influence of variables governing behavioural change in terms of governing assumptions (double loop) and governing values (triple loop). I further developed this basic concept to depict the dynamics of governance systems as multi-level and multi-loop learning processes where actors experiment with innovation. In doing so they may encounter structural constraints which they try to overcome by operating in niches and by initiating transformative change (Pahl-Wostl 2009). Single loop learning is intended to improve established routines within the structural constraints imposed by the governance system. Structural characteristics are not called into question. Structural characteristics are not called into question. Double-loop learning embraces a broadening of the interpretation and a questioning of structural constraints. Triple-loop learning embraces the transformation of structural constraints.

I derived a succession of changes that would be expected for the different stages of learning in the structural elements of a governance system: institutions, actors, governance modes, multi-level interactions (Pahl-Wostl 2009). Such a conceptualization can provide the foundation for a theory of the dynamics of governance systems. Furthermore, it can also serve as a basis for developing a measure for the adaptive and transformative capacity of a governance system based on the premise that the presence of higher levels of learning indicates a higher capacity for change. These developments is discussed in more depth in subsequent chapters.

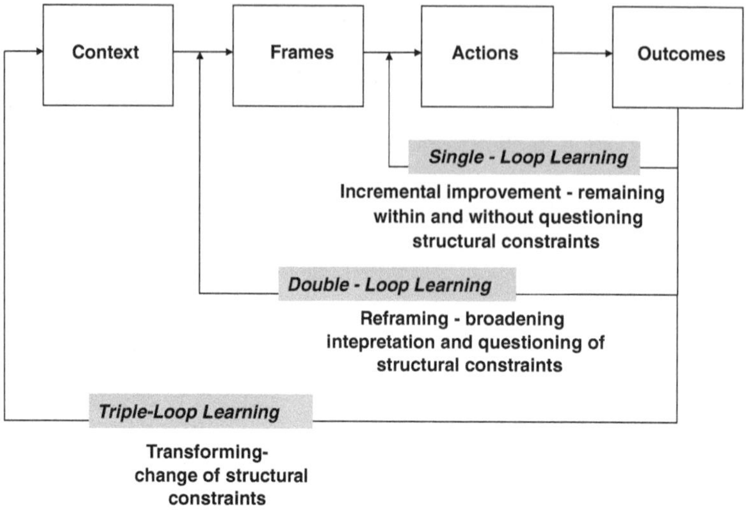

Fig. 3.3 Triple loop learning—an evolutionary, learning-based model of change in governance systems (Modified from Pahl-Wostl 2009)

3.4 Conceptualizing and Operationalising Multi-level Governance Processes

The analysis of the nature of interactions and the dynamics of governance systems requires a systematic and standardized way of representing multi-level governance processes. The approach presented here is based on the Management and Transition Framework (MTF).

3.4.1 Building Blocks of Social Interactions

The MTF was developed to allow a coherent representation of multi-level water governance and management systems and to support structured analyses through the provision of standardized definitions of variables and their relationships (Pahl-Wostl et al. 2007b, 2010). It was designed to analyse the requirements for adaptive water governance and management and the characteristics of social learning processes and whole system transformations.

Figure 3.4 shows an Action Situation, the core building block of governance processes which are depicted in the MTF as sequences of action situations.

The notion of an Action Situation (AS) was initially introduced by Elinor Ostrom as a core concept of the institutional analysis and development (IAD) framework to depict a collective choice situation in a common pool resource game (Ostrom 2005). The notion of an AS was further developed and broadened for application in the MTF

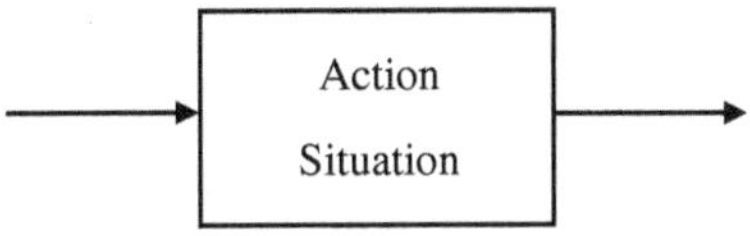

Fig. 3.4 An action situation as elementary building block of a process description

(Pahl-Wostl et al. 2010; Knieper et al. 2010). It describes, in general, a structured social interactions context that leads to identifiable outcomes. In comparison to the IAD framework, the MTF does not focus just on institutions but also gives equal emphasis to actors. The MTF broadens the behavioural model by embracing aspects of social constructivism. Furthermore, ASs are also used at higher levels of aggregation to analyse processes (e.g. policy development) over time. In such applications, the focus is not on what happens within an AS but rather on how ASs are connected and on the interactions across levels. This is explained in Sect. 3.4.2.

Figure 3.5 shows the major elements of social interaction contexts as represented in the MTF. The diagram uses UML (Unified Modelling language), a formalized language for object oriented design. Not all classes of the MTF are shown. The highest class in which all other classes including ASs are embedded is the '**Water System**'. To simplify the representation this class is only indicated in the figure without proper UML notation. The '**Water System**' comprises all environmental and human components and includes also the factors defining the context in which the governance system is embedded. As already noted an '**Action Situation**' constitutes the core building block of governance processes by representing a structured social interaction context that leads to specific outcomes. Such outcomes can be '**Institutions**' or '**Knowledge**' or '**Operational Outcomes**' which can for their part influence other ASs. '**Institutions**' may be formal or informal such as regulatory frameworks and societal norms. It is unlikely though that a societal norm can be meaningfully represented as an outcome of an AS. For example, the emergence of a societal norm is far more difficult to capture and to attribute to a social interaction context than the passing of a formal regulation. However, ASs may be influenced by societal norms, and a voluntary agreement of how a societal norm is interpreted and applied to a specific water-related issue could well be an outcome of an AS. '**Operational Outcomes**' refers to the results of direct physical interventions such as infrastructure for flood management, irrigation or a drop in the groundwater table. Another group of operational outcomes refers to changes in societal characteristics such as increased societal awareness for environmental problems.

'**Actors**' are individuals or collectives (e.g. governments, organized interest groups, companies) who take part in 'action situations' in which they hold certain 'roles' which entitle them to perform certain actions. '**Roles**' are based on a shared understanding of their meaning and function. A 'role' is held by an 'actor' during an 'action situation'. Roles belong thus to the relationship 'actor'—'action situation' and not to the 'actor' in general. Roles such as leadership may also emerge in the context of interaction. By performing certain actions, an actor may shape and

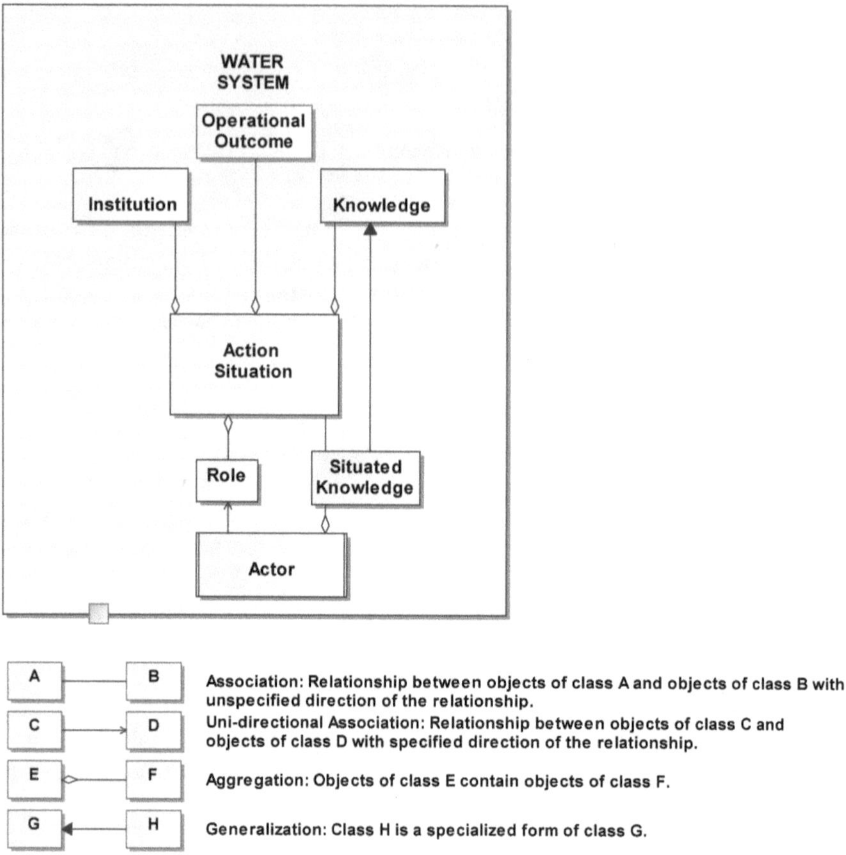

Fig. 3.5 Elementary building blocks of a social interaction context as represented in the MTF in UML notation (Unified Modelling Language). UML is a general purpose modelling language which allows representing object oriented hierarchies (www.uml.org). Each *box* represents a class. Classes may have different kinds of relationships. Situated knowledge is for example a special kind of knowledge. Actors have situated knowledge which they activate in the context of an AS. General knowledge is available at the level of an AS to all actors. Not all classes of the MTF are shown. The highest class in which all other classes are embedded is the 'Water System'. To simplify the representation this class is only indicated without proper UML notation

adopt a certain role. In contrast, in the IAD, the whole setting of an AS is defined by different kinds of rules. Position rules assign positions (which are comparable to roles) to actors. Roles as used in the MTF resonate with Ervin Goffman's interactionist approach (Goffman 1959; Smith 1999).

'**Knowledge**' refers to meaningful information and experience which can be externalized and made publicly accessible. '**Knowledge**' may be available in an 'action situation' to either all or selected actors. It is assumed that actors activate '**Situated Knowledge**' within the context of a specific AS. '**Situated Knowledge**' refers to a situation-specific interpretation of available knowledge. Situated knowledge captures the importance of framing and reframing and the embeddedness of knowledge in a social context. Such an assumption deviates considerably from the rational choice approach that underlies the IAD framework. A more in depth discussion about actors and theories on individual and collective behaviour is provided in Chap. 4.

This conceptualization of social interaction contexts can now be used to depict policy trajectories and analyse the characteristics of governance processes as outlined in subsequent sections.

3.4.2 Representing Multi-level Policy Trajectories

Governance processes can be represented as sequences of connected ASs. Figure 3.6 illustrates such a representation of a multi-level governance process. ASs

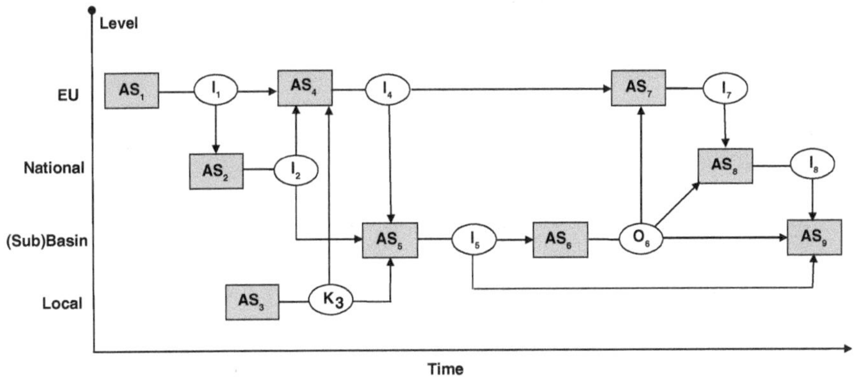

Fig. 3.6 Schematic representation of a chronological mapping of a multilevel governance process in the MTF. *AS* refers to action situation, *I* to institution, *K* to knowledge and *O* to operational outcome. The numbering of the ASs reflects their succession in time. The representation is inspired by the implementation of European water policy. It does not aim though at representing a specific process at specific locations. AS_1 could represent the passing of legislation (I_1) at the European level. AS_2 represents translation into national law. AS_3 stands for pilot studies at the local level to test new methods (e.g. monitoring, stakeholder involvement). Such pilot studies produce knowledge that feeds into AS_4, the development of guidance documents (I_4) for the implementation process at the EU level and directly into AS_5, the development of basin management plans (I_5) at the level of national (sub)basins. AS_6 represents the implementation of river basin management plans which leads to operational outcomes (O_6) such as improved water quality. Effectiveness of implementation and compliance with regulations is monitored at the European level (AS_7). Lack of compliance could lead to some formal request and/or sanctions (I_7). Experience from the first round of implementation of basins plans, in combination with the formal request from the EU level, may lead, to an amendment of national regulations (AS_8). Obviously, AS_8 would also be influenced by the present national regulation (I_2) and presumably by the river basin management plans (I_5). To avoid excessive complexity, these links have not been included. AS_9 would then represent a new round of development of river basin management plans

and their connections are represented as a function of time and governance level. The representation assumes that each AS can be characterized by a dominant administrative or spatial level. This implies that social interactions occur in a spatial context. The level of an AS may be determined by the dominant administrative level in a formal policy process (e.g. development of a drought policy at national level). A community-led initiative for restoring a wetland would be assigned to the local level.

Such a representation of policy trajectories allows the analysis of cross-level interactions and vertical integration. The network of influence can be much more complex than depicted in the highly simplified scheme in Fig. 3.6 and can involve multiple pathways and multiple processes. ASs may have more than one outcome and may be influenced by more than one input and thus by different ASs.

The process represented in Fig. 3.6 is a chronological mapping which follows the development of a process. The schematic representation was inspired by the implementation of European water policy to render the example less abstract. But it does not aim at representing a specific process at specific locations. Some may argue that such a representation is far too simple and mechanistic and does not capture the politics and multitude of influences in real world policy processes. Indeed, even in a quite well structured and reasonably transparent political process such as the implementation of European policy, influences on the process across levels are far more complex and presumably less well organized than can be captured in a graphical representation. However, I argue that such a structured analytical approach is essential for a meaningful and, in particular, comparative analysis and to avoid being overwhelmed by the complexity of real world policy processes.

To better understand the overall logic of a governance process, it may also be of interest to introduce mapping which represents a logical progression of ASs. Given the guiding assumption that governance is a social function and serves a certain purpose, one should expect to detect some logic in problem solving within specific policy processes. In political science, the policy cycle has a long tradition of being used as a tool to analyse the development of public policy. The policy cycle makes a distinction between different phases of a policy process which comprise problem definition, agenda setting, policy formulation, implementation and evaluation. The policy cycle has been used as a normative approach (in particular Bridgman and Davis (1998) suggesting that "good" processes result in good policies; a "good" process is characterized by a logical progression of phases as depicted in the policy cycle model). This approach is highly controversial. In reality, the different phases cannot often be clearly distinguished; they overlap or run in parallel. The policy cycle has also been used as an analytical device, as a heuristic rather than as a normative model. I adhere to this approach when introducing different phases to structure complex governance processes.

The MTF introduces stylized policy processes which comprise, as set of characteristic phases, Strategic goal setting, Assess current state, Policy formation, Developing operational goals, Developing measures, Implementation, and Monitoring. Figure 3.7 is a representation of the process depicted in Fig. 3.6 but

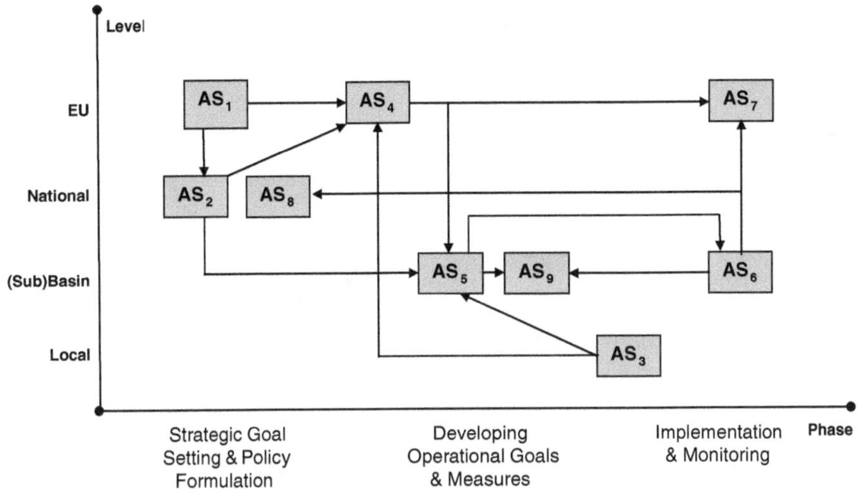

Fig. 3.7 Schematic representation of the multilevel governance process of Fig. 3.6 mapped to different phases of the policy cycle. The *labels* for institutions, knowledge and operational outcomes, respectively, are omitted to simplify the representation

mapped to the different phases of the policy cycle. Each AS was either associated with one phase or between the different phases. This representation shows the feedbacks and the iterative nature of policy processes.

Furthermore, the MTF makes an analytical distinction between formal policy and management processes on the one hand and learning processes that are at least partly informal on the other (Pahl-Wostl et al. 2010). This is schematically depicted in Fig. 3.8. The policy cycle includes the development of a policy framework which may form the basis for new formal institutions. Policy implementation includes not only the coming into force of a legal framework but also operational implementation and the design of management actions. These refer to the development and implementation of operational measures such as direct technical intervention in the environment or the implementation of pricing policies to achieve a specific goal.

The analytical distinction between learning and policy cycle does not imply that learning is impossible within a formal policy process. However, it is assumed that within the formal policy cycle, learning refers mainly to the incremental improvement of established routines which may also be called singe loop learning (Pahl-Wostl 2009). Informal settings are required to stimulate higher levels of learning. Adaptive learning refers mainly to the operational implementation and monitoring phases, to single or in part double-loop learning (the beginning of reframing and of calling into question established institutions) that does not yet change the reigning paradigm and the whole structural context (e.g., institutions. technical infrastructure) settings. Transformative learning implies a change in strategic goals and policy formulation, which means triple-loop learning and a change in the reigning paradigm, in regulatory frameworks, and prevailing codified

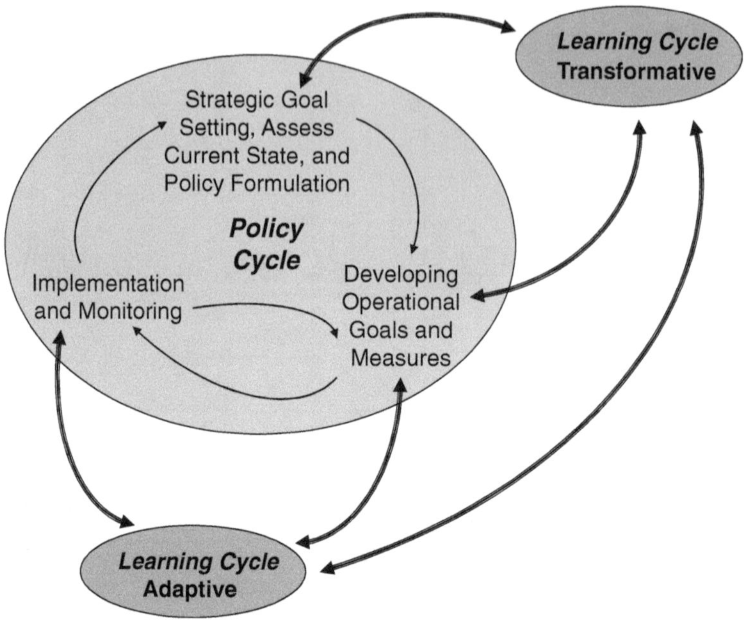

Fig. 3.8 Schematic representation of links between formal policy and informal learning cycles. The policy cycle is depicted in three aggregated phases rather than the seven individual phases distinguished by the MTF: Strategic goal setting—assess current state—policy formation—developing operational goals—developing measures—implementation—monitoring. *Bold arrows* denote links between learning and policy cycles (Reproduced from Fig. 1 in Pahl-Wostl et al. 2013a)

practices (e.g., technical design principles). In the long term, such change has an influence on operational goals and measures, and on implementation and monitoring.

The policy cycle may not follow a sequential logic and a progression of phases but can be much more iterative. Nevertheless, one may obtain some insights into policy failures if certain phases are disconnected or missing, in particular since the cycle as used in the MTF comprises policy and management. Management without measurable goals or an evaluation if goals are achieved is rather meaningless. However, the phases may overlap, run in parallel, or not follow a strictly sequential order.

A multi-level process can now be represented schematically (Fig. 3.9) as a sequence of ASs which are mapped to phases of a stylized process. Each circle denotes an AS. Black circles refer to formal policy and management and light-shaded circles to informal learning processes. The arrows denote influence between ASs. The temporal sequence can be inferred from the pathways of influence as denoted by the direction of the arrows. An AS can only influence another AS if it occurs before the AS to be influenced.

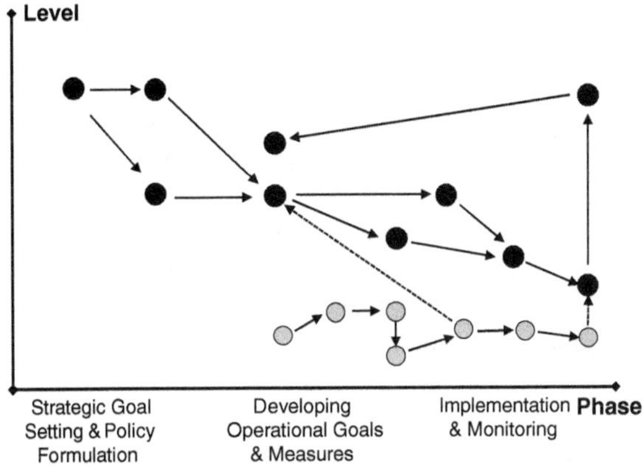

Fig. 3.9 Schematic representation of interconnected formal policy (*black circles*) and informal social learning (*light shaded circles*) processes as a function of stylized process phases and levels. Each *circle* denotes an AS. The *arrows* denote pathways of influence. Since real processes rarely follow the linear logic of a stylized process, links may also appear in the reverse direction

The at least partial informality of learning processes is assumed to be essential for supporting higher levels of learning. However, without connections to formal policy and management processes, informal learning processes will most likely have little effect on the water governance regime (Pahl-Wostl 2009; Sendzimir et al. 2010). Hence it is of major interest to analyse the relationship between formal policy/management and learning processes to understand which connections may be effective. Effective implies that learning has an enduring influence on formal policy without being constrained. Learning cycles may be connected to the formal policy cycle/management by formal connection (e.g., formal mandate, institutions regulating learning processes), an overlap in actors or the mutual transfer of knowledge. Higher levels of learning may be supported by an openness of learning cycles in terms of who participates and the kind of issues to be addressed. Learning cycles may be limited to double-loop learning and be thus mainly concerned with reframing and searching for innovation within the constraints provided by the structural context. Or, they may explicitly address issues related to the transformation of the structural context (e.g., integration of sectors, changes in water rights, formal rights for stakeholder participation) and thus lead to triple-loop learning deemed to be essential for an effective implementation of innovative forms of water management and governance.

3.4.3 Functional Analysis of Governance Systems

Another important characteristic of governance systems is their functional organization. A functional analysis of water governance systems is based on what may be called governance sub-functions. Each sub-function can be represented by one or more ASs that capture the characteristics of how a sub-function is realized and how sub-functions are connected in a certain governance system. Such analysis could be seen as a bridge between configuration-based and process analysis (cf. Fig. 3.2).

A similar kind of approach was suggested by McGinnis (2011) in his analysis of adjacent action situations in polycentric resource governance systems. Given that he used the IAD, the focus was predominantly on rules as outcomes linking ASs, and on developing a systematic approach to the construction of more elaborate models of complex policy networks in which overlapping sets of actors have the ability to influence the rules under which their strategic interactions take place. He identified what he calls core governance tasks in public policy: production, provision, consumption, financing, coordination, dispute-resolution, and rule-making. The core governance tasks identified by McGinnis correspond, in principle, to what I call governance sub-functions. One difference lies in the fact that what I refer to as sub-functions does not only refer to public policy. The identification of such sub-functions is based on assumptions about the requirements needed to fulfil the governance function in complex and uncertain environments. The distinction made here is based on the approach developed by Pahl-Wostl et al. (2013b) for the functional analysis of policy trajectories to compare processes in global water governance. This framework includes the following as key elements of governance sub-functions: policy framing, conflict resolution, rulemaking, knowledge generation and resource mobilization. One can note the overlap with the governance tasks identified by McGinnis (2011) with respect to resource mobilization, conflict resolution, and rulemaking. Since water governance is not only about production and consumption, these tasks are not included in the framework for the functional analysis used here. However, compared to the framework presented in Pahl-Wostl et al. (2013b), one additional sub-function was added: monitoring and evaluation. The resulting framework that identifies the key sub-functions and critical properties of these sub-functions is shown in Fig. 3.10. The figure presents the interdependent

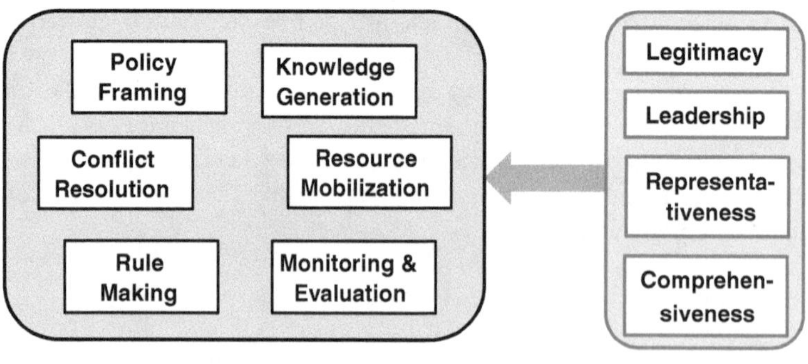

Fig. 3.10 Governance sub-functions (*left box*) and critical properties characterizing their performance (*right box*). *Note* The figure presents sub-functions assumed to be necessary for dealing with a complex policy problem, without assuming a sequential logic among them. Hence, arrows linking sub-functions were omitted. Interdependencies between sub-functions may be complex and recursive

sub-functions which are assumed to be necessary for dealing with a complex policy problem, without assuming a sequential logic among them. Hence, arrows linking sub-functions were omitted. Interdependencies between sub-functions may be complex and recursive.

Knowledge Generation: Knowledge Generation may encompass the collection of new information and/or the integration of available, fragmented evidence from different sources. A key step in this process is the translation of that information into validated and legitimized ‘knowledge’ that is sufficiently stable to have shared meaning for the various actors involved (albeit in different ways and to varying extents) (Jasanoff 2004).

Policy Framing: The framing of the problem is of key importance in shaping a policy, particularly in its initial stages. Framing involves identifying the nature of a problem, potential causes, and solutions. Framing identifies not simply ends but also the range of plausible means to an end, thereby focusing, but also limiting the imagination of actors and the feasibility of various forms of action. For this reason, frames that take a pluralistic approach, allowing for different world views, are important for dealing with complex issues in conflictive settings (Conca 2006; Pahl-Wostl et al. 2007a).

Rule-making: Rule-making occurs at many stages, but in particular as actors move from deliberation and learning towards more formal commitments (Young 1998; Pahl-Wostl et al. 2007a; Pahl-Wostl 2009). Rule-making is an essential ingredient for enabling informal social learning processes to structure interactions and support progress towards tangible outcomes (Mostert et al. 2007; Pahl-Wostl et al. 2007a).

Resource mobilization: Governance requires resources, including funding, expertise, and the political resources that generate support for policy implementation at different levels. Difficulties in mobilizing resources may come from several sources: inability of actors to agree or prioritize actions; classic barriers to collective action (Ostrom 1990) such as uncertainty, mistrust, transaction costs, and coordination barriers; or a lack of leadership or stewardship.

Conflict resolution: Conflicts are endemic to governance processes and to water governance in particular given the resource's multiple uses, irreplaceability, unpredictability, and strategic value as a productive resource, a constituent of critical ecosystems, and an anchor of local livelihoods and cultures (Conca 2006). Unresolved conflicts may jeopardize the implementation of a policy or marginalize certain actors, who may in turn seek to block action through coercive, extra-institutional, or even violent means (Gray 2004).

Monitoring and evaluation: Monitoring and evaluation are essential prerequisites for learning for any adaptive governance and management approach. This implies setting tangible short-term targets for assessing success or failure and implementing transparent processes with respect to who decides on which kind of evidence is required for the adjustment of policies and/or measures.

In order to support effective governance, it is assumed that governance sub-functions require a particular set of properties as indicated in Fig. 3.9.

Legitimacy: Legitimacy refers to the validity and broad-based acceptance of the authority of an actor or event, making it possible for those actors and events to play an influential role in the overall process. Legitimacy may derive from the way authority was conferred on an event or group (for example, through a democratic, open and inclusive process). Legitimacy may also be gained by generating outcomes that are endorsed by many participants of the overall process (so-called 'performance legitimacy'). A lack of legitimacy may lead to opposition, resistance, or loss of commitment. Legitimacy is frequently contested in complex, multi-level governance settings given the involvement of a multitude of actors and their often poorly defined roles.

Representativeness: Representativeness refers to the adequate involvement of all stakeholder groups. The active involvement of not only powerful actors but also affected stakeholder groups has proven crucial for ensuring that a process is perceived as legitimate and for reducing the likelihood of the process being jeopardized by narrow interests (Plummer et al. 2013). Broader participation may also enhance effectiveness through learning mechanisms or the generation of new information.

Leadership: Complex governance processes are characterized by self-organization and emergence. However, self-organization without leadership may fail to produce tangible outcomes. This may be leadership of a governmental body based on formal regulations. It may also be the emergent leadership that develops from an actor's influential role in a network (Pahl-Wostl 2009). Two recent, comprehensive studies confirmed the importance of this kind of emergent, forward-looking leadership for sustainable resource governance at local and regional levels (Gutiérrez et al. 2011; Kenward et al. 2011).

Comprehensiveness: As highlighted previously, the issues of water governance must be addressed from an integrated perspective. Problems often arise due to the involvement of different policy fields and a lack of coordination across them. Again, the water arena has (partially in theory and partially in practice) been at the forefront of developing and "testing" concepts of integrated resource management. Some have criticized such approaches as unrealistic abstractions requiring rigid bureaucratic structures for their implementation (Biswas 2004). While such critiques may raise important concerns about the need for flexible, adaptive policy systems under conditions of uncertainty, they may also be rooted in a technocratic paradigm that fails to recognize the need for the integrated governance of interdependent policy fields (Pahl-Wostl et al. 2011). Comprehensive-ness is more likely to be achieved in open and flexible governance settings (Galaz et al. 2012; Pahl-Wostl 2009).

These functional categorizations can be used to analyse and compare the realization or absence of governance sub-functions, their interdependence, and their properties in different governance systems. Such analyses can form the basis for an improved understanding of the relationship between functional organization and functional performance of governance systems.

3.5 Concluding Remarks

After laying out the foundations of a conceptual and methodological approach to analysing the structure, dynamics and performance of governance systems, the subsequent chapters elaborate in more depth on important dimensions that were introduced: institutions and actors, governance modes and multi-level interactions.

References

Argyris, C., & Schön, D. (1978). *Organizational learning: A theory of action perspective*. Reading, Mass., USA: Addison Wesley.

Benz, A. (2004). Governance—Modebegriff oder nützliches sozialwissenschaftliches Konzept? In A. Benz (Ed.), *Governance—Regieren in komplexen Regelsystemen: Eine Einführung* (pp. 11–28). Wiesbaden, Germany: Verlag für Sozialwissenschaften.

Berkes, F., Colding, J., & Folke, C. (Eds.). (2003). *Navigating social-ecological systems. Building resilience for complexity and change*. Cambridge: Cambridge University Press.

Berkes, F., & Folke, C. (2002). Back to the future: Ecosystem dynamics and local knowledge. In L. H. Gunderson & C. S. Holling (Eds.), *Panarchy. Understanding Transformations in Human and Natural Systems* (pp. 121–146). Washington, Covelo, London: Island Press.

Biswas, A. K. (2004). Integrated water resources management: A reassessment. *Water International, 29*(2), 248–256. doi:10.1080/02508060408691775

Borowski, I., Le Bourhis, J., Pahl-Wostl, C., & Barraqué, B. (2008). Spatial misfit in participatory river basin management: Effects on social learning. A comparative analysis of German and French case studies. *Ecology and Society, 13*(1), 7. URL: http://www.ecologyandsociety.org/vol13/iss11/art17/

Bridgman, P., & Davis, G. (1998). *Australian policy handbook*. Sydney: Allen & Unwin.

Casti, J. (1997). *Reality rules I & II. Picturing the world in mathematics: The fundamentals, the frontier*. Chichester: Wiley.

Conca, K. (2006). *Governing water: Contentious transnational politics and global institution building*. Cambridge, MA, USA: MIT Press.

Flood, R., & Romm, N. (1996). *Diversity management: Triple loop learning*. Chichester, UK: Wiley.

Galaz, V., Crona, B., Österblom, H., Olsson, P., & Folke, C. (2012). Polycentric systems and interacting planetary boundaries—Emerging governance of climate change–ocean acidification–marine biodiversity. *Ecological Economics, 81*, 21–32.

Goffman, E. (1959). *The presentation of self in erveryday life*. New York: Anchor.

Gray, B. (2004). Strong opposition: Frame-based resistance to collaboration. *Journal of Community & Applied Social Psychology, 14*, 166–176.

Gutiérrez, N., Hilborn, R., & Defeo, O. (2011). Leadership, social capital and incentives promote successful fisheries. *Nature, 470*, 386–389.

Helmke, G., & Levitsky, S. (2004). Informal institutions and comparative politics. *Perspectives on Politics, 2*, 725–740.

Héritier, A. (2002). New modes of governance in Europe: Policy-making without legislating? In A. Héritier (Ed.), *Common goods: Reinventing European and international governance* (pp. 185–206). Lanham, Md.: Rowman & Littlefield Publishers.

Holton, R. J. (1998). *Globalization and the nation-state*. Basingstoke, UK: Palgrave.

Jasanoff, S. (Ed.). (2004). *States of knowledge: The co-production of science and social order*. London, UK: Routledge.

Kauffman, S. (1995). *At home in the universe: The search for the laws of self-organization and complexity*. New York: Oxford University Press.

Kenward, R. E., Whittingham, M. J., Arampatzis, S., Manos, B. D., Hahn, T., Terry, A., et al. (2011). Identifying governance strategies that effectively support ecosystem services, resource sustainability, and biodiversity. *Proceedings of the National Academy of Sciences, 108*(13), 5308–5312. doi:10.1073/pnas.1007933108

Knieper, C., Holtz, G., Kastens, B., & Pahl-Wostl, C. (2010). Analysing water governance in heterogeneous case studies—Experiences with a database approach. *Environmental Science & Policy, 13*(7), 592–603. doi:10.1016/j.envsci.2010.09.002

Kohler-Koch, B. (1999). The Evolution and Transformation of European Governance. In B. Kohler-Koch & R. Eising (Eds.), *The transformation of governance in the European Union* (pp. 14–35). London, UK: Routledge.

Kooiman, J. (2000). Societal governance: Levels, modes and orders of social-political interaction. In J. Pierre (Ed.), *Debating governance: Authority, steering, and democracy* (pp. 138–164). Oxford UK: Oxford University Press.

Kooiman, J. (2003). *Governing as governance*. London: SAGE.

Krysanova, V., Dickens, C., Timmerman, J., Varela-Ortega, C., Schlüter, M., Roest, K., et al. (2010). Cross-comparison of climate change adaptation strategies across large river basins in Europe, Africa and Asia. *Water Resources Management, 24*(14), 4121–4160. doi:10.1007/s11269-010-9650-8

Leadley, P., Proença, V., Fernández-Manjarrés, J., Pereira, H. M., Alkemade, R., Biggs, R., et al. (2014). Interacting regional-scale regime shifts for biodiversity and ecosystem services. *BioScience*,. doi:10.1093/biosci/biu093

Mayntz, R. (2004a). Governance Theory als fortentwickelte Steuerungstheorie? *MPIfG Working Papers*. Cologne, Germany Max-Planck-Institute for the Study of Societies.

Mayntz, R. (2004b). Mechanisms in the analysis of social macro-phenomena. *Philosophy of the Social Sciences, 34*, 237–259.

McGinnis, M. D. (2011). Networks of adjacent action situations in polycentric governance. *Policy Studies Journal, 39*(1), 51–78. doi:10.1111/j.1541-0072.2010.00396.x

Moss, T. (2003). Solving problems of 'fit' at the expense of problems of 'interplay'? The spatial reorganisation of water management following the EU water framework directive. In H. Breit, A. Engels, T. Moss, & M. Troja (Eds.), *How institutions change. Perspectives on social learning in global and local environmental contexts* (pp. 85–121). Opladen: Leske + Budrich.

Mostert, E., Pahl-Wostl, C., Rees, Y., Searle, B., Tàbara, D., & Tippett, J. (2007). Social Learning in European river-basin management: Barriers and fostering mechanisms from 10 river basins. *Ecology and Society, 12*(1).

Newig, J., Pahl-Wostl, C., & Sigel, K. (2005). The role of public participation in managing uncertainty in the implementation of the water framework directive. *European Environment, 15*(6), 333–343.

North, D. C. (1990). *Institutions, institutional change and economic performance*. Cambridge: Cambridge University Press.

Ostrom, E. (1990). *Governing the commons. The evolution of institutions for collective action* (Political Economy of Institutions and Decisions). Cambridge: Cambridge University Press.

Ostrom, E. (2005). *Understanding institutional diversity*. Princeton: Princeton University Press.

Ostrom, E. (2007). A diagnostic approach for going beyond Panaceas. *Proceedings of the National Academy of Sciences of the United States of America, 104*, 15181–15187.

Pahl-Wostl, C. (1995). *The dynamic nature of ecosystems: Chaos and order entwined*. Chichester: Wiley & Sons.

Pahl-Wostl, C. (2007a). The implications of complexity for integrated resources management. *Environmental Modelling & Software, 22*, 561–569.

Pahl-Wostl, C. (2007b). Requirements for adaptive water management. In C. Pahl-Wostl, P. Kabat, & J. Möltgen (Eds.), *Adaptive and integrated water management. Coping with complexity and uncertainty* (pp. 1–22). Heidelberg: Springer.

Pahl-Wostl, C. (2009). A conceptual framework for analysing adaptive capacity and multi-level learning processes in resource governance regimes. *Global Environmental Change, 19*, 354–365.

Pahl-Wostl, C., Becker, G., Sendzimir, J., & Knieper, C. (2013a). How multilevel societal learning processes facilitate transformative change: A comparative case study analysis on flood management. *Ecology and Society, 18*(4), 58. doi:10.5751/ES-05779-180458

Pahl-Wostl, C., Conca, K., Kramer, A., Maestu, J., & Schmidt, F. (2013b). Missing links in global water governance: A processes-oriented analysis. *Ecology and Society, 18*(2), 33. doi:10.5751/ES-05554-180233

Pahl-Wostl, C., Craps, M., Dewulf, A., Mostert, E., Tabara, D., & Taillieu., T. (2007a). Social learning and water resources management. *Ecology and Society, 12*(2), 5. URL: http://www.ecologyandsociety.org/vol12/iss12/art15/

Pahl-Wostl, C., Holtz, G., Kastens, B., & Knieper, C. (2010). Analyzing complex water governance regimes: The management and transition framework. *Environmental Science & Policy, 13*(7), 571–581. doi:10.1016/j.envsci.2010.08.006

Pahl-Wostl, C., Jeffrey, P., Isendahl, N., & Brugnach, M. (2011). Maturing the new water management paradigm: Progressing from aspiration to practice. *Water Resources Management, 25*(3), 837–856. doi:10.1007/s11269-010-9729-2

Pahl-Wostl, C., & Knieper, C. (2014). The capacity of water governance to deal with the climate change adaptation challenge: Using fuzzy set Qualitative Comparative Analysis to distinguish between polycentric, fragmented and centralized regimes. *Global Environmental Change, 29*, 139–154.

Pahl-Wostl, C., Lebel, L., Knieper, C., & Nikitina, E. (2012). From applying panaceas to mastering complexity: Toward adaptive water governance in river basins. *Environmental Science & Policy, 23*, 24–34.

Pahl-Wostl, C., Möltgen, J., Ebenhöh, E., & Holtz, G. (2007b). The NeWater management and transition framework—State and development process. In C. Pahl-Wostl, P. Kabat, & J. Möltgen (Eds.), *Adaptive and integrated water management. Coping with complexity and uncertainty* (pp. 75–96). Heidelberg: Springer.

Pahl-Wostl, C., Sendzimir, J., Jeffrey, P., Aerts, J., Berkamp, G., & Cross, K. (2007c). Managing change toward adaptive water management through social learning. *Ecology and Society, 12* (2), 30. URL: http://www.ecologyandsociety.org/vol12/iss32/art30/

Plummer, R., Armitage, D. R., & de Loe, R. C. (2013). Adaptive Comanagement and Its Relationship to Environmental Governance. *Ecology and Society, 18*(1), doi:10.5751/ES-05383-180121

Ragin, C. (1987). *The comparative method. Moving beyond qualitative and quantitative strategies.* Berkeley, Los Angeles and London: University of California Press.

Rogers, P., & Hall, A. W. (2003). Effective water governance. *TEC Background Papers Nr. 7*: Global Water Partnership Technical Committee (TEC).

Rosenau, J. N. (1995). Governance in the 21st century. *Global Governance, 1*, 13–43.

Scheffer, M. (2009). *Critical transitions in nature and society*. Princeton and Oxford: Princeton University Press.

Schuppert, G. F. (2006). Governance im Spiegel der Wissenschaftsdisziplinen. In G. F. Schuppert (Ed.), *Governance Forschung: Vergewisserung über Stand und Entwicklungslinien* (pp. 371–469). Baden-Baden, Germany: Nomos Verlag.

Scott, R. W. (2008). *Institutions and organizations: Ideas and interests* (3rd edn.). Thousand Oaks: Sage Publications.

Sendzimir, J., Flachner, Z., Pahl-Wostl, C., & Knieper, C. (2010). Stalled regime transition in the upper Tisza River Basin: The dynamics of linked action situations. *Environmental Science & Policy, 13*(7), 604–619.

Smith, G. (Ed.). (1999). *Goffman and social organization: Studies of a sociological legacy*. New York: Routledge.

Thompson, G., Frances, J., Levacic, R., & Mitchell, J. (Eds.). (1991). *Markets, hierarchies and networks: The co-ordination of social life*. London, UK: Sage Publishers.

Treib, O., Bähr, H., & Falkner, G. (2007). Modes of governance: Towards a conceptual clarification. *Journal of European Public Policy, 14*(1), 1–20. doi:10.1080/13501760601071406

UNESCAP. (2009). *What is good governance?* Bangkok, Thailand: United Nations Economic and Social Commission for Asia and the Pacific.

Wilensky, U. (1999). *NetLogo traffic basic model.* http://ccl.northwestern.edu/netlogo/models/TrafficBasic. Evanston, IL: Center for Connected Learning and Computer-Based Modeling, Northwestern Institute on Complex Systems, Northwestern University.

Young, O. R. (1998). *Creating regimes: Arctic accords and international governance.* Ithaca, NY, USA: Cornell University Press.

Young, O. R. (2002). Usable knowledge: Design principles and institutional diagnostics. In O. R. Young (Ed.), *The institutional dimensions of environmental change: Fit, interplay and scale* (pp. 165–190). Cambridge, MA: MIT Press.

Young, O. R. (2013). Sugaring off: Enduring insights from long-term research on environmental governance. *International Environmental Agreements: Politics, Law and Economics, 13*(1), 87–105. doi:10.1007/s10784-012-9204-z

Chapter 4
The Role of Institutions, Actors and Social Networks in Societal Change

In Chap. 3, I defined adaptive capacity as the ability of a governance system to alter processes and to adapt its structural elements in response to current or anticipated changes in the social or natural environment. Transformative capacity was defined as the ability of a governance system to first adapt and if required transform structural elements in response to current or anticipated changes in the social or natural environment. The selected phrasing does not imply that a system as a whole is assumed to have agency. It is the collective behaviour of actors in the system which leads to change or which may also prevent change. Understanding what determines adaptive and transformative capacity requires an understanding of the relationship and the interplay between structure and agency. This chapter describes the role of institutions and actors in processes of change and their role with respect to the adaptive and transformative capacity of governance systems.

4.1 Determinants of Institutional Change

Institutions have been a major topic of research in various social science disciplines—in particular economics, sociology and political science—with diverse and often divergent interpretations. Despite significant research efforts, the dynamics of institutional change is not yet well understood. The conclusion of Young (2008) "*Knowledge regarding the nature of change in the institutional dimensions of socio-ecological systems remains relatively underdeveloped*" (ibid, p. 140), is still valid. A comprehensive understanding of the complex nature and central role of institutions in governance systems, and in particular of institutional change requires crossing disciplinary boundaries. Unfortunately such scholarship is the exception rather than the rule. One pioneer in institutional analysis and environmental governance, Elinor Ostrom, was both a political scientist and an institutional economist. Her work paved the way for recognizing the importance of self-organization, cooperation and collective action with respect to common pool resources. Her focus though was on the requirements for and the results of self-organization rather than the process of change itself. But important insights have arisen from her work that are of relevance for understanding learning and change from a rational choice perspective.

C. Pahl-Wostl, *Water Governance in the Face of Global Change*,
Water Governance—Concepts, Methods, and Practice,
DOI 10.1007/978-3-319-21855-7_4

4.1.1 Institutional Change in Governing the Commons—Rational Choice Approach

Elinor Ostrom's work is based on a rational choice approach albeit not adhering to the narrow interpretation of neo-classical economics. Ostrom (2005, p. 3) referred to institutions as "*the prescriptions that humans use to organise all forms of repetitive and structured interactions including those within families, neighbourhoods, markets, firms, sports leagues, churches, private associations, and governments at all scales. Individuals interacting within rule-structured situations face choices regarding the actions and strategies they take, leading to consequences for themselves and for others*". This broad definition of institutions reflects the complexity and diversity of the role of institutions in any kind of social interaction. It is not, however, highly conducive to analysis. Crawford and Ostrom (1995) introduced in their grammar of institutions a formalized syntax to classify the wide range of rule-based interactions. The syntax is based on a definition of rules as "*shared understandings by participants about en-forced prescriptions concerning what actions (or outcomes) are required, prohibited, or permitted.*" (Ostrom 2005, p. 18). Social interactions are conceptualized as games where alternative actions can be evaluated by their expected pay-offs. Being the rules of the games, institutions are essential to developing such expectations. Taken together, the different types of rules fully describe a social interaction context in an action situation as conceptualized in the IAD (Institutional Analysis and Development) framework introduced in Chap. 3 (Ostrom 2005): e.g., position rules specify a set of positions actors may hold and how many actors hold one; boundary rules specify how actors are chosen to enter or leave these positions; choice rules specify which actions are assigned to an actor in a position; pay-off rules specify how pay-offs are to be distributed to actors in certain positions.

The identification of these rule types was based on empirical evidence from a large number of case studies on local resource governance systems. Furthermore, Ostrom derived a set of design principles that could be detected in common-pool resource governance systems sustained over longer periods of time. They encompass the presence of clearly defined boundaries, proportional equivalence between benefits and costs, collective-choice arrangements, monitoring, graduated sanctions, conflict-resolution mechanisms, minimal recognition of the right to organize and nested enterprises (Ostrom 2005, p. 259).

One of the design principles refers to the importance of external sanctions to enforce compliance with rules. External enforcement is to some extent at odds with self-organization. Highlighting the importance of sanctions does not imply though that Ostrom did not take into consideration different mechanisms of compliance. The grammar of institutions (Crawford and Ostrom 1995; Ostrom 2005) makes the distinction between formal sanctions on the one hand and moral and emotion-based factors on the other. Institutions are classified according to strategies, norms and rules that are distinguished by the kind of sanctioning associated with them. Rules have material sanctions, norms have sanctions based on moral values or social pressures, and strategies have no sanctions associated with them. The grammar makes no distinction between formal (i.e. legal prescriptions) and informal institutions. The distinction between rules and norms is based on the presence or absence of clearly defined sanctions. According to this syntax a law would be called a norm if it does not specify a sanction. And it does not make a distinction between rules agreed to by local communities and rules prescribed by formal law as long as sanctions are clearly specified. Particularly, at the local level such an approach may make a great deal of sense since, in the end, only rules in use matter for the governance of a resource. In this book I maintain the distinction between formal (in the sense of jurisdiction) and informal institutions since it is important to analyse the role of government and the effectiveness of governmental bureaucracies and legal procedures as well as the relationship between formal and informal governance processes to understand institutional change and the role of various change agents.

The distinction between different pathways of rule compliance and the role of internal (to the individual) and external factors is increasingly recognised as being essential to understanding human behaviour in general and collective action in particular (Mantzavinos et al. 2004; Ostrom 2010). Ostrom's work provided evidence that trust, reciprocity and reputation are essential to support cooperation among resource users. Actors seem to develop and strengthen or weaken internal norms of trust and fairness based on experience in social interactions. This implies that cooperation can be stabilized or destabilized in actor groups. The emergence of and compliance with a shared norm of cooperation is dependent upon the learning of individuals in repeated social interactions. In experiments with public goods, players introduced punishment for non-cooperative behaviour to prevent the detrimental effect of free-riding by sanction (Fehr and Gächter 2000, 2002). Learning behaviour is assumed to be influenced by two kinds of variables—micro-situational which describe the decision context and broader contextual variables. The framing of the decision context (micro-situational variables) and contextual variables were found to affect levels of trust and cooperation in experimental economics games and case studies on local resource governance (Ostrom 2005, 2010; Ebenhöh and Pahl-Wostl 2008).

Regarding individual and collective learning Ostrom was inspired by the work of Douglas North on the role of mental models (Ostrom 2005, pp. 106–109). Already in his early work North pointed out the importance of mental models and beliefs and criticized standard neo-classical assumptions for their static and unrealistic approaches (North 1990, 1994; Denzau and North 1994). He emphasizes even more strongly in his later work that institutional performance and in particular institutional change cannot be understood without the input from cognitive sciences (North 2005; Mantzavinos et al. 2004). As pointed out in Mantzavinos et al. (2004, p. 76): "*A mental model can best be understood as the final prediction that the mind makes or expectation that it has regarding the environment before getting feedback from it. .....When environmental feedback confirms the same mental model many times, it becomes stabilized, in a way. We call this relatively crystallized mental model a "belief"; and we call the interconnection of beliefs (which can be either consistent or inconsistent) a "belief system*"". Collective beliefs are referred to as 'shared mental models'. Institutional change and the formation of institutions proceed thus in a kind of feedback loop: Interaction with reality generates belief systems that lead to design or emergence of institutions. Institutions shape policies and behavioural patterns which lead to some outcomes such as reduced water consumption and thus altered realities. The outcomes are evaluated and may or may not reinforce existing beliefs and satisfaction with prevailing institutional frameworks.

This description of a learning sequence incorporates the essential ingredients of rational choice despite considerable deviations from the neo-classical model. However, moving towards a theory of institutional change requires an in depth understanding of the underlying processes. North and colleagues identified what they considered important elements and pointed out as well major knowledge gaps: the interaction and mutual dependencies between individual and social learning, between internal (shared mental models) and external (shared behavioural regularities) perspectives, between formal and informal institutions. To deal with these knowledge gaps implies moving beyond cognition and individual learning. One might conceive of a hierarchy of beliefs that change according to time scale. Ostrom (2005) introduced a hierarchy of rules in her work: operational rules, collective choice and constitutional rules. She assumed that efforts and costs associated with change increase with the more hierarchical the system which implies the rate of change is increasing from operational rules to constitutional rules. However, it remains an open question as to how multi-level institutional change of this nature takes place. As already pointed out, Ostrom had a stronger focus on the outcomes of processes of self-organization rather than on the dynamics of the processes themselves.

Furthermore, Elinor Ostrom's work had a strong focus on the local level. It is still not clear to which extent her insights and approaches can be applied at higher governance levels (Araral 2014). Some of the design principles may not be implementable at higher levels and further design principles might be needed to capture the complex multi-level dynamics characterizing contemporary environmental governance. In this context one needs also to consider if Ostrom's design principles should be understood as necessary conditions that prescribe a deterministic blueprint for effective resource governance systems. In their review of the extensive literature on application and testing of the design principles Cox et al. (2010) argue in favour of a probabilistic rather than deterministic approach in the application of these principles. This means that they can inform and guide a structured approach to institutional diagnosis. However, one must keep in mind that the relative importance of such design principles is influenced by contextual conditions and the resource problem under consideration. When understood in this way, the principles are, in my view, a useful and flexible tool for institutional diagnosis.

Huntjens et al. (2012) used Ostrom's institutional design principles as point of departure for analysing governance of adaptation to climate change in river basins. Based on our analyses we concluded that there is a need to distinguish between design principles for sustaining enduring common pool resource systems on a local scale and design principles for adaptation to climate change in complex, cross-boundary and large-scale resource systems where levels of complexity and uncertainty related to the policy problem for larger jurisdictional and geographical scales are high. Climate change adaptation is a dynamic and complex process which requires responsiveness and high capacities for learning. In these situations more attention is needed on institutions that facilitate systemic learning processes. Design principles are required that address the need for robust and flexible processes and for policy learning through the exploration of uncertainties, the deliberation of alternatives and the reframing of problems and solutions. Systemic learning also needs to address the problem of governance and management systems that are often shaped by a kind of systemic logic, a paradigm. To develop an improved understanding how such systemic path-dependence and inertia can be overcome a broader understanding of institutions as provided by the categorization of institutions introduced by Scott (2008) is needed (Pahl-Wostl 2009).

4.1.2 A Broader Understanding of Institutional Change

Richard Scott, an organizational sociologist, introduced a broad definition of institutions to capture essential assumptions within the diverse approaches pursued by scholars in institutional analysis and to support integration rather than controversial discussions: "*Institutions are comprised of regulative, normative and cultural-cognitive elements that, together with associated activities and resources, provide stability and meaning to social life*" (Scott 2008, p. 48). This definition encompasses not only institutions as symbolic elements (i.e. rules, norms, cultural-cognitive beliefs) but highlights as well the role of social activities and material resources which are required to give symbolic elements meaning. Scott makes a distinction between three kinds of institutions (Scott 2008): regulative (what is formally allowed and what is not allowed), normative (what is appropriate and what is not appropriate as judged by societal standards), and cultural-cognitive (what is conceivable and what is inconceivable).

Regulative institutions normally have formal legal structures, regulatory frameworks, and formalized professional rules of good practice typically codified in professional handbooks. Such institutions are products of purposeful design. They follow an instrumental logic. Compliance is based on expedience and an assessment of sanctions related to rule violation. Institutionalization, processes that stabilize and sustain institutions, is driven by increasing returns, in other words by processes of positive feedback. However, the introduction of new regulative institutions is associated with high transaction costs. Hence a broader interpretation of existing institutions will most likely be the first approach to widening the scope of existing regulatory frameworks rather than introducing new institutions [e.g., analysis of institutional change in EU budgetary politics by Lindner (2003)].

Normative institutions correspond to informal societal norms, with shared but non-codified rules of good practice. Normative institutions reflect societal value structures. They determine what is considered to be appropriate in a certain societal context. Compliance is based on the desire to meet social obligations, to fulfil expectations of stereotypical roles and repertoires of context-specific actions. Contrary to regulative institutions, normative institutions emerge from human interaction, and change is not based on negotiations and formal agreements but is more gradual. Institutionalization emerges from increasing commitments.

Cultural-cognitive institutions correspond to paradigms, in other words, dominant world-views that strongly influence meaning, understanding and perception of reality and of problem situations, how boundaries are delineated, and how the space for identifying problems and developing solutions is determined (Pahl-Wostl et al. 2007). Cultural-cognitive institutions reflect a shared understanding with involved actors and instil a sense of certainty and confidence among them. Similar to normative institutions, change is not negotiated but enacted in shared practices. Institutionalization is supported by increased objectification and shared beliefs. Paradigms in water management are a typical example. The command and control approach has prevailed for decades and is engrained in artefacts (e.g., technical infrastructure for flood protection) and shared practices and management strategies (e.g., urban development in floodplains and deserts largely ignoring potential constraints imposed by environmental conditions).

The three types of institutions are not independent. Regulative institutions are influenced by normative and cultural-cognitive institutions. Regulation may also influence and reinforce normative institutions. Polluting the environment may, for example, at first be prohibited by law but over time become unacceptable from a normative point of view. Flood protection provides another a good example of interdependence and co-evolution of these different kinds of institutions (Pahl-Wostl 2006; Pahl-Wostl et al. 2013). Regulating rivers allowed development on floodplains. Inhabitants of these former floodplains take it for granted that they are safe and protected. Legal regulations prescribe a certain protection level. Best practice in building dikes is codified in professional scripts, engineering education and experience. Understanding and supporting institutional change in such an intertwined institutional setting requires an innovative and encompassing concept of institutional change.

An attentive reader might now suspect that the author will next offer a solution to the numerous scientific challenges identified. Indeed I do so, but first I discuss another recent stream of research advancing a theory of gradual institutional change in the context of political economies.

4.1.3 Patterns of Institutional Change in Political Contexts

Prevailing theories in political science and sociology assume path dependence and institutional inertia and correspondingly the necessity of discontinuous change and abrupt shocks as causes for the breakdown of a dominant regime and a subsequent regime shift. A collection of in-depth empirical case studies on institutional change in advanced political economies counters arguments of prevailing theories by providing evidence for the widespread presence of gradual institutional change (Streek and Thelen 2005). The volume presents a conceptual framework which emphasizes the importance of social negotiation processes for the interpretation and enactment of institutions. Gradual institutional change may result from the process of enactment of institutions and negotiations on their meaning. Institutions may be enforced by calling in a third party if what are considered legitimate, normative expectations have not been fulfilled. Such a third party may be a jurisdictional court depending on the kind of institution under consideration. The interpretation of rules may thus change over time and different actors may hold different interpretations.

Streek and Thelen (2005) identified five modes of gradual but nevertheless transformative change in the range of case studies analysed. They refer to these modes as Displacement (gradual replacement of dominant by subordinate institution), Layering (attachment of new elements), Drift (change in meaning due to negligent practice), Conversion (reinterpretation of old institutions for new purposes) and Exhaustion (gradual breakdown over time).

Mahoney and Thelen (2009) further developed this conceptual framework into what they call a theory of gradual institutional change. Their theory emphasizes the distribution of power effects of institutions. They introduced a framework which conceptualizes characteristics of the political context and of the institution in question together as drivers of institutional change. The influence of political context is determined by the veto-power available to defenders of the status-quo. The decisive characteristic of the institution is the extent of discretion in its interpretation and enforcement. Mahoney and Thelen assume that political context and institutional form shape the type of dominant change agent likely to emerge in any specific institutional context as well as the kinds of strategies change agents might pursue to effect change. Drift and conversion is more likely to occur if the extent of discretion in interpretation and enforcement of an institution is high. The veto-power of defenders of the status-quo favours layering and drift as strategies for change. In this model, transformative change need not be a purposeful goal of actors involved in the process. It can be a largely unintended by-product of social negotiation and distributional struggles.

It is interesting to note that Young (2010) identified comparable modes of institutional change in his work on the dynamics international environmental regimes. Young distinguishes five different kinds of institutional change (ibid, pp. 9–12): Progressive Development (gradual increase in the capacity of the regime to deal with the problem it is designed to address), Punctuated Equilibrium (periodic stresses challenge the capacity of regimes to adjust while also triggering instances of progressive regime building), Arrested Development (after a phase of unobstructed development regimes encounter obstacles that impede further development), Diversion (the original purpose of a regime is changed to such an extent that if may even run counter to the original purpose) and Collapse (after a phase of operation a regime may be seriously challenged which either leads to a termination of the regime in formal terms or to a severe decline).

As Streck, Thelen and Mahoney, Young emphasizes the importance of a more or less strategic reinterpretation of existing institutions and gradual institutional change. He also highlights the interaction between exogenous and endogenous factors as a determinant of the nature of change by positing the endogenous—exogenous alignment thesis: "*patterns of change occurring in individual regimes are determined by interactions between endogenous, or regime-specific, factors and exogenous factors, or forces operative in the biophysical and socioeconomic settings in which regimes are located*" (ibid, p. 14). Young's definition of exogenous embraces a broader set and different factors which reflect his focus on international regimes with a clearly defined purpose such as the Climate Regime or the Regime on Stratospheric Ozone. Regarding the characteristics of the institutional setting he makes a distinction between rigid and flexible regimes as determinants of the type of expected institutional change. This distinction is comparable to the level of discretion introduced by Mahoney and Thelen.

What are the conclusions that can be drawn from these approaches to conceptualizing institutional change in more formal and political settings? Institutional settings are not static and gradual change is ubiquitous. Ambiguity, the degree to which institutions allow broader interpretation, is a central characteristic of institutions which influences institutional change. Interpretation hinges on agency, the more or less purposeful behaviour of actors in generating such change, and the distribution of power and how it influences and is influenced by institutional change. These approaches do not give much insight though and did not focus on the question of the extent to which such institutional change can be governed.

A point I would still like to stress is that all authors ground their theories in empirical evidence and formulate their theories such that they can be tested by further empirical studies. This is highly commendable and a requirement for cumulative knowledge generation.

4.1.4 Institutional Change as Multi-loop and Multi-level Learning

Governance scholars are facing a novel challenge as the transformation towards sustainability is deemed a necessity. The governance of transformation implies that transformative change is not an unintended by-product of negotiations but a collective goal in itself. Institutional settings should thus support a negotiation process among the actors involved about the direction of change and reflection on the process of change itself.

Gradual institutional change seems to be an appropriate type of change to strive for in any intentional transformation in water governance systems. A rapid breakdown and subsequent replacement of institutions is hardly a desirable path for water governance reform and a paradigm shift in water management. But what does desirable mean and whom does it serve? Gradual change must not imply that change is only proceeding smoothly in incremental steps at a slow pace. Transformative change in complex governance systems may and most likely will always include instances of non-linear, fast development alternating with phases verging on stasis. But a model of gradual change implies that systems must not break down to overcome path-dependence and institutional inertia. A breakdown is usually the result of the complete failure of the prevailing system and/or irreconcilable conflicts among supporters of different governance systems. Breakdown entails the danger of loss of functionality—at least during a transition period. In countries with reasonably well functioning governance systems—as is the case in most developed countries—this is clearly a development that should be avoided. The situation may present itself differently in developing countries characterised by dysfunctional governance in all domains of public live. In such cases, the transformation of water governance systems can hardly be detached from broader societal development. And societal development may not always be gradual and smooth. A breakdown in a national governance system, resulting from regime change for example, may offer a window of opportunity for a radical shift in water governance as well (Herrfahrdt-Pähle and Pahl-Wostl 2012). After the abolishment of apartheid and a radical change in the political system, South Africa introduced one of the world's most innovative regulatory frameworks on water. Unfortunately, the limited capacity to implement change has meant that the lofty goals of the new water legislation have not been met. That is, implementation has been slow. In a contrasting case, that of Uzbekistan, the breakdown of the Communist regime did not trigger much change in water governance. Water management remains unsustainable from an environmental, social and economic point of view. Building capacity for and sustaining fundamental change takes time.

Understanding and supporting purposive transformative change requires an understanding of social and societal learning. To develop such an understanding, I further refined the concept of triple-loop learning (cf. Fig. 3.3) to explain social learning and transformative change from an evolutionary perspective (Pahl-Wostl 2009). Table 4.1 summarizes processes of institutional change within the three loops of learning for institutions in general and for the three categories of institutions as introduced by Scott in particular.

Single-loop learning refers to an incremental improvement of established action strategies without questioning the underlying assumptions. When dealing with floods due to climate change, for example, this might imply the construction of taller dikes as a safety margin. But prevailing regional planning and flood risk management practices are not contested. Regulatory frameworks which prescribe safety margins persist even when the statistical foundations on which they are based may be discredited by new scientific findings (Milly et al. 2008). In water scarce regions, attempts are made to re-direct rivers but prevailing land-use practices are not scrutinised.

Double-loop learning refers to a revisiting of assumptions (e.g., about cause-effect relationships) within a normative framework. This stage is characterized by a

Table 4.1 Characteristics of institutional change for the three different levels of learning (Modified from Pahl-Wostl 2009)

	Single loop	Double loop	Triple loop
Institutions general	Reproduction and no contesting of established institutions, signs of unilateral reinterpretation	Reinterpretation and contesting of established institutions by many parties	Established institutions changed or abolished and new institutions implemented
Regulative institutions	Existing regulations are strictly adhered to and used to justify established routines Introduction of new by-laws and interpretations of existing law to accommodate exceptions	Regulatory frameworks identified as major constraints for innovation Juridical conflicts about interpretation of rules Exemptions allowing innovative approaches and experimentation	Substantial formal changes in regulatory frameworks, new policies implemented Institutions reflect logic of a new paradigm
Normative Institutions	Established norms are used to justify prevailing system Relying on codes of good practice	Established norms and routines are contested	Change which can be identified in public discourse and new practices
Cultural-cognitive Institutions	Discourse remains in established paradigms that are refined Radical alternatives to reigning paradigm clearly dismissed	New ideas emerge beyond isolated groups Discourse (media, political debate, public hearings, scientific conferences) embrace new paradigm Strong arguments on alternative views —"ideological" debates	Discourse dominated by new paradigm Powerful representatives of "main-stream" support the new paradigm

reframing and discourse which may be controversial. For example, flood policy discourse may start with an innovation in flood management paradigms; 'living-with-water' or 'more-space-for-the-river' which take into account ecosystem services and aim at the design of resilient landscapes. Actors who want to introduce innovative flood risk approaches (e.g., real time forecasting) may encounter regulatory constraints. Local initiatives aimed at experimenting with more integrated flood management practices may conflict with powerful stakeholder groups (Sendzimir et al. 2010). In water scarce regions, one may question current land

and water-use practices and discuss the needs for shifting from supply to demand-management.

In triple-loop learning, underlying values and beliefs and world views are reconsidered if the assumptions of a world view do not hold anymore. This is the stage where the real transformation takes place, when institutions are replaced. Given the complexity of such transformations and the interdependence of many system elements, it is impossible to design and implement a comprehensive blueprint for a new governance and management approach. Thus, an important element of such transformation includes exploratory search and innovation processes where actors experiment with innovative ideas and try to overcome constraints imposed by the prevailing institutional settings. Layering (adding additional elements to existing institutions) and conversion (reinterpretation of existing institutions) may be important in doing so. Moving from discourse to action is the difficult and critical stage where the political support and willingness to act is needed. Furthermore, supportive regulatory frameworks are required which give actors greater freedom to experiment with innovations and which allow broader interpretation of formal regulations without jeopardizing the enforceability of the overall regulatory goals.

Supporting transformative change and sustaining the adaptive capacity of water governance systems requires flexible institutions which leave room for interpretation, for tailoring institutions to specific circumstances, and for responding to developments which impact water governance (e.g., climate change, population change). Ambiguities—the degree to which institutions allow broader interpretations—have always been inherent in institutional settings. Institutions cannot regulate any and every circumstance under which they are applied. Whereas in the past this was perceived as a nuisance rather than a desirable property, in particular for regulatory frameworks, flexibility has increasingly become an element of purposeful design (Garmestani and Benson 2013; Green et al. 2013). The more flexible the regulatory framework, the more responsibility is delegated to actors during policy implementation. Administrators do not only implement rules but also make decisions on how specific rules under certain circumstances may be interpreted. This may require a change in regulatory culture if administrators have been accustomed to merely implementing prescriptions rather than acting with some degree of autonomy. On the other hand, flexibility may open the door for powerful actors to try to impose their vested interests. The grammar of institutions introduced by Crawford and Ostrom (1995) may be useful for operationalizing the degree of ambiguity inherent in rules and for analysing the implications for the performance of governance systems (see also Chap. 9).

In summary, the concept of triple-loop learning captures different levels of institutional change. Single-loop learning refers to reproduction, double-loop learning to (re)interpretation, and triple-loop learning to a transformation of institutions. While there is already doubt about whether the effectiveness of institutions can be understood without giving due credit to agency, it is entirely impossible to understand institutional change without doing so. It does not seem to be meaningful to ask if institutions determine actor behaviour or the converse. Institutions are not static. They are continuously enacted and reinterpreted in social interactions which are themselves shaped and constrained by existent institutions. This mutual feedback is a characteristic of complex systems. In his theory on structuration, Giddens (1984) highlighted the interdependence between autonomous individual action and constraints imposed by the social structure, by institutions. Three decades later, there is still an urgent need to develop an improved understanding of this mutual feedback and interdependence, of processes of institutional performance and change which takes into account both the characteristics of institutions and the role of agency.

4.2 Actors as Agents of Change

Human action—agency—is essential to both the reproduction of established governance structures and the transformation of the governance system. Through their activities, actors give meaning to and interpret institutions or other social artefacts. Understanding the interplay between governance structure/institutions and agency is strongly influenced by underlying theories of human behaviour and learning processes. This became evident in the preceding sections in the discussions of different approaches in institutional theory and analysis. Theories differ in their basic assumptions of the determinants of human behaviour and the role of institutions. Furthermore, theories need to be able to capture the behaviour of both individual and collective actors. In particular because collective actors play a central role at higher levels of governance. This does not imply that an identical behavioural model must be applicable across all levels of aggregation. This is often the case in rational choice and game theoretical approaches where players may represent individuals, companies or nation states. In the following I use the notion of 'actor' to refer to both individual and collective actors. If statements apply to individual actors only, I make that explicit.

4.2.1 Theories and Models of the Behaviour of Individuals and Collectives

The essence of this section is the age-old agency/structure problematic in philosophy and the social sciences. Conceived as a spectrum, there is, on the one hand, an unrestricted free will and, on the other, a structural determination That is, the individual has little discretion with respect to the interpretation of rules and norms. Social reality falls somewhere within that spectrum.

A variety of theories, sometimes competing, are to be found in the social sciences to explain the behaviour of human actors. To cope with such bewildering diversity, the classification of theories according to shared characteristics is useful. One quite coarse but effective approach is to make a distinction between purposeful, strategic behaviour on the one hand and social determination on the other. The most prominent and often criticized representative of the first category is the neo-classical model of a 'homo economicus' maximizing self-interest and being isolated from any kind of social interactions. A more realistic model in this category, 'homo reciprocans', goes beyond the neo-classical model of a 'homo economicus' by recognizing the importance of intrinsic norms of cooperation and fairness that may be strengthened or weakened in social interactions (Dohmen et al. 2009). Occupying the other end of the spectrum, away from the emphasis on the behaviour of individuals, is 'homo sociologicus'. Originating in sociology, this model argues that actor behaviour is determined by social context and the fulfilment of social roles.

Such a broad distinction between strategic and socially determined behaviour has been used by several scholars to categorize the wide range of approaches which exist for explaining (collective) actor behaviour in a social (societal) context. In their seminal work Hall and Taylor (1996) introduced the distinction between three different kinds of institutionalism: rational choice, sociological and historical institutionalism. Rational choice theorists perceive institutions as being crafted by actors to solve collective action problems to maximize their benefits. The behaviour of actors is based on a logic of calculus. On the other hand, the proponents of sociological institutionalism assume that a logic of appropriateness is the basis of behaviour. That is, actors internalize collective scripts regulating human behaviour. Finally, historical institutionalism emphasizes the importance of the evolution of institutional forms and how they shape the behaviour of actors. It makes less stringent assumptions about the logic of behaviour.

With respect to international environmental regimes, Young (2001) classified the diverse theoretical approaches found in the literature as collective-action models and social-practice models. Collective-action models assume the utility maximizing behaviour of self-interested actors. Social-practice models view the behaviour and identities of actors to be strongly influenced by the social environment in which the actors are embedded. Furthermore, and as an example of reflexivity, the behaviour of actors itself shapes the social environment. Cleaver (2012) derived a similar categorization in her review of institutional development in natural resource governance at the community level. She made a distinction between 'Mainstream Institutionalism' and 'Critical Institutionalism'. Mainstream Institutionalism emphasizes instrumental rationality and the importance of institutional design. Instrumental rationality is goal directed and guides actors to find the most suitable means to an end. Critical Institutionalism views agency as relational, with identities being shaped by social interactions and institutions being pieced together through practice and adaptation, a process referred to by Cleaver as 'bricolage'. Both Young and Cleaver note that what they called the schools of thought of 'social practice' and 'critical institutionalism', respectively, embrace a quite heterogeneous collection of approaches with little exchange among the various disciplines. Scott and Young as well as Cleaver point out that the two ways of understanding human behaviour should not be seen as mutually exclusive but that research should search for a better understanding how they can be integrated.

In the remainder of this section, I discuss the basic assumptions of the two broad classes for understanding human behaviour, purposeful action and social determination, and their implications for the understanding of institutional change and agency. Furthermore, I introduce as a third category, 'routine behaviour', which I consider to fall between the two.

According to a utilitarian rational choice model, actors comply with rules if the expected costs of rule violation exceed the expected benefits. For example, companies pollute the environment if fines are lower than the costs associated with avoiding pollution. Actors evaluate alternative options and either optimize or satisfice (to act in such a way as to satisfy the minimum requirements for achieving a particular result). This evaluation occurs under conditions of bounded rationality which take into account limitations in availability and processing capacities of information. Collective action can be represented as actors playing a game with well-defined rules and pay-offs. Classical game theory is based on a complete knowledge of the game (rules and pay-offs) and shared beliefs among all players. Such games terminate in an equilibrium state, a rational expectations equilibrium where each player's move is a best response. Agency is limited to the choice of strategies in a certain game. However, there is no room for agency which aims at changing the rules of the game (Greif and Latin 2004). Changes may be introduced by exogenous influence such as factors which change the pay-off structure. Evolutionary game theory relaxes assumptions that action is only possible if expectations converge to a rational-expectations-equilibrium (Gintis 2000). Actors may learn and change expectations based on experience. Learning is often represented as Bayesian learning where actors update their subjective probabilities on expected outcomes. In recent years more attention has been devoted to the willingness of actors to cooperate, to reciprocate cooperative behaviour rather than to abuse cooperation to maximize their own utility. However, this behaviour is not altruistic because it is linked to expectations of reciprocity. Actors are willing to punish behaviour that is considered to be unfair (Fehr and Gächter 2000, 2002). Despite numerous modifications and extensions, rational choice theory retains as its core the notion of instrumental rationality. One may call this type of behaviour 'strategic and self-interested'. Regarding the role of agency in the context of institutional change, actors may reinterpret or violate rules if required to pursue their interests. Such agency is associated with the willingness to take risks if the returns are expected to be high.

Approaches rooted in sociology assume that an actor's role is defined by societal schemata and that this role determines actions. Compliance with rules and social norms is based on the desire to fulfil societal expectations and to conform to stereotypes. Stereotypes may refer to societal groups, professions or nationalities. In political processes, the identities of actors may be (re)shaped in processes of negotiation and interpretation. Behaviour is relational and the meaning of institutions depends on social context. Countries differ in the overall willingness of citizens to comply with formal rules (e.g., traffic rules). In Germany, most citizens adhere to the law not only since they fear sanctions but since this is a strong societal norm. One may call this behaviour 'social expectations conforming'. With respect to the role of agency in the context of institutional change, actors may deviate from standards and norms based on a new interpretations of roles and rules. Such agency requires willingness to not comply with social expectations and to not be susceptible to social pressure.

I would like to explicitly distinguish a third category, 'routine behaviour'. Routine behaviour posits that actors use heuristics and routines in situations which are assumed to be similar. Situational cues determine if a situation is judged to be similar to those already encountered (Gigerenzer and Selten 2001; Ebenhöh and Pahl-Wostl 2006). Actors may use established routines since they perceive it as too costly to evaluate and try new actions every time. Costs are associated with searching for and processing information. Actors may thus adopt satisficing behaviour and stick to routines as long as goals are met (Simon 1982). Furthermore, resources are required to test and achieve good performance with new types of behaviour. Routine behaviour is thus compatible with a rational choice approach taking into account bounded rationality. Routine behaviour may also derive from social influences. Social costs may be accrued since deviating from socially established and accepted routines may result in the sanctioning of social norms (e.g., peer-group pressure imposing group-conforming behaviour, routines linked to stereotypical roles). Finally, actors may stick to routines since underlying premises are taken for granted. Professional routines follow established scripts which conform to dominant cultural-cognitive institutions. Neither of these explanations is exclusive. They are complementary and all these factors may play a role in shaping routine behaviour albeit to varying extents depending on context. Regarding the role of agency in the context of institutional change, actors may call established behavioural routines into question if achieved or expected outcomes are judged as unsatisfactory or as complete failures. A lack of satisfaction may also be linked to a change in societal and peer-group recognition with respect to compliance with a certain norm. New insights may undermine dominant cultural-cognitive institutions and some actors may promote different world-views and new paradigms. Some understanding of the role of agency can also be found in the work of the sociologist Goffman (1959) and Smith (1999). According to Goffman actors engage in symbolic interaction. They are not passive objects controlled by societal rules that determine their behaviour. Instead they deliberately decide on the kind of impression they want to convey in social interactions. In most cases that will imply that actors reproduce behavioural routines and what they consider socially appropriate in a given situation. However, they may also make deliberate choices to adopt different behavioural patterns to attract attention or to point out what they consider flaws in behavioural routines.

4.2.2 Endogenous Innovation as One Driver of Transformative Change

Moving towards more sustainable water governance and management requires profound structural transformation and systemic innovations. Established governance structures and reigning paradigms need to be challenged. A broader interpretation of existing institutions and experimenting with innovative approaches are steps towards transformative change. Treading new paths entails uncertainty and risks and potential conflicts with supporters of established routines. Prevailing institutions may sanction innovative behaviour. Individual actors differ in their desire to conform to societal expectations and in their willingness to take risks. Work on the diffusion of innovations has shown that the majority of a population follow the crowd whereas only a small minority of a few percent are pioneers and early adopters (Rogers 2003). Such findings have also important implications for understanding institutional change and the role of innovation. The majority of actors do not search for innovations and change but rather tend to search for stability and the confirmation of established principles. They remain in the stage of single-loop learning.

In the face of the need for major structural transformation, such inertia may be perceived as being a negative property of social systems. However, it would be quite stressful to live in a world where conventions and guiding principles would be contested all the time. The stability of governance structures is the basis for the existence of functioning nation states and of societal organization. The backdrop of such inertia is indeed that a social system may not respond to challenges and not recognize the need for change. Actors may be caught in a socially-constructed reality where they search for confirmation of prevailing world views, of prevailing paradigms and of the appropriateness of established behavioural routines.

Selective information processing may prevent learning and the adaptation to a changing environment; this applies for individuals, for enterprises or for scientific organizations. Mental models play an important role in such processes. Or phrased more appropriately: the concept of mental models may be highly useful for developing an improved understanding of social constructions of reality and the role of cultural-cognitive institutions (viz. Pahl-Wostl et al. 2011 for a discussion of the relevant paradigms). People make sense of the world in different ways. Individuals cannot and do not pay equal attention or attribute equal value to all information available in a specific situation. Instead, they selectively process information according to their interest and concern (see e.g., Denzau and North 1994; Doyle and Ford 1998, 1999). As a consequence, they value and make use of only those considerations which are most relevant to them while others remain unrecognized or overlooked. Figure 4.1 represents the role of mental models in such selective processing of information. The figure sketches important elements and processes in the structuring of a decision situation by an individual actor in a social context. The actor needs to process information about the decision situation derived from the social and physical environment to be able to classify the situation and choose appropriate strategies for acting.

I use mental models to refer to "*a relatively enduring internal abstraction of an external system to aid and govern activity*" (after Doyle and Ford 1998, p. 17). Mental models are required to translate information into knowledge which is meaningful to an actor. They are not static but may undergo changes over time. One can distinguish among three kinds of factors that shape mental models: cognitive biases, individual experience and social norms. These factors may interact and reinforce or weaken each other.

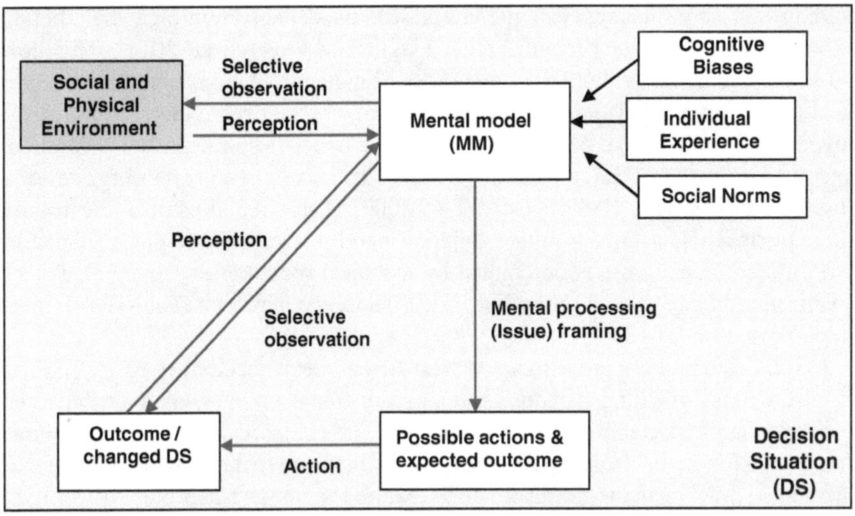

Fig. 4.1 The role of mental models in the selective processing of information and structuring of decision situations in a social context (based on Fig. 1 Pahl-Wostl et al. 2011, p. 852)

Cognitive biases result from heuristics which are shortcuts or simplifications for actions that allow human beings to survive and act in a highly complex and partly unpredictable world without deeply analysing and calculating every detail (Gigerenzer 2002). Individual experiences may help in the construction of a context from few pieces of information, to draw analogies to previous situations and to select a type of response and behaviour that is deemed to be appropriate based on past experience. A common form of bias is 'confirmation bias', which posits that information confirming one's beliefs receives more weight than contradicting evidence. Social norms, or rather their interpretation by individual actors and the role actors hold, influence mental models. They influence what is considered to be important and appropriate in a given decision context.

Mental models shape the selective processing of information. Processes of selective observation of the social and physical environment lead to a certain perception of the decision context. Such filtered information is translated into strategies to deal with situations (mental processing/issue framing as represented in Fig. 4.1). The way people act is influenced by how they frame a certain situation, an issue or relationships with other actors (see Dewulf et al. 2009 for further details). A frame refers to the wider context into which an actor embeds a mental model and which gives sense and meaning to it. For example, a decision situation may be judged to be conflictual. Under the influence of a paradigm, individual actors hold mental models significant parts of which they share and according to which they frame the situations which they seek to influence and manage. In terms of a water management paradigm, the respective epistemic community of actors possesses a shared mental model with respect to the nature of the system to be managed, the management goals and the way the goals may be achieved. A group of actors holding the same management paradigm will most likely reinforce their beliefs through their interactions (confirmation bias) (Pahl-Wostl et al. 2011). This does not imply that actors all hold the same types of motives for doing so and the degree of intentionality may also differ. The adherence to certain behavioural routines and governance and management settings may be linked to strategic considerations since transformative change leads to changes in power structures and the roles of different actor groups (Pahl-Wostl 2006, 2009). Table 4.2 illustrates the role of different stakeholder groups in two different flood management regimes shaped by two different paradigms: flood control by technical measures and integrated flood (plain) management based on an integrated landscape approach (Pahl-Wostl 2006; Pahl-Wostl et al. 2013).

Governance regimes are associated with stereotypical social roles partly codified by institutions—reigning paradigms, regulatory frameworks—which attribute certain tasks and expectations to certain groups. The comparison in Table 4.2 shows that a transformation from one regime to another is associated with fundamental changes in roles and the corresponding required skills and identities of different groups. Such changes may be major barriers and sources for conflicts. Engineers will have to extend their skills and share responsibility and influence with ecologists and landscape architects. Homeowners will have to take more responsibility in dealing with flood risk and some homeowners may face a loss in the value of their property if certain areas currently protected from flooding are assigned to temporally flooded zones. Some actors from the business sector will face losses, and others will benefit. Roles and identities of actors become less clearly defined, in particular during transition phases.

Table 4.2 Comparison of the roles of various stakeholder groups in two flood management regimes operating under different paradigms (Derived from Pahl-Wostl 2006)

	Technical flood control—regulated and controlled rivers	Integrated flood management—multi-functional dynamic landscapes
Stakeholder groups and their role and interests	• Authorities as regulators in a highly controlled environment • Engineers as technical experts and implementers of regulations who construct and operate dams, reservoirs and levees • Insurance companies selling insurances against flood damage • Homeowners largely ignorant of living on floodplains, and if aware, expecting to be protected from flooding • Agricultural use of land in the vicinity of rivers • Shipping industry interested in functioning water-ways	• Authorities as regulators and facilitators of adaptive management processes with shared responsibilities • Engineers as participants in implementation teams cooperate with ecologists and landscape architects • Homeowners responsible for taking appropriate precautions against temporary flooding • Tourism industry and tourists using the floodplains for recreation

This implies major changes in generic governance sub-functions (cf. Chap. 3, Sect. 3.4.3). In particular, in the transition phase from one regime to the other, policy framing and knowledge generation are of prime importance. Given the complexity and potential for conflicts of the new integrated management paradigm, the legitimacy and comprehensiveness of policy framing and knowledge generation require the involvement of a wide range of actors representing diverse interests and perspectives. Processes of social learning and governance in actor networks become increasingly important.

4.2.3 *Processes of Social Learning*

The increasing awareness of complexity and the need to render water governance and management more flexible has led to an increased interest in network governance and in engaging different stakeholder groups in participatory management. Social learning is considered to be an essential process for building capacity for dealing with complex resource challenges. As a consequence, one can now find multiple applications and interpretations of social learning in resources management (Reed et al. 2010; Rodela 2011). The original concept of social learning as introduced by Bandura (1977) referred to the learning of individuals in a social environment by the observation and imitation of others. By focusing on the cognitive processes of individuals in a social context, the original concept did not consider group processes such as the development of shared meanings and values. Concepts which focus on the learning of social entities as a whole can be mainly found in work on organizational learning (Argyris and Schön 1978, 1996). Jean Lave, an anthropologist, and Etienne Wenger, originally trained in computer science and artificial intelligence, coined the concept of 'Community of Practice (CoP)'. This concept emphasizes the importance of situated cognition and the development of shared meanings based on shared practices (Lave and Wenger 1991). Wenger developed comprehensive theoretical foundations for CoPs drawing on a range of social science disciplines (Wenger 1998, 2000). However, CoPs have become highly popular among practitioners more than in academic circles. Over the past decades, the concept has turned into a normative approach with design principles about how CoPs can be established,[1] and its application in resource governance and management (Dionnet et al. 2013).

The concept of social learning that we developed in the context of the European project HarmoniCOP (Harmonizing Collaborative Planning) was partly inspired by Wenger's work. The concept is characterized by a broad understanding of social learning that is rooted in the more interpretative strands of the social sciences (Pahl-Wostl et al. 2007). It was originally developed to explain and analyse the

[1]See for example the resource guide on the homepage of E. Wenger: http://wenger-trayner.com/map-of-resources/.

processes in multi-party platforms dealing with complex problem situations in water management. Figure 4.2 represents the social learning framework which is structured as context, process and outcomes. The process concept refers to multi-party interactions in actor networks having two pillars which emphasize the different aspects of cognition and learning. One pillar relates to the processing of factual information about a problem (content management) and the other to engaging in processes of social exchange (social involvement). Relational practices which are task-oriented actions with relational qualities of reciprocity and reflexivity facilitate the integration of these pillars. Social involvement refers to engaging in processes of major relevance for problem solving. These processes include the framing of the problem (e.g., what is important, causes and potential solutions), the management of the boundaries between different stakeholder groups, the type of ground rules applied and negotiation strategies chosen or the role of leadership in the process. The social learning concept builds on the central hypothesis that the processing of factual knowledge cannot be separated from social interactions, an interdependence which has often been neglected in the technocratic tradition of water management.

The overall process has two kinds of potential outcomes. Technical qualities refer to the improvement of the direct target of the problem solving process such as an improved water allocation scheme or the state of the environment. Relational qualities reflect the development of the capacity for collective action and the build-up of social capital. Improved communication pathways and the establishment of trustful relations increase the capacity of a stakeholder group to manage a problem and/or institutional change. This social learning concept does not assume that consensus is a necessary condition for joint action. Social learning develops the capacity to recognize and accept differences (e.g., in problem framing) and to deal with them constructively rather than being caught in intractable conflicts (Gray 2004).

Social learning in the context of dealing with resource governance and management problems can thus be defined as a process of multiparty interactions where actors engage in relational practices to assess and generate knowledge about a problem domain. Social learning manifests itself at the levels of participating individuals and the group as a whole in the change in, and development of, shared practices, change in individual mental models, individual and collective perceptions or cultural-cognitive institutions, and in the development of trust and the capacity for collective action.

The feedback loop from outcomes to context takes into account potential change and learning processes at wider societal scales. Such feedback may, for example, be induced when participants in such multi-party processes act as representatives of larger constituencies. They may spread innovative knowledge and practices in their home organizations. Furthermore, actors groups who encounter externally imposed constraints in a learning process may try to actively alter them. For example, stringent regulations may prevent experimentation with innovative management approaches.

Social learning as introduced here in the context of resource governance and management has a positive connotation. It is assumed that it enhances the capacity of a collective of actors to deal with complex resource governance and management problems and to find solutions that are deemed satisfactory by the majority of actors

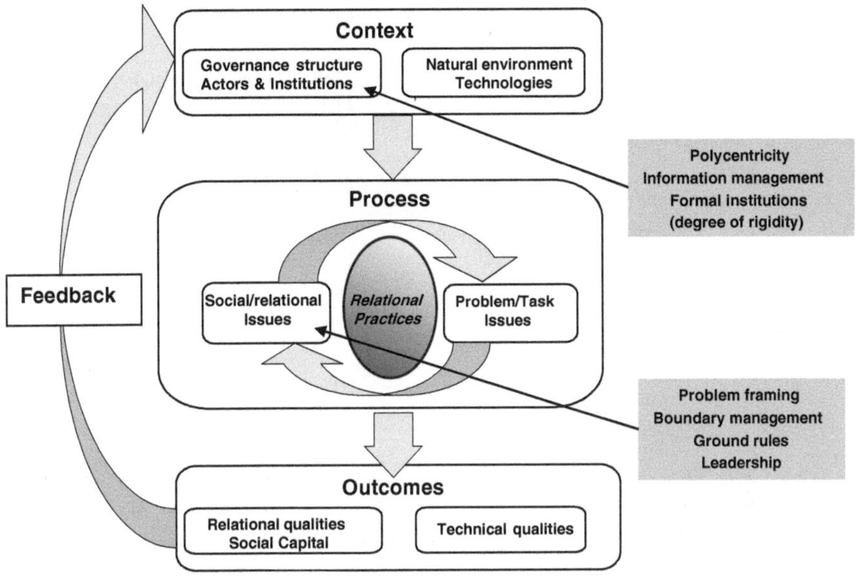

Fig. 4.2 Framework for social learning (based on Fig. 4 in Pahl-Wostl et al. 2007)

involved in the process. Such processes may be jeopardized by powerful actors with vested interests. Furthermore, not all actors may have equal access to a process due to resource constraints and power constellations. Such a development would detract from the quality and success of a social learning process. Therefore, the agreement on ground rules and boundary management are emphasized as important process conditions in the social learning concept (cf. Fig. 4.2).

4.2.4 The Importance of Actor Networks on Social Learning

It is evident that actor networks have a strong influence on social learning and innovation processes. The composition and the degree of openness of actor networks influence the diversity of mental models. Innovative ideas may be cultivated in localized sub-networks first and later diffuse to other networks or be actively disseminated to other networks.

Newig et al. (2010) elaborated the characteristics of networks that foster social learning. To characterize network structure, we considered network measures based on Social Network Analysis. Table 4.3 summarizes the hypothesized influence of several network characteristics on information transmission, single-loop learning and higher (double and triple) loop learning. In this section, I describe briefly the network characteristics that we have chosen. For a more in depth treatment of Social Network Analysis and network measures readers are advised to consult among others Borgatti et al. (2013) and Scott (2000).

Table 4.3 Hypothesized influence of several network characteristics on learning (extracted from Table 2 Newig et al. 2010)

Network function (network characteristic)	Information transmission	Single-loop learning	Higher levels of learning
Homophily (average)	+	+	−/+/− (concave curve)
Relation of weak to strong ties	+	○	+ (−)
Network size	+	+	−/+/− (convex curve)
Density	++	+	−
Cohesion/absence of structural holes	+	+	−
Centralization	+	+	+

Homophily describes the degree of similarity of actors who are connected in a network. It is measured by characterizing actors by certain attributes and measuring the degree of homogeneity with respect to these attributes. Regarding the influence on learning in the context of water governance and management attributes of relevance would, for example, be profession, environmental consciousness or adherence to a certain management paradigm.

The strength of network ties is determined by the intensity and quality of a relationship. In most cases, actors have more weak ties than they do strong ones. Weak ties are more flexible than strong ties. They may provide links to other groups and may be activated to search for or disseminate information.

"Network size" is defined by the number of actors in a network. "Network density" is defined as the number of realized relations in a network divided by the number of possible relations. Network density decreases with network size, since the number of possible connections increases approximately with the square of the number of nodes in the network.[2] Hence, density is a relative rather than an absolute measure and size must be taken into account when densities of networks are compared.

"Cohesion" is a measure of the degree of embeddedness of actors in the network. It is defined by a high degree of network closure and the absence of "structural holes". "Structural holes" refers to the absence of connections between clusters or cohesive subgroups in a network.

"Network centralization" is a measure of how "uneven" centrality is distributed in a network. Centrality is an actor-related measure that describes the "importance" or "power" of an actor in a network. There are different ways of measuring centrality, including the number of connections of an actor. A star-like network with one central node has the highest degree of centralization whereas a network shaped like a circle (all nodes with equal centrality) has the lowest degree of centralization.

[2]The number of possible undirected connections equals 0.5 * N * (N − 1) where N is the number of nodes.

Let us now turn to the hypothesized relationships summarized in Table 4.3. Information transmission and single loop learning are assumed to be supported by high degrees of similarity, and large, cohesive and dense networks with centralized structures. In such networks, transaction costs related to information transmission are low since actors share similarities and a similar normative framework and engage in regular exchanges. But such network characteristics are not necessarily beneficial for higher levels of learning. Dense and cohesive networks support network closure and group think. Actors reinforce themselves in their beliefs and are resistant to innovation. The −/+/− symbol for homophily with respect to higher levels of learning indicates that neither a very high nor a very low degree of homophily is beneficial to higher levels of learning. If groups share no similarity at all, communication and the development of shared meaning and experimentation with innovative ideas is unlikely to occur. The same applies to network size. Networks that are neither too small nor too big support higher levels of learning. Medium sized networks with a certain degree of heterogeneity, with a modular structure and some redundancy are assumed to support higher levels of learning. The presence of structural holes offers opportunities for actors to establish new connections and act as bridging nodes. Overall such networks encourage and offer more opportunities for innovative agency.

The reflections of the previous sections demonstrate that individual and social learning are closely entwined. Identities of actors and the nature of governance as societal function are socially constructed. Individual agency is the nucleus of innovation. Actor networks provide the social interaction context for innovations to be further developed and spread. However, actor networks can also block change. Table 4.4 summarizes the characteristics of agency and changes in actor networks expected for single, double and triple-loop learning.

Table 4.4 Characteristics of agency and changes in actor networks for the three different levels of learning (slightly modified from Pahl-Wostl 2009)

	Single loop	Double loop	Triple loop
Actors—Agency and Networks	Actors remain mainly within their networks—communities of practice Search for advice/opinion mainly within established peer groups Established roles and identities are not challenged	Explicit search for advise/opinion from actors outside of established network (e.g., invitation to meetings) New roles become important (e.g., facilitators in participatory processes) Arguments about identity frames (e.g., what does it mean to be an "engineer") Boundary spanners start to connect different networks—communities of practice Emergent leadership of increasing importance	Changes in network boundaries and connections New actors groups and roles become established Changes in power structure (formal power, centrality—new actors in centre) Identity frames/roles get blurred and become less important, emergence of joint approaches instead of isolated performance according to one's role

In the following sections I elaborate how institutions, actors and actor networks are represented and operationalized in the MTF (Management and Transition Framework). These sections build on the introduction of the MTF in Chap. 3.

4.3 Representation of Actors and Institutions in the MTF

The representation of actors and institutions in the MTF integrates a variety of schools of thought. It is compatible with both rational choice and sociological theories. The key building blocks of social interactions as represented in the MTF were introduced in Sect. 3.4.1. The important elements of an actor's direct and wider social interaction contexts are captured in Fig. 4.3 which is a refinement of Fig. 3.5. Action Situations (ASs) and Actors are embedded in an 'Action Arena'. An Action Arena is an issue-specific political arena focused on a societal function such as flood protection or water supply. It is characterized by a dominant issue framing linked to a dominant 'Management Paradigm'. A Management Paradigm is a cultural-cognitive institution which determines, for example, the preference for certain kinds of solutions (e.g., technical) or strategies for managing risks. A discourse of the reframing of guiding assumptions underlying the current

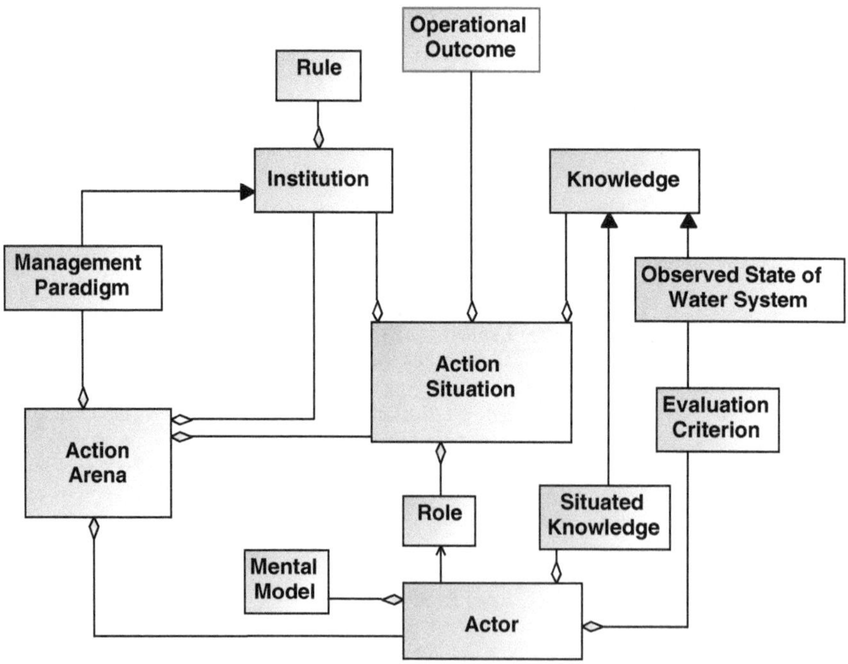

Fig. 4.3 Elementary building blocks of an Action Arena as represented in the MTF in UML notation (Unified Modelling Language). (Expansion of Fig. 3.5—see also for explanation of UML notation)

management policy is a sign of a weakening of the dominant paradigm (double-loop learning). An AS may be influenced by and/or produce 'Institutions', 'Operational outcomes' or 'Knowledge'. The class 'Observed State of Water System' is a specification of knowledge held by an actor and is used to evaluate the state of the 'water system' based on an actor's 'Evaluation Criteria'. The choice of factors and criteria used reflect the perception of 'actors' about what is important and meaningful for them. However, introducing evaluation criteria does not imply utility maximizing behaviour. Evaluation criteria determine the degree of satisfaction of actors with the current state of the water system. For example, actors who attribute more importance to environmental rather than economic considerations may come to different conclusions regarding the state of the water system and the need for action than actors who attribute little importance to the environment and give priority to economic concerns. The whole evaluation process is also influenced by an actor's 'Mental Model'. In the MTF, Mental Model is assumed to refer to individual actors only. Though the mental model concept has been used to refer to 'shared' mental models (Denzau and North 1994). I consider that this is an inapt use of the word. Mental models refer to cognitive constructs and are often determined by cognitive maps (Doyle and Ford 1998; Sterman 2000). Hence, I prefer to make a clear distinction between mental models at the level of an individual actor and social-cultural institutions and paradigms at the level of the collective. If individuals adhere to the same paradigm and are living in the same cultural environment, it is quite likely that their mental model share similarities.

Actors also hold 'Situated Knowledge' which is activated in a specific interaction context. Situated Knowledge results from the selective processing and integrating of different kinds of information and knowledge (subjective experiential knowledge, publicly available general knowledge). The selective processing of information is influenced by the setting of the AS, subjective experience and societal context as reflected in an actor's mental model (c.f. Fig. 4.1). The behaviour of actors is also influenced by the role actors hold in an AS. 'Roles' are based on a shared understanding of their meaning and function. A Role as defined in the MTF is held by an 'Actor' during an 'Action Situation'. Roles belong thus to the relation 'Actor'—'Action Situation' and not to the 'Actor' in general. In general, actors can be individual or collective actors. However, some classes (e.g., mental model, situated knowledge) specifying factors essential for the behaviour of actors are meaningful for individuals only. For other classes, in particular for roles, this is more complicated. Collective actors may hold a certain role in a social context which, in practice, is exercised by an individual acting as representative of the collective actor.

4.3.1 Role of Actors in Social Interactions

A role of an actor in the MTF is only defined in a social context and is linked to expectations about behaviour. It is useful to make a distinction between two kinds of roles: comparatively static roles in generic functional ASs shaped by a prevailing governance and management regime (e.g., water service provider, monitoring authority) and changing, partly emergent roles in negotiation and learning processes. In the following I elaborate in particular on this latter category.

Actors may hold different roles in dynamic learning and transformation processes. The MTF makes an analytical distinction between formal policy and management processes and learning processes that are at least partly informal (cf. Sect. 3.3.2) (Pahl-Wostl et al. 2010). Both kinds of processes have in common the aim of engaging a multitude of actors in collaboration and joint problem solving, and possibly innovation. An actor may take more than one role during a process. Roles may be assigned by rules (formally documented or informally decided upon). Such assignment of roles by rules is usually explicit in that it is based on a decision making process but it might also be implicitly accepted by the group as result of an emerging process. One may identify different kinds of roles in collaborative multi-party processes.

Lead/Convener: The lead of a formal reform process is, in general, explicitly assigned to one actor and may change during different phases of a process. The lead could be taken by a collective actor (e.g., ministry) in a process of water policy reform. It could be an individual for a collaborative process at the local level such as forming and implementing village irrigation boards. At the local level, the convening role usually includes a highly visible public discussion of community issues. These discussions are often related to data-gathering or studies which provide information intended to highlight a common understanding of the issues at hand. Such discussions are important prerequisites for collaborative community problem-solving. Depending on the openness of the boundaries (who is allowed to participate, which topics can be addressed, etc.), the convenor decides or at least prepares the process structure and agenda, the nature of participation and implementation, and has the legitimacy and capacity to gather the parties concerned.

Participant: This would appear to be the most obvious role in a collaborative process but the way that this role is played greatly affects the quality of the collaborative process and the likely outcomes of its activities. Hence it is useful to distinguish between active and passive participants:

Active Participant: Actively engages in the content and structure of a process and shapes its outcomes. Participants, who are entitled to join the process, share the risks, responsibilities, resources and rewards in the collaborative efforts. They may establish mutually respectful, trusting relationships; take the time to understand each other's motivations and hoped-for accomplishments, and state problems 'in a manner that provides opportunities for others to share in their solutions. Active participants may also try to jeopardize a process and bargain for their own advantage.

Passive Participant: More a "by-stander" and not really taking part in the exchange of knowledge and views. Such passiveness may be caused by a lack of resources. It could also derive from a lack of interest in active engagement despite the possibility of doing so. However, this role may also be relevant. For example, in being perceived as an observer by the other participants. A passive participant is not identical to actors who are entirely excluded from the process.

Facilitator/Mediator: A facilitator helps a multi-party group work collaboratively by focussing on the process of how participants work together. Mediators are required to resolve persistent or emergent conflicts. Facilitators apply their expertise in leading the process but they are not participants/partners, have no authority to impose any action on the group, and have no vested interest in the outcome. Facilitation increases the legitimacy of a process. This role is usually explicitly designated to one external person (or external facilitator team).

Technical/Scientific Expert: This role is to bring technical/scientific expertise on an issue to a process. A technical/scientific expert may be independent or be linked to one of the participating organizations. Independent experts can be characterized by not supporting one of the interest groups in the process related to the issue under consideration. The power of one group can also derive from its monopoly on giving technical/scientific advice. Instead of simply bringing himself/herself in as a "technology advocate" and/or promoting arguments supporting the interests of one group, the technical expert ideally appears as the "honest broker" of technical/ scientific information.

The following are the typically *emergent roles* due to the activities of individuals or groups of individuals—not collective actors:

Leadership (emergent): This role is an emergent element during a multi-party process. An actor assuming leadership connects people by, for example, building trust among the participants and supporting the convergence of opinions. Moreover, actors in a leadership role connect people by ensuring the engagement of all actors, spanning boundaries and linking key actors who operate in different policy arenas and/or different policy levels. As well, they reconcile and integrate different understandings of an issue or problem. Leadership might also become apparent by bringing in new perspectives, creating and communicating visions to deal with the issue at hand, pursuing alternatives ways of the process' management and promoting the agenda.

Policy entrepreneur: Policy entrepreneurs are individuals or groups of individuals who actively try to promote innovative policy ideas (Kingdon 1984). According to Huitema and Meijerink (2010) they may in particular make use of the following strategies: develop new ideas; build coalitions and sell ideas; recognize and exploit windows of opportunities; recognize, exploit, create, and/or manipulate the multiple venues in modern societies; orchestrate and manage networks. These strategies combine the generation of knowledge with an active shaping of processes of change.

4.3.2 Zooming in Social Learning Processes

Up to now the MTF has been used to describe the characteristics of processes and the dynamics at what may be called the macro-level—as sequences of linked ASs. The MTF can also be used to represent the dynamics at the micro-level. In principle the relevant classes are already represented in Fig. 4.4. However, for more specific kinds of interactions it may be required to introduce a refinement of the existing classes. The representation of actors in the MTF has, for example, been extended to better represent the details of social learning processes facilitated by participatory methods (Scholz et al. 2013).

Building on and further developing this extension of Fig. 4.4 shows additional sub-classes of the MTF and which are useful for capturing the important characteristics of social learning processes in general. This extension builds on the concepts of social learning and collaborative governance developed by Pahl-Wostl et al. (2007) and as summarized in Sect. 4.2.3. These concepts highlight the dual nature of social learning entailing cognitive and relational components. Also taken into account is the processing of factual information and the engagement in social exchange processes. Thus the substantive and relational outcomes are closely intertwined.

Collective practice was introduced as a subclass of action situation. Collective practice denotes social interactions targeted at addressing a complex, often ill-defined problem situation. Such situations typically require social learning to develop the capacity of an actor group to effectively deal with complex, at times conflictual, issues. Collective practice could, for example, embrace deliberations on factual evidence, group model building exercises or scenario planning.

Two subclasses were introduced for mental models: a substantive and a relational mental model. The substantive model includes as major components interdependencies and the causal relationships of human-technology-environment systems. The relational model includes representations of other actors—including a memory of prior encounters -, expectations about future actions (e.g., willingness of other actors to cooperate), and of skills and knowledge.

Two subclasses were also introduced for operational outcomes: a substantive and a relational outcome. Outcomes are defined at the level of the AS as a whole—they are collective outcomes. Substantive outcomes refer, for example, to new insights and agreements on substantive issues. Relational outcomes refer, for example, to trust among group members, to mutual understanding, and to agreement on procedural rules or establishment of group identities.

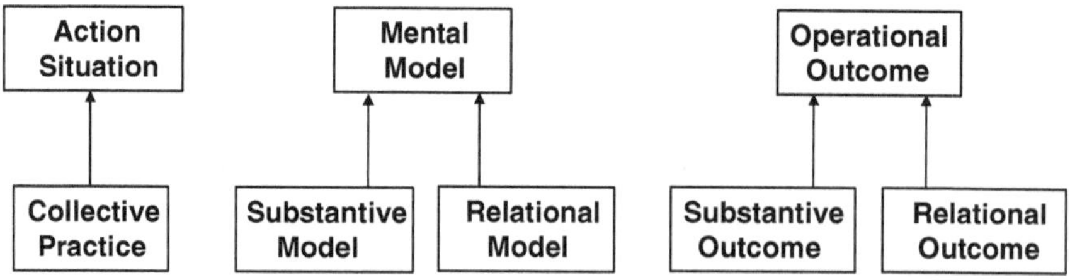

Fig. 4.4 Additional subclasses introduced in the MTF to represent social learning process. The *arrow* denotes an inheritance relationship. Collective practice is a kind of action situation, substantive and relational models are kind of mental models, substantive and relational outcomes are kind of operational outcomes

Figure 4.5 sketches the different kinds of interactions in such a social learning process. It includes, as well, a direct comparison with the social learning framework depicted in Fig. 4.2 to illustrate the analogy of the representations.

An actor participates with a certain role in a collective practice. Only one actor is shown in the diagram to simplify the representation. An actor holds a mental model with relational and substantive components. This mental model shapes the perception of the actor with respect to the social interaction context (cf. also Fig. 4.1). Collective Practice leads to Substantive and Relational Outcomes at the level of the AS as a whole. These Outcomes influence the mental model of the actor—possibly leading to a convergence of certain assumptions in the group as a whole. The actor is also influenced by individual experience such a direct interaction with other actors. Relational Outcome may also influence the role of the actor. A leadership role of individual actors might emerge but agreement on certain procedural rules could also change the role of all participants in a collective practice.

Figure 4.5 captures essential elements of the interplay between individual and social learning. It is not trivial to operationalize such a description for empirical analysis (Scholz 2014). Getting access to data is difficult. Chapter 11 elaborates on the potential of agent based simulation models as a promising and complementary approach to empirical analyses to test the plausibility of assumptions and inspire and guide empirical studies.

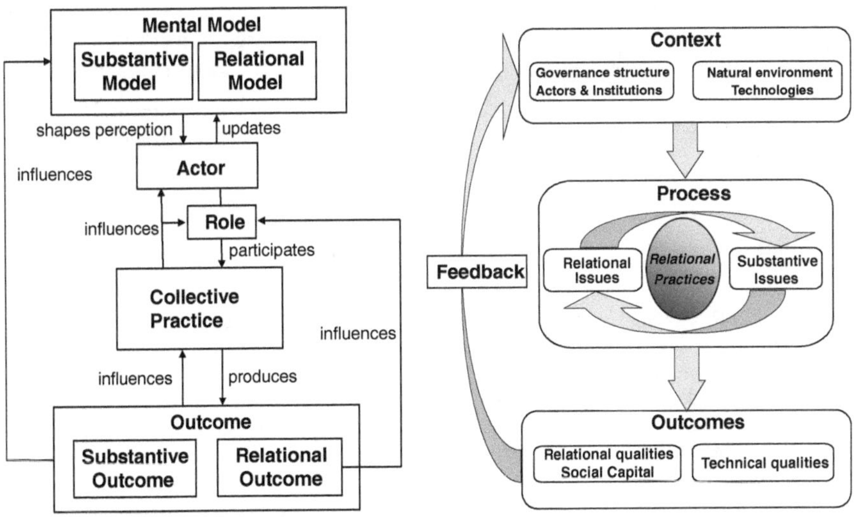

Fig. 4.5 *Left side* Representation of different kinds of interactions in a social learning process. *Arrows* denote the kind of influence between system elements. This figure does not use UML notation. UML is used to represent a hierarchy of concepts, of knowledge objects. It does not capture functional, process-based relationships. *Right side* a direct comparison of the social learning framework from Fig. 4.2 illustrates the analogy of the representations

4.4 Conclusions

Conceptualizing transformative change requires an improved understanding of the interplay between structure and agency. Governance structure is never static but is continuously interpreted and reproduced. The interdependence of various structural elements of governance systems and the selective power of a logic shaped by a dominant paradigm stabilize an established governance regime and impose barriers to change. However, positive feedback may also support the stabilization of a new governance regime and paradigm. The process of transformative change is conceptualized as an evolutionary search process since such complex interdependencies cannot be changed in a planned, purposeful process.

Three kinds of activities related to agency and transformative change were identified:

- new/broader interpretation of existing institutions;
- new interpretation of existing roles and the shaping of new roles; and
- engaging in changing and shaping actor networks.

Transformative change requires informal contexts—ambiguity and space for innovation. At the same time, converting innovation into collective practice and stabilizing a new regime require the codification of new institutions. Furthermore, the importance of power constellations and politics cannot be ignored. Phases of change may also open up space for powerful actors to try to impose their vested interests. Transformation processes are particularly challenging with respect to good governance. Such complex, multi-faceted dynamics can only result from the interplay between various governance modes—hierarchies, networks and markets. The next chapter elaborates on the current state and requirements for an improved understanding of the various governance modes, their interplay and interdependence.

References

Araral, E. (2014). Ostrom, Hardin and the commons: A critical appreciation and a revisionist view. *Environmental Science and Policy, 36*, 11–23. doi:10.1016/j.envsci.2013.07.011

Argyris, C., & Schön, D. (1978). *Organizational learning: A theory of action perspective*. Reading, MA, USA: Addison Wesley.

Argyris, C., & Schön, D. (1996). *Organizational learning II: Theory, method, and practice*. Reading, MA, USA: Addison Wesley.

Bandura, A. (1977). *Social learning theory*. Ehglewood Cliffs, NJ: Prentice-Hall.

Borgatti, S. P., Everett, M. G., & Johnson, J. C. (2013). *Analyzing social networks*. London: Sage.

Cleaver, F. (2012). *Development through Bricolage: Rethinking Institutions for Natural Resource Management*. London: Routledge.

Cox, M., Arnold, G., & Tomás S. V. (2010). A review of design principles for community-based natural resource management. *Ecology and Society, 15*(4), 38. [online] URL:http://www.ecologyandsociety.org/vol15/iss34/art38/

Crawford, S. E. S., & Ostrom, E. (1995). A grammar of institutions. *American Political Science Review, 89*(3), 582–600.

Denzau, A. T., & North, D. C. (1994). Shared mental models: Ideologies and institutions. *Kyklos, 47*(1), 3–31. doi:10.1111/j.1467-6435.1994.tb02246.x

Dewulf, A., Gray, B., Lewicki, R., Putnam, L., Aarts, N., Bouwen, R., et al. (2009). Disentangling approaches to framing in conflict and negotiation research: A meta-paradigmatic perspective. *Human Relations, 69*(2), 155–194.

Dionnet, M., Daniell, K. A., Imache, A., von Korff, Y., Bouarfa, S., Garin, P., et al. (2013). Improving participatory processes through collective simulation: Use of a community of practice. *Ecology and Society, 18*(1), 36. doi:10.5751/ES-05244-180136

Dohmen, T., Falk, A., Huffman, D., & Sunde, U. (2009). Homo reciprocans: Survey evidence on behavioural outcomes*. *The Economic Journal, 119*(536), 592–612. doi:10.1111/j.1468-0297.2008.02242.x

Doyle, J. K., & Ford, D. N. (1998). Mental models concepts for system dynamics research. *System Dynamics Review, 14*(1), 3–29. doi:10.1002/(SICI)1099-1727(199821)14:1<3:AID-SDR140>3.0.CO;2-K

Doyle, J. K., & Ford, D. N. (1999). Mental models concepts revisited: Some clarifications and a reply to Lane. *System Dynamics Review, 15*(4), 411–415.

Ebenhöh, E., & Pahl-Wostl, C. (2006). Agent-based modelling with boundedly rational agents. In J.-P. Rennard (Ed.), *Handbook of research on nature inspired computing for economy and management* (Vol. 2, pp. 225–245). Hershey, PA, USA: Idea Group Reference.

Ebenhöh, E., & Pahl-Wostl, C. (2008). Agent behavior between maximization and cooperation. *Rationality and Society, 20*(2), 227–252. doi:10.1177/1043463108089546

Fehr, E., & Gächter, S. (2000). Cooperation and punishment in public good experiments. *The American Economic Review, 90*(4), 980–994.

Fehr, E., & Gächter, S. (2002). Altruistic punishment in humans. *Nature, 415*, 137–140.

Garmestani, A. S., & Benson, M. H. (2013). A framework for resilience-based governance of social-ecological systems. *Ecology and Society, 18*(1), 9. doi:10.5751/ES-05180-180109

Giddens, A. (1984). *The constitution of society: Outline of the theory of structuration*. Cambridge: Polity Press.

Gigerenzer, G. (2002). *Adaptive thinking: Rationality in the real world*. New York: Oxford University Press.

Gigerenzer, G., & Selten, R. (Eds.). (2001). *Bounded rationality: The adaptive toolbox*. Cambridge, MA: MIT Press.

Gintis, H. (2000). *Game theory evolving: A problem-centered introduction to modeling strategic interaction Princeton*. New Jersey: Princeton University Press.

Goffman, E. (1959). *The presentation of self in everyday life*. New York: Anchor.

Gray, B. (2004). Strong opposition: Frame-based resistance to collaboration. *Journal of Community and Applied Social Psychology, 14*, 166–176.

Green, O. O., Garmestani, A. S., van Rijswick, H. F., & Keessen, A. M. (2013). EU water governance: Striking the right balance between regulatory flexibility and enforcement? *Ecology and Society, 18*(2), 10. doi:10.5751/ES-05357-180210

Greif, A., & Latin, D. (2004). A theory of endogenous institutional change. *American Political Science Review, 98*(04), 633–652. doi:10.1017/S0003055404041395

Hall, P. A., & Taylor, R. C. R. (1996). Political science and the Three New Institutionalisms*. *Political Studies, 44*(5), 936–957. doi:10.1111/j.1467-9248.1996.tb00343.x

Herrfahrdt-Pähle, E., & Pahl-Wostl, C. (2012). Continuity and change in social-ecological systems: The role of institutional resilience. *Ecology and Society, 17*(2), 8. doi:10.5751/ES-04565-170208

Huitema, D., & Meijerink, S. (2010). Realizing water transitions: The role of policy entrepreneurs in water policy change. *Ecology and Society, 15*(2), 26.

Huntjens, P., Lebel, L., Pahl-Wostl, C., Camkin, J., Schulze, R., & Kranz, N. (2012). Institutional design propositions for the governance of adaptation to climate change in the water sector. *Global Environmental Change, 22*(1), 67–81. doi:10.1016/j.gloenvcha.2011.09.015

Kingdon, J. (1984). *Agendas, alternatives and public policies*. New York, USA: Harper Collins.

Lave, J., & Wenger, E. (1991). *Situated learning: Legitimate peripheral participation*. Cambridge, MA: Cambridge University Press.

Lindner, J. (2003). Institutional stability and change: Two sides of the same coin. *Journal of European Public Policy, 10*(6), 912–935.

Mahoney, J., & Thelen, K. (2009). A theory of gradual institutional change. In J. Mahoney & K. Thelen (Eds.), *Explaining Institutional change ambiguity, agency, and power* (pp. 1–37). Cambridge: Cambridge University Press.

Mantzavinos, C., North, D. C., & Shariq, S. (2004). Learning, institutions, and economic performance. *Perspectives on Politics, 2*(01), 75–84. doi:10.1017/S1537592704000635

Milly, P. C. D., Betancourt, J., Falkenmark, M., Hirsch, R. M., Kundzewicz, Z. W., Lettenmaier, D. P., et al. (2008). Stationarity is dead: Whither water management? *Science, 319*(5863), 573–574.

Newig, J., Günther, D., & Pahl-Wostl, C. (2010). Synapses in the network: Learning in governance networks in the context of environmental management. *Ecology and Society, 15* (4), 24.

North, D. C. (1990). *Institutions, institutional change and economic performance*. Cambridge: Cambridge University Press.

North, D. C. (1994). Economic performance through time. *The American Economic Review, 84*(3), 359–368. doi:10.2307/2118057

North, D. C. (2005). *Understanding the process of economic change*. Princeton, USA: Princeton University Press.

Ostrom, E. (2005). *Understanding institutional diversity*. Princeton: Princeton University Press.

Ostrom, E. (2010). Beyond markets and states: Polycentric governance of complex economic systems. *American Economic Review, 100*, 641–672. doi:10.1257/aer.100.3.641

Pahl-Wostl, C. (2006). The importance of social learning in restoring the multifunctionality of rivers and floodplains. *Ecology and Society, 11*(1), 10. [online] URL: http://www.ecologyandsociety.org/vol11/iss11/art10/

Pahl-Wostl, C. (2009). A conceptual framework for analysing adaptive capacity and multi-level learning processes in resource governance regimes. *Global Environmental Change, 19*, 354–365.

Pahl-Wostl, C., Craps, M., Dewulf, A., Mostert, E., Tabara, D., & Taillieu., T. (2007). Social learning and water resources management. *Ecology and Society, 12*(2), 5. [online] URL: http://www.ecologyandsociety.org/vol12/iss12/art15/

Pahl-Wostl, C., Holtz, G., Kastens, B., & Knieper, C. (2010). Analyzing complex water governance regimes: The management and transition framework. *Environmental Science and Policy, 13*(7), 571–581. doi:10.1016/j.envsci.2010.08.006

Pahl-Wostl, C., Jeffrey, P., Isendahl, N., & Brugnach, M. (2011). Maturing the new water management paradigm: Progressing from aspiration to practice. *Water Resources Management, 25*(3), 837–856. doi:10.1007/s11269-010-9729-2

Pahl-Wostl, C., Becker, G., Sendzimir, J., & Knieper, C. (2013). How multilevel societal learning processes facilitate transformative change: A comparative case study analysis on flood management. *Ecology and Society, 18*(4), 58. doi:10.5751/ES-05779-180458

Reed, M. S., Evely, A. C., Cundill, G., Fazey, I., Glass, J., Laing, A., et al. (2010). What is social learning? *Ecology and Society, 15*(4):r1.

Rodela, R. (2011). Social learning and natural resource management: The emergence of three research perspectives. *Ecology and Society, 16*(4), 30. doi:10.5751/ES-04554-160430

Rogers, E. M. (2003). Innovativeness and adopter categories. In E. M. Rogers, (Ed.), *Diffusion of innovations* (5 ed, pp. 267–299). New York: Free Press, Simon & Schuster Inc.

Scholz, G. (2014). *How participatory methods facilitate social learning in natural resources management*. Osnabrück: Osnabrück University.

Scholz, G., Dewulf, A., & Pahl-Wostl, C. (2013). An analytical framework of social learning facilitated by participatory methods. *Systemic Practice and Action Research*, 1–17, doi:10.1007/s11213-013-9310-z

Scott, J. (2000). *Social network analysis: A handbook* (2nd ed.). London, UK: Sage.

Scott, R. W. (2008). *Institutions and organizations: Ideas and interests* (3ed.). Thousand Oaks: Sage Publications.

Sendzimir, J., Flachner, Z., Pahl-Wostl, C., & Knieper, C. (2010). Stalled regime transition in the upper Tisza River Basin: The dynamics of linked action situations. *Environmental Science and Policy, 13*(7), 604–619. doi:10.1016/j.envsci.2010.09.005

Simon, H. (1982). *Models of bounded rationality*. Cambridge, MA, USA: MIT Press.

Smith, G. (Ed.). (1999). *Goffman and social organization: Studies of a sociological legacy*. New York: Routledge.

Sterman, J. D. (2000). *Business dynamics: System thinking and modeling for a complex world*. USA: McGraw-Hill.

Streek, W., & Thelen, K. (2005). Introduction: Institutional change in advanced political economies. In W. Streek & K. Thelen (Eds.), *Beyond continuity: Institutional change in advanced political economies* (pp. 3–39). Oxford: Oxford University Press.

Wenger, E. (1998). *Communities of practice; learning, meaning, and identity*. Cambridge, UK: Cambridge University Press.

Wenger, E. (2000). Communities of practice and social learning systems. *Organization, 7*, 225–246.

Young, O. R. (2001). The behavioral effects of environmental regimes: Collective-action vs. social-practice models. *International Environmental Agreements, 1*(1), 9–29. doi:10.1023/A:1010181007938

Young, O. R. (2008). Building regimes for socioecological systems: Institutional diagnostics. In O. R. Young, L. A. King, & H. Schroeder (Eds.), *Institutions and environmental change: Principal findings, applications, and research frontiers* (pp. 115–144). Cambridge, MA: MIT Press.

Young, O. R. (2010). *Institutional dynamics—Emergent patters in international environmental governance*. Cambridge, MA: The MIT Press.

Chapter 5
Governance Modes

The concept of governance aims at capturing the complexity of real world policy processes. The distinction between those who govern and those who are governed has become increasingly blurred. Governance processes take place at the interface between state, market and civil society and may take various forms. These different forms are called modes of governance. They differ in terms of the kind of actors involved and their roles, and in terms of the nature and logic of interactions. Depending on the governance challenge, a particular governance mode or a combination of modes may be most effective in addressing the challenge. In this chapter I review the notion of governance modes and how it has been defined and applied by various governance scholars. I argue in favour of using the classical distinction of bureaucratic hierarchies, networks and markets as major governance modes. Through analysis the role of diverse hybrid forms among these three modes needs to be identified, in particular in the context of governance of transformation and institutional change.

5.1 Governance Modes—Conceptualizations

When public policy was dominated by hierarchical government steering there was no need to introduce a discussion about other governance modes. In democracies with strong governmental control the responsibility for collectively binding decisions is entirely delegated to elected representatives in legislatures and governments. Decisions are then implemented by the public administration using a top-down approach. With the discourse on the shift 'from government to governance' discussions about 'new' modes of governance have come to the fore as a hallmark of this shift.

There is no overall agreement across social science disciplines on how to define governance. Hence it is not too surprising that a shared understanding of the meaning of governance modes does not exist either. The various approaches all broach the issues of new modes of coordination and steering and the increasing importance of non-governmental actors. The approaches differ though in their conceptualization of governance and the logic selected to delineate governance

C. Pahl-Wostl, *Water Governance in the Face of Global Change*,
Water Governance—Concepts, Methods, and Practice,
DOI 10.1007/978-3-319-21855-7_5

modes. Furthermore, a stream of literature focuses mainly on governance modes within public administration whereas others analyse governance modes within public policy in general or at an even broader level in domain-specific collective action targeted at dealing with societal problems and achieving collective goals.

Kooiman (2000, 2003) was one of the first to try to capture more systematically the change in the nature of interactions in modern governance and the diversity, dynamics and complexity of societal situations. He distinguished three governance modes: self, co-, and hierarchical governance which differ mainly in the role of governmental and non-governmental actors. Hierarchical governance refers to the classical mode of governmental steering and top-down control. The other end of the spectrum is represented by self-governance which refers to situations in which actors take care of themselves, outside the realm of governmental control. Co-governance refers to organized forms of governance interaction where different actors (public and private) coordinate and communicate to deal with the issues at stake without a central governing actor. Kooiman argues that most governance-related societal interactions can be expressed by these modes of governance which may often occur in combination. Empirical analyses that were based on this conceptualization provided evidence that governance interactions had indeed become increasingly complex and diverse (Kooiman et al. 2008). This was also confirmed by Arnouts et al. (2012) who introduced a distinction between closed and open co-governance. This extension of Kooiman's categorization allows a finer distinction to be made regarding the role of actors, the distribution of power and interaction rules. In their analyses of historical changes in governance in the Dutch forestry sector, Arnouts et al. (2012) could also detect a reversal of the expected trend from government to governance. Government took again a stronger and more powerful role. More nuanced and comparable analyses would appear to be required in order to capture different developments under different contextual conditions.

To structure their review of the literature on governance modes, Treib et al. (2007) introduce a broad categorization for conceptions of modes of governance according to the emphasis on state intervention versus societal autonomy. Furthermore they distinguish approaches according to the extent to which these highlight elements of the politics (actors and political processes), polity (kind of institutions) and policy (policy content) dimensions. In the politics dimension the main emphasis is on the presence of public versus private actors. In the polity dimension state intervention is associated with hierarchy, a central locus of authority and institutionalized interactions. Conversely, societal autonomy is associated with markets, dispersed loci of authority and non-institutionalized interactions. The policy dimension covers a broader range of different approaches. State intervention relies on legal bindingness, rigid approaches to implementation, the presence of sanctions, material regulation and fixed norms. Conversely, governance modes rooted in societal autonomy emphasize soft law, flexible approaches to implementation, the absence of sanctions, procedural regulation and malleable norms. Modern governance increasingly combines approaches which emphasize state intervention with those that emphasize societal autonomy. Treib et al. (2007) argue in favour of using the three dimensions politics, polity and policy as point of departure to develop typologies for governance modes. To illustrate this, they developed a typology of modes of governance in the policy dimension which builds on the analysis of EU policies by Knill and Lenschow (2004). This 2-dimensional typology makes a distinction between binding and non-binding legal instruments and between rigid and flexible implementation. EU-policies increasingly move towards framework regulation with binding legal instruments and flexible implementation replacing the traditional mode of coercion with binding legal instruments and rigid implementation (cf. also Chap. 2). Voluntarism with non-binding legal instruments and flexible implementation is only rarely encountered. The development of such typologies aims at mapping and comparing changes in modes of governance in public policies. The ambition is thus to provide an analytical rather than a normative framework.

By contrast, Lange et al. (2013) take a normative stance in their conceptualizing of governance modes for the governance of sustainability. They deplore the prevailing lack of conceptual clarity in the field of governance in general and governance modes in particular. This lack of clarity is identified as significant obstacle to making advances in our theoretical understanding of how best to govern a process towards sustainability. They explicitly dismiss using the distinction between 'hierarchy', 'market' and 'network' for the categorization of governance modes since they consider that understanding real-world governance arrangements (and their relationship to sustainable development) means going beyond what they consider highly abstract, aggregated ideal types. As did Treib et al. (2007), they use the distinction between politics, polity and policy which they consider to be particularly useful for analysing the shift from government to governance. Lange et al. (2013) tested the meta-framework they developed by applying it to the comparison of various frameworks that conceptualized governance modes. This led them to the conclusion that their meta-framework provides a comprehensive basis for theorizing and empirical analysis. However, they have not yet come to any substantive conclusions regarding the role and performance of different governance modes in governing towards sustainability.

In conclusion we observe that increasing efforts are devoted to the development of shared approaches and common conceptualizations of governance modes. Frameworks are used to analyse changes in governance style and to assess if policy developments confirm an overall shift from government to governance and a fundamental transformation in the nature of public policy. Generally valid insights on the implications of governance shifts for the functional performance of governance systems are not yet available. As in many areas of governance research individual cases studies abound and systematic comparative analyses are lacking.

5.2 Ideal Types—Hierarchies, Markets, Networks

An often employed conceptualization of idealized governance modes makes the distinction between 'Hierarchies', 'Markets' and 'Networks' (Thorelli 1986; Thompson et al. 1991; Thompson 2003; Lowndes and Skelcher 1998). Arguments were put forward against the usefulness of such abstract, aggregated ideal-types for meaningful analysis and in favour of more detailed conceptualizations (Driessen et al. 2012; Lange et al. 2013; Treib et al. 2007). I agree that it might be quite a futile endeavour to try classifying governance modes in empirical analyses as either hierarchies or markets or networks. However, I argue for using these ideal types as points of departure for more refined analyses.

Hierarchies, markets and networks denote different ways of coordinating collective action and operate under different logics. Hence such an ideal-typical configuration has a strong explanatory power in terms of logical coherence as well as potential conflict if governance modes are combined. Furthermore, hierarchies, networks and markets already encompass politics, polity and policy dimensions combined to operate under a certain logic. In their analyses of modes of governance Treib et al. (2007) made the distinction between state intervention and societal autonomy to group the modes identified in the politics, polity and policy dimension in two major categories following a different logic. To some extent this reflects the distinction between hierarchical and network governance.

As represented in Fig. 5.1 the three governance modes hierarchies, markets and networks differ along the dimensions of the degree of formality of institutions and the role of state versus non-state actors (Pahl-Wostl 2009). In bureaucratic hierarchies regulatory processes are based on formal institutions and governmental actors play the dominant role. Markets are based on a combination of formal (i.e. property rights) and informal institutions and non-state actors dominate. Networks are largely governed by informal institutions and both state and non-state actors may participate even though the latter are generally in the majority.

Each governance mode is associated with different kinds of processes to confer power to actors: formal power in a hierarchical order, centrality of an actor's role in a network and economic resources in markets. This also implies that actors possess different strategies to accumulate and to exercise power and control.

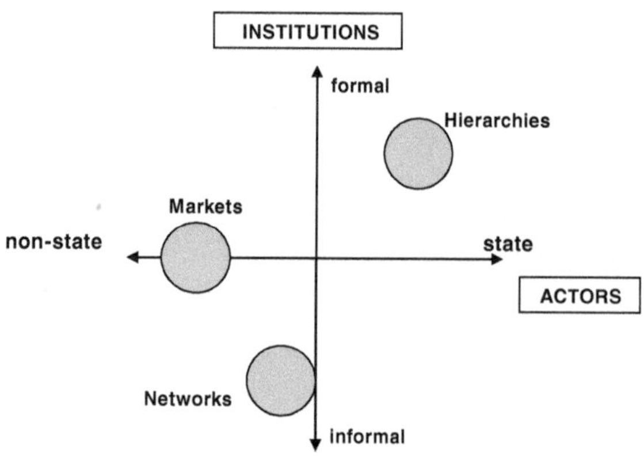

Fig. 5.1 Difference in governance modes of bureaucratic hierarchies, markets and networks with regard to the formality of institutions and the importance of state and non-state actors (Reproduced from Fig. 1 in Pahl-Wostl 2009, p. 358, with permission)

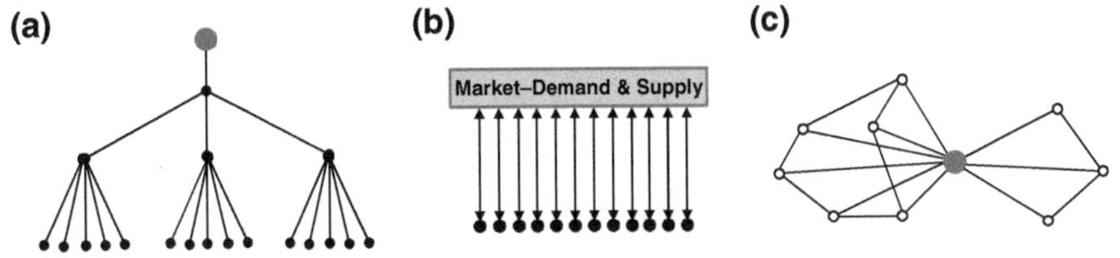

Fig. 5.2 Different structures of interactions between actors and the locus of power and control in **a** hierarchies, **b** markets and **c** networks. *Red colour* and *size* or *shading*, respectively, denote the locus of power

Figure 5.2 presents different idealized modes of interaction in hierarchies, markets and networks respectively. In hierarchical governance, most of the power is conferred by the hierarchical system to those at the top of a hierarchical pyramid. Hierarchical steering can be depicted by a technical systems metaphor. Being on top of a hierarchy is like being in the driver seat of a car. It does not necessarily mean that such a powerful position is used for the benefit of the collective, i.e., the system as a whole. The "benevolent dictator" may be more the exception than the rule. Introducing elections to select those who govern can be a mechanism to introduce some feedback control.

In idealized markets individual actors do not have any powerful position. There is no direct interaction between actors—they only interact via the institution of the market accessible and visible to all. The market is steered by Adam Smith's 'invisible hand'—an optimal collective outcome (Pareto optimum) in terms of the allocation of a scarce resource is achieved by competition and the balance between demand and supply. The 'power' of steering resides in the institution of the market itself. Real markets are not ideal and some actors are more powerful than others. They exercise control by interfering with the market, for example by controlling access or prices.

In networks power is linked to the centrality of actors. The central actor node labelled dark red in the network in Fig. 5.2 has a pronounced position. It is connected to all other nodes and it connects two subgroups within the network. An actor with such a position can control information flows, has a high visibility and potential influence on other actors. Such network structures are in general the result of self-organization and emergence. Some actors may strategically shape networks and their position (cf. Chap. 4) and use this to steer the network in a certain direction. Also networks are not immune to the abuse of power by those in such a central position.

As already discussed, the various governance styles have their strengths and weaknesses. Hybrid forms of governance should try to combine the strengths in a complementary way rather than enhancing the weaknesses. In his book 'Public Management and the Metagovernance of Hierarchies, Networks and Markets' Meuleman (2008a) analysed potential and limitations of the purposeful combination of governance styles which he refers to as meta-governance. Based on a comprehensive review of literature he identified a number of differences among

Table 5.1 Differences among the three governance styles derived from (Meuleman 2008b, pp. 45–48)

	Hierarchical style	Network style	Market style	Dimension
Key concept	Public goods	Public value	Public choice	Vision/strategy polity
Common motive	Reliable	Great discretion, flexible	Cost-driven	Vision/strategy polity
Motive of sub-ordinate actor	Fear of punishment	Belonging to group	Material benefit	Vision/strategy politics
Roles of government	Government rules society	Government is partner in a network society	Government delivers services to society	Vision/strategy politics
Choice of actors	Controlled by written rules	Free, ruled by trust and reciprocity	Free, ruled by price and negotiation	Orientation policy
Aim of stock-taking of actors	Anticipating protest/obstruction	Involving stakeholders for better results and acceptance	Finding reliable contract partners	Orientation politics
Power (added indicator)	Position in formal hierarchy	Centrality of role in network	Degree of wealth, market share	Not available politics
Control	Authority	Trust	Price	Structure policy
Coordination	Imperatives; ex ante coordination	Diplomacy; self-organized coordination	Competition; ex-post coordination	Structure politics
Flexibility	Low	Medium	High	Structure policy
Roles of knowledge	Expertise for effectiveness of ruling	Knowledge as shared good	Knowledge for competitive advantage	Structure politics
Leadership	Command and control	Coaching and supporting	Delegating, enabling	People politics
Relations	Dependent	Interdependent	Independent	People politics

The final column shows the categorization according to Meuleman and an alternative mapping of the polity, politics and policy dimensions

hierarchical, network and market-style governance in public administration. He introduced yet another categorization to support his analysis of meta-governance and the development of strategies for what he refers to as "meta-governors" in public administration. Differences between governance styles are classified as belonging to the dimensions vision and strategy, orientation, structure, people and results, respectively. The differences identified by Meuleman for the different dimensions can also be roughly mapped to the polity, politics and policy domains.

Table 5.1 summarizes a selection of differences among governance styles which I consider of particular relevance for highlighting the difference among a hierarchical, network or market logic in relation to the three domains.

A hierarchical style sees government as ruling society with a command and control approach. Choices are constrained and prescribed by written rules. Control builds on authority of those who govern and the fear of punishment of sub-ordinate actors. Relationships between actors are mainly built upon dependencies determined by formal rules. Such a governance style is supposed to be and aims at being predictable and reliable. It is characterized by a technocratic approach to knowledge as expertise supporting those who govern.

In a network style government interacts as a partner with other societal actors. Relations are interdependent and built on trust. Actors engage in ongoing interactions and develop relationships as a result of their feeling of belonging to a group. Knowledge is co-produced and shared. Network governance is more flexible than hierarchical governance which is controlled by rigid rules and structures. But networks of interdependence and strong identification with one group can also lead to some rigidity and network closure.

In a market style of governance, government is seen as service provider. Control is exercised through price-setting. Water is seen as a commodity. Actors engage in interactions since they expect material benefits. Knowledge production and use is instrumental with the goal of providing a competitive advantage. In its idealized form it is characterized by independence of actors, and market governance is thus the most flexible of all governance modes.

Different governance styles also lead to different governance sub-functions and a different interpretation of the governance properties introduced in Chap. 3. Based on the characterization of governance modes I developed stylized representations as summarized in Table 5.2. Depending on their overall orientation different actor groups may hold different views on how a governance sub-function should be realized and they may use different criteria for evaluating properties of governance functions. Governmental authorities may typically argue in favour of a hierarchical style, NGOs a network style and enterprises a market style.

Most real governance settings do not correspond to such stereotypical configurations but are characterized by hybrid forms of governance styles. However, a stereotypical approach clarifies and highlights the different assumptions and logic by which the different governance styles operate. It helps to flesh out incoherence and potentially incompatible combinations of approaches that may lead to conflict and governance failure.

5.2.1 Potential Conflicts Between Governance Styles

As different governance styles operate according to different logic the combining of these styles is by no means straightforward. Incompatibilities and contradictions may lead to ineffective and inefficient approaches and even to severe conflicts rather

Table 5.2 Governance sub-functions and governance properties (cf. Fig. 3.8) in the three governance styles

Governance sub-functions	Hierarchical style	Network style	Market style
Policy framing	Prescribed by regulation; Expert judgement of problem identification; Focus on prescriptions and command and control instruments	Broad process on problem identification encompassing different perspectives; Focus on voluntary agreements	Problem identification based on profitability, cost consideration, market failure; Focus on pricing and market based instruments
Knowledge generation	Technocratic focus; Only technical experts involved	Knowledge generation as part of group building process; Different types of knowledge acknowledged; Broad sharing of knowledge	Knowledge serves to increase competitive advantage
Resource mobilization	Engage actors with political power; Tax, governmental budgets for financing	Mobilize broad stakeholder support; Voluntary financing	Engage actors with market power; Investment
Conflict resolution	Jurisdiction; Legal procedures	Mediation; Aim for consensus	Survival of the fittest; Compensation payments
Rule making	Political parliamentary process; Jurisdiction and formal procedures for rule extension if needed	Broad negotiation of and deliberations on rules; Malleable rules open to renegotiation	Negotiations on prices; As few rules as possible
Monitoring and evaluation	Compliance with regulation and quantifiable standards Rigid in terms of learning	Participatory; Reflection on agreed goals Openness to adaptive approaches—change negotiated	Cost-benefit calculations; Rapid changes in individual strategies if needed to increase profitability

(continued)

Table 5.2 (continued)

Governance properties	Hierarchical style	Network style	Market style
Legitimacy	Legitimacy as representation; Democratic elections of governments; Constitutional rules as the basis for authorities; Output legitimacy[a]	Legitimacy as participation; Process-based procedural arguments; Input legitimacy[a]	Profit counts; Input (efficiency) and output (effectiveness) legitimacy combined[a]
Leadership	Prescribed by formal rules; Command and control	Often emergent in a process; Coaching and supporting	Determined; Delegating and enabling
Representativeness	Elected representatives; Technical experts on problem domain	All voices heard, openness of process; Those affected participate in decision-making	Access for all market players
Comprehensiveness	Technocratic integration of relevant issues	Participatory integration of perspectives	Integration of all relevant costs and benefits

[a]Input and output legitimacy refers to different ways of legitimizing agency. Legitimization by output assesses legitimacy by the product of an action. Legitimization by input assesses legitimacy by the process by which actors acquire particular roles and how an outcome is derived

than expected synergies. Water authorities have, for example, operated for decades in a, by now, deeply ingrained technocratic and hierarchical mode. The current trend towards reliance on participatory approaches poses serious challenges to a hierarchical governance tradition in terms of lack of skills, trust in expertise and institutional inertia (Allan and Curtis 2005; Roth and Warner 2007; Pahl-Wostl et al. 2011). Organizational culture clashes with changes in the policy landscape. Innovative water policy frameworks such as the European Water Framework Directive (WFD) include more elements derived from market and network governance. The WFD prescribes, for example the involvement of stakeholders in the development of river basin management plans but without specifying the rights of those participating or procedural rules. Authorities often include participation to comply with legal prescriptions but without deliberate consideration of the potential benefits of a network governance approach. Participatory processes need agreement on the ground rules, for example, how decisions are made based upon the recommendations developed in a participatory process. Trust is destroyed if consensual agreements achieved in participatory processes are overridden by top-down decisions in a hierarchical mode (Mostert et al. 2007).

Problems of accountability may arise if hierarchical control by the state is increasingly replaced by cooperative arrangements in complex networks (Papadopoulos 2003). Who is accountable for failed governance notions in diffuse networks? This poses problems for democratic accountability since those who shape policies are not necessarily responsible to an electorate. Problems may also arise with legal accountability if rules for attributing accountability do not match the complexity of participatory and decentralized decision making processes (Lemos and Agrawal 2006).

A classical conflict arises between the freedom of markets and demands for decentralization versus centralized hierarchical control. What can and should be left to the market in public policy? And to which extent can and should governmental control and market-based approaches be combined? Different attitudes may exist though regarding the role of government if profits or costs are involved. Market failure arises if profits are private and environmental costs are public. The privatization debate has shown that it is by no means trivial to design hybrid forms and implement public-private partnerships (cf. Sect. 2.2). But it has demonstrated as well that such hybrid forms seem to be superior to purely market or government-based approaches.

Market-based approaches assign more responsibility for risk management to individuals. Innovation is not possible without accepting some risks when trying something new. Water management though has been dominated by a risk-averse conservative strategy. 'Safety first' has for a long time been the guiding principle for many public service provisions. This applies, for example, to drinking water supply where costs for the provision of services were not taken into account in the past (Tillman et al. 2005). Demand management and water pricing have been introduced and even prescribed by law in many places. This may lead to the counter-intuitive consequence that in some places water-saving leads to an increase in water price since the price is determined by sunk costs rather than by the amount of water consumed (Tillman et al. 2005). Public investment in large-scale infrastructure generates legacies for decades.

Similar problems are encountered in policy changes to address flood risk. People settle in former floodplains encouraged by improved flood protection. In the event of high levels of damage due to extreme flooding events, governmental support has been provided even when not prescribed by law. Changes in flood risk management reflect a shift towards more individual responsibility—where managing risks are linked to individual decisions. This generates conflicts, for example, with land owners who are suddenly faced with potential costs for insurance and a loss of the value of their property.

In dealing with risks, water security problems arise because the diverse aspects of water security are approached with different governance styles operating on the basis of a different reasoning (Pahl-Wostl et al. 2013). Based on the definition of Grey and Sadoff (2007, p. 545) "*Water security refers to the availability of an acceptable quantity and quality of water for health, livelihoods, ecosystems and production, coupled with an acceptable level of water-related risks to people, environments and economies*" we distinguished four approaches to defining what is acceptable. These differ in the kind of knowledge used, in the institutional setting of and in the actors involved in the process in the following ways:

- Scientific analysis and expert judgement supporting hierarchical governmental control of risks;
- Invoking widely shared societal norms in a network governance approach but also a basis for rule-making in a hierarchical governance approach;
- Economic cost-benefit types of analysis providing decision-support for market-based risk governance;
- Place-based assessment of perceptions of concerned stakeholders feeding into participatory deliberations in network governance.

We demonstrated that different approaches and thus different governance styles dominate the four domains (health, livelihoods, ecosystems, production) identified in the water security definition, and that this rendered an integrative approach to negotiating trade-offs and resolving conflicts problematic (Pahl-Wostl et al. 2013). These findings highlight the need for integrative institutional arrangements supporting negotiations and transparent and evidence-based decisions about trade-offs between and among the various water security dimensions.

Given the potential for conflicts between different governance modes an essential part of the design of governance systems should be an explicit consideration of how to avoid these conflicts and how to generate synergies instead. In his analysis of sustainable land use and property right systems Young (2011) concluded that successful governance systems are characterized by hybrid forms of governance and typically involve regulatory top-down strategies, normative bottom-up strategies and some combination of the two. There is a clear need for meta-governance which aims at achieving an effective combination of governance styles and addresses potential conflicts.

5.3 Governance Modes—Hybrid Forms and Meta-Governance: Potential and Limitations

Let us recapitulate. The increasing complexity and interdependencies of modern societies and the need for more coordination of and collaboration among societal actors in order to address complex governance problems has led to the emergence of new and diverse forms of governance. I argue in favour of using the classical distinction between hierarchies, networks and markets as ideal-typical governance modes to analyse complex and hybrid governance settings. As substantiated by Grande (2012), the universe of discourse of governance research more or less embraces the interface between hierarchical governmental control, self-organization of civil society and competitive markets. Opinions diverge as to the extent to which the interactions at this interface can be influenced by purposeful design and if any kind of design principles can be derived. Meuleman (2008a), for example, explicitly argues in favour of purposeful design—at least in public administration. He recommends the development of tool boxes which allow public managers to act as "meta-governors" combining instruments of hierarchical, network or market governance styles, respectively, as required by particular governance problems and process-specific conditions. Kooiman and Jentoft (2009) promote a less instrumental understanding of meta-governance. For them meta-governance, the governance of governance, refers to reflection on values, norms and principles which pertain to governance system issues, for instance, their institutional design. In terms of Ostrom's IAD framework meta-governance is located at the level of metaconstitutional action situations which determine the constitutional rules-in-use (Ostrom 2005, p. 59). They set the context within which other governance activities unfold. However, a game-theoretical, rational choice approach may be too constraining to capture what I consider to be the essence of meta-governance. I define '**meta-governance** as a reflexive process of societal learning to develop, to evaluate and to adapt governance approaches with the purpose of addressing complex societal challenges'. Defined as such meta-governance is an important element of triple loop learning. However, societal capacities for such reflexive governance and discourse are not well developed and are less supported by solid empirical knowledge on the performance of different governance arrangements and the role of combinations of governance modes.

5.4 Concluding Comments

The concept of governance modes is a promising approach for capturing the complexity of the many diverse forms of governance encountered in resources management. The distinction between ideal typical modes of hierarchical, market and network governance allows the most important logic underlying the organization of governance to be described. In real world governance systems hybrid forms prevail. It is remains an open question as to the extent to which such hybrid forms can and should be the product of purposeful design. Different governance modes also encompass different modes of coordination across scales and levels which are addressed in the next chapter.

References

Allan, C., & Curtis, A. (2005). Nipped in the bud: Why regional scale adaptive management is not blooming. *Environmental Management, 36*(3), 414–425. doi:10.1007/s00267-004-0244-1

Arnouts, R., van der Zouwen, M., & Arts, B. (2012). Analysing governance modes and shifts—Governance arrangements in Dutch nature policy. *Forest Policy and Economics, 16*, 43–50. doi:10.1016/j.forpol.2011.04.001

Driessen, P. P. J., Dieperink, C., van Laerhoven, F., Runhaar, H. A. C., & Vermeulen, W. J. V. (2012). Towards a conceptual framework for the study of shifts in modes of environmental governance—Experiences from the Netherlands. *Environmental Policy and Governance, 22* (3), 143–160. doi:10.1002/eet.1580

Grande, E. (2012). Governance Forschung in der Governance Falle. *Politische Vierteljahresschrift, 4*, 565–592.

Grey, D., & Sadoff, C. W. (2007). Sink or swim? Water security for growth and development. *Water Policy, 9*, 545–571.

Knill, C., & Lenschow, A. (2004). Modes of regulation in the governance of the European Union: Towards a comprehensive evaluation. In J. Jordana, & D. Levi-Faur (Eds.), *The politics of regulation: Institutions and regulatory reforms for the age of governance* (pp. 218–245).

Kooiman, J. (2000). Societal governance: Levels, modes and orders of social-political interaction. In J. Pierre (Ed.), *Debating governance: Authority, steering, and democracy* (pp. 138–164). Oxford UK: Oxford University Press.

Kooiman, J. (2003). *Governing as governance*. London: SAGE.

Kooiman, J., Bavinck, M., Chuenpagdee, R., Mahon, R., & Pullin, R. (2008). Interactive governance and governability: An introduction. *The Journal of Transdisciplinary Environmental Studies, 7*(1), 1–11.

Kooiman, J., & Jentoft, S. (2009). Meta-governance: Values, norms and principles, and the making of hard choices. *Public Administration, 87*(4), 818–836. doi:10.1111/j.1467-9299.2009.01780.x

Lange, P., Driessen, P. P. J., Sauer, A., Bornemann, B., & Burger, P. (2013). Governing towards sustainability—Conceptualizing modes of governance. *Journal of Environmental Policy & Planning, 15*(3), 403–425. doi:10.1080/1523908X.2013.769414

Lemos, M. C., & Agrawal, A. (2006). Environmental governance. *Annual Review of Environment and Resources, 31*(1), 297–325. doi:10.1146/annurev.energy.31.042605.135621

Lowndes, V., & Skelcher, C. (1998). The dynamics of multi-organizational partnerships: An analysis of changing modes of governance. *Public Administration, 76*(2), 313–333. doi:10.1111/1467-9299.00103

Meuleman, L. (2008a). *Public management and the metagovernance of hierarchies, networks and markets*. Heidelberg: Physica-Verlag.

Meuleman, L. (2008b). Theoretical framework. In L. Meuleman (Ed.), *Public management and the metagovernance of hierarchies, networks and markets*. Heidelberg, Germany: Physica-Verlag.

Mostert, E., Pahl-Wostl, C., Rees, Y., Searle, B., Tàbara, D., & Tippett, J. (2007). Social learning in European river-basin management: Barriers and fostering mechanisms from 10 river basins. *Ecology and Society, 12*(1).

Ostrom, E. (2005). *Understanding institutional diversity*. Princeton: Princeton University Press.

Pahl-Wostl, C. (2009). A conceptual framework for analysing adaptive capacity and multi-level learning processes in resource governance regimes. *Global Environmental Change, 19*, 354–365.

Pahl-Wostl, C., Jeffrey, P., Isendahl, N., & Brugnach, M. (2011). Maturing the new water management paradigm: Progressing from aspiration to practice. *Water Resources Management, 25*(3), 837–856. doi:10.1007/s11269-010-9729-2

Pahl-Wostl, C., Palmer, M. A., & Richards, K. (2013). Enhancing water security for the benefits of humans and nature—the role of governance. *Current Opinion in Environmental Sustainability* (Vol. 5, pp. 676–684).

Papadopoulos, Y. (2003). Cooperative forms of governance: Problems of democratic accountability in complex environments. *European Journal of Political Research, 42*(4), 473–501. doi:10.1111/1475-6765.00093

Roth, D. I. K., & Warner, J. (2007). Flood risk, uncertainty and changing river protection policy in The Netherlands: The case of 'calamity polders'. *Tijdschrift voor economische en sociale geografie, 98*(4), 519–525. doi:10.1111/j.1467-9663.2007.00419.x

Thompson, G. (2003). *Between hierarchies and markets: The logic and limits of network forms of organization*. Oxford: Oxford University Press.

Thompson, G., Frances, J., Levacic, R., & Mitchell, J. (Eds.). (1991). *Markets, hierarchies and networks: The co-ordination of social life*. London: Sage.

Thorelli, H. B. (1986). Networks: Between markets and hierarchies. *Strategic Management Journal, 7*(1), 37–51.

Tillman, D. E., Larsen, T. A., Pahl-Wostl, C., & Gujer, W. (2005). Simulating development strategies for water supply system. *Journal of Hydroinformatic, 7*, 41–51.
Treib, O., Bähr, H., & Falkner, G. (2007). Modes of governance: Towards a conceptual clarification. *Journal of European Public Policy, 14*(1), 1–20. doi:10.1080/13501760606l071406
Young, O. R. (2011). Land use, environmental change, and sustainable development: The role of institutional diagnostics. *International Journal of the Commons, 5*(1).

Chapter 6
Multi-level and Cross-Scale Governance

Governing water implies governance of a complex social-ecological system at and across different scales in space and time. The spatial scale selected for governance has implications for both the biophysical boundaries that are taken into consideration and the administrative level of the actors involved in a governance system. This chapter summarizes the development of research and policy discourses on the ideal level at which to focus water governance. Systems analysis has been used to determine an appropriate scale for dealing with a governance problem. The scale could also be the subject of political discourse since actors may have different reasons for up- or down-scaling a particular water-related issue. On the basis of current scientific understanding, multi-level and cross-scale water governance is promoted. Water issues can rarely be dealt with at one scale only. I therefore argue in favour of polycentric governance combining decentralization with effective coordination of decision centres as the normative governance model. Finally the chapter elaborates on insights from and methods for structured analyses of modes of cross-scale coordination.

6.1 The Spatial Dimension of Water Governance

6.1.1 Evolution from Local to Global Water Governance

Water governance has a long history and what was perceived as the preferred spatial scale at which to govern and manage water resources has changed considerably over time (Pahl-Wostl et al. 2008). In particular the importance of the global scale was contested for a long time.

During the first joint Conference "Challenges of a Changing Earth" of the Global Change Research Community in Amsterdam 2002 we organized a session on global water issues (Pahl-Wostl et al. 2002). The relevance of the global dimension of water governance was a controversial issue at that time. A participant of the panel discussion entirely dismissed the necessity for any governance arrangements at the global level and asked if we intended to tow and trade icebergs. The argument on the global dimension illustrates that water was not perceived as a commodity in

C. Pahl-Wostl, *Water Governance in the Face of Global Change*,
Water Governance—Concepts, Methods, and Practice,
DOI 10.1007/978-3-319-21855-7_6

global trade and that water problems were perceived to be regional or even local issues. Diverse arguments can and have been put forward for favouring different governance levels and spatial scales and awareness of the complex multi-level and cross-scale nature of water governance has increased over the past decade.

Arguments in favour of the local level stress the importance of understanding local rights, needs, and stakeholders in order to effectively address governance issues. The driving forces behind such arguments are the notions of decentralization and subsidiarity. Water problems were largely perceived as being local and correspondingly they should be handled at the lowest appropriate governance level. Issues related to local access to water and sanitation services, factors affecting local demand or local vector-borne diseases, for example, can best be addressed by local authorities and communities because local residents have the local knowledge needed and a greater incentive to address their own problems than those at other levels of governance. The work of Elinor Ostrom provided ample empirical evidence that local communities can develop effective rules for sustainable resource governance of common pool resources (Ostrom 1990, 2005).

There also exist strong arguments in favour of water governance at national levels to overcome the potential bias of local interests and to guarantee generally-applicable rules based on formal legislation. All national water and resource laws, for example, are based on this concept. From this perspective, water is a public good, a national resource that should be governed for the benefit of the national economy and society as a whole: domestic interests come first. Elected national governments are the legitimate actors to do so. To overcome the problem of massive overexploitation of groundwater resources the Spanish government, for example, introduced a law (Spanish Water Act 1985) shifting water resources from private to public goods and imposing groundwater use restrictions (Costejà et al. 2002). Not all land owners accepted this law though since they considered it to be an illegitimate intrusion on their private rights. This has generated ongoing conflicts (Knüppe and Pahl-Wostl 2013). Furthermore, the superiority of national water governance is threatened, if the willingness of the state to promote the welfare of the people within it is questioned (Lemos and Agrawal 2006). Nevertheless, certain rules such as water quality standards should be set at higher level rather than being subject to negotiations at local scale.

An approach to governance of water that has dominated the past decades focuses on the *basin level*. Arguments in favour of this approach posit that water-related problems and conflicts are best dealt with within the natural sphere of the system—that is, the hydrologically defined basin, catchment, or watershed (Newson 1997; Global Water Partnership GWP 2000; Hooper 2005; Molle 2009). This concept combines notions of efficiency with a hydrological systems approach. As an attempt to overcome problems of spatial 'misfit' between boundaries of natural resources and administrative boundaries (Young 2002), functionally-specific governance institutions have been increasingly implemented at a scale that corresponds with the hydrological boundaries of water resource issues within a basin. It allows comprehensive problem analysis and the internalization of otherwise externalized problems as they arise, for instance, from up- and downstream activities. As summarized by Vogel (2012) governance and management at the river basin scale leads to a more holistic and integrated approach in terms of management goals and areas, the distribution of benefits and decision making that produces balanced management.

However, river basin governance cannot operate in isolation and must operate within a complex institutional landscape. Problems of fit between natural resources and administrative boundaries may be solved at the expense of problems arising from the interplay between different administrative levels (Moss 2003). Interplay refers to interactions between different governance regimes and different governance levels (Young 2002). Integrated management crossing sectoral boundaries may encounter bureaucratic hurdles since other sectors such as agriculture are governed at administrative scales. Cohen and Davidson (2011) identified a number of challenges associated with governing water at the basin scale: boundary choice, accountability, public participation, and the asymmetries of watersheds' with 'problem-sheds' and 'policy-sheds'. Defining the boundaries of a watershed is by no means trivial and controversial debates may arise. The boundaries of a watershed do not necessarily coincide with boundaries of other environmental systems such as groundwater aquifers or ecosystems. The setting of hydrological boundaries is associated with uncertainties in the scientific knowledge base even when data availability is extensive which, inevitably results in the need for political bargaining. Accountability may become blurred when new actors enter the scene and governmental actors have to make decisions at scales which do not coincide with jurisdictional boundaries. Cohen and Davidson (2011) also address the problem of legitimacy that is often closely related to accountability and a shift from legitimacy as representation to legitimacy as participation. They note that basin management has been conflated with a broader, participatory and integrated governance framework. This renders the assessment of governing at the basin scale difficult. The basin concept which was originally introduced based on scientific and technical considerations encounters challenges which can only be addressed in a political process and not by technocratic expertise. Such a requirement often does not get sufficient awareness and attention in research and practice (Jeffrey and Gearey 2006).

The basin level approach is also at the root of the equitable management of transboundary rivers (Bernauer and Siegfried 2008; Dombrowsky 2008). International cooperation in managing the Rhine basin is often praised as the success story in the history of international basin management organizations. But transboundary water governance is quite prone to being connected or even dominated by political considerations and agendas. A case in point is the securitization[1] and the influence of geo-political factors on water policy in the Middle East (Lowi 1993; Katz and Fischhendler 2011). Further tension on transboundary river basin

[1]Securitization of water policy refers to a process whereby political actors transform water issues into matters of security. It is a politicization of water governance that is not primarily rooted in material but in political arguments. Securitization of water policy has been prevalent in the Middle East (Fischhendler 2015).

management arises from the ever increasing global boom in dam construction (Zarfl et al. 2015) caused by the interest in hydropower and by the need for additional storage as a result of climate-induced retreat of glaciers.

The emergence of a *global* perspective on governance of water is quite recent. The global dimension of water-related problems has for a long time been neglected or even disputed (Pahl-Wostl et al. 2002; Vörösmarty et al. 2013). As already pointed out, it does not make sense to trade water over large distances. However, enormous flows of virtual water are associated with food trade and sophisticated accounting schemes have been developed to quantify such water transfers (Yang et al. 2013). Many water-related environmental and societal problems as well as water use related conflicts elude appropriate solutions at the local level or within national or basin boundaries. The regional manifestations of global climate change have severe implications for water resource governance and management. In all these cases, it is important to address issues at a global level. Thus, growing attention is being given to multilateralism in the international politics of water (Gleick and Lane 2005; Conca 2006; Varady and Iles-Shih 2011) and to the recognition that local, national, and basin-level water issues are interlinked within a global water system. Regional multilateralism is also extended to water policy. The prime example here is the European Union and European environmental legislation, but also the Southern African Development Community (SADC) countries have, for example, developed a joint water policy.

The UNECE Water Convention[2] is an interesting example of the evolution of a global water convention. This Convention on the Protection and Use of Transboundary Watercourses and International Lakes was adopted in Helsinki in 1992 and entered into force in 1996. Initially the convention was limited to countries in the region of the United Nations Economic Commission for Europe (UNECE). In 2013 an amendment entered into force that allows accession by countries outside the UNECE region, thus turning the Water Convention into a legal framework for transboundary water cooperation worldwide. The Convention is a framework agreement which does not replace bilateral and multilateral agreements for specific basins or aquifers. It fosters cooperation between countries and provides guidance for the establishment and implementation of specific agreements.

In 2010 the United Nations General Assembly adopted a resolution which explicitly recognized the human right to water and sanitation: "*The Resolution calls upon States and international organisations to provide financial resources, help capacity-building and technology transfer to help countries, in particular developing countries, to provide safe, clean, accessible and affordable drinking water and sanitation for all.*" (http://www.un.org/waterforlifedecade/human_right_to_water.shtml). The Resolution is clearly linked to the Millennium Development Goals on water which resulted in improvements but did not meet their ambitious targets (Pahl-Wostl et al. 2013b).

---

[2] http://www.unece.org/env/water/text/text.html.

Table 6.1 Examples of arguments for the different governance level (partly derived from Gupta and Pahl-Wostl 2013)

Level	System-analytic argumentation	Political reasons
Global	Global feedback cycles—virtual water, teleconnections	Prevent free-riding; policy coherence; share information/experiences; transfer technologies and resources
Fluvial/transboundary	Management scale boundaries of the resource—connection via hydrology	Prevent free riding; develop common principles of water sharing and water pollution; create a level playing field; jointly solve common problems
National	Coherence with other policy domains governed at national level Major boundary of societal system	Prevent free riding: State is negotiating unit in international relations; domestic regulatory functions rest with the state; state can empower other actors to take a role
Local	Local problems do not need higher level intervention	Laboratories of policymaking; capable of own initiative; ownership of problem and hence solution

The different approaches to water governance—local, national, basin, and global—are not mutually exclusive. They indicate that different water issues are dealt with at different levels and that historically diverse perspectives have dominated. All approaches encounter challenges which can only be dealt with by addressing problems at their appropriate scale rather than promoting panaceas.

What is then an appropriate spatial scale—and administrative level—to deal with a specific water governance problem? Thus far, I have mainly elaborated on scholarly arguments that adopt a problem solving perspective. However, actors may have many different reasons for selecting a certain scale for dealing with a water issue. Table 6.1 lists both system-analytical arguments and political reasons for choosing a certain level. The reasons for the choice of a certain spatial scale or level are manifold and may be rooted in problem assessment as much as in political processes and power games. In the following I elaborate in more depth on different approaches to selecting what is considered an appropriate governance level for addressing governance challenges.

6.1.2 *Characterization of Spatial and Temporal Dimensions of Water Resource Problems*

A technocratic approach assumes that there exists an appropriate governance level to deal with a governance problem which can be derived from external, objective analysis based on factual knowledge. This implies that the boundaries of a problem (issues of concern, scales in space and time) as well as criteria for assessing the appropriateness of a strategy for dealing with it can be derived from such analysis. The Driver, Pressure, State, Impact, Response (DPSIR) framework has been used as an approach to analyse the cause-effect chains in environmental problems (Smeets and Weterings 1999; Borja et al. 2006). Drivers are societal characteristics such as economic development and regulatory frameworks. These drivers influence pressures which are the processes directly affecting ecosystems (e.g. water use for irrigation in agriculture) and changing their state (e.g. groundwater tables). Alterations in state lead to impacts on socio-ecological systems (e.g. degradation of wetlands, increased droughts). It is assumed that an improved understanding of these causal relationships allows the development and assessment of appropriate and effective response strategies and measures.

The DPSIR framework has been used in particular in Europe to inform policy processes and support policy implementation (EEA 2010; Impress 2003). It can also be used to analyse the scale characteristics of environmental problems (Pahl-Wostl 2002)—an approach which has not yet found much application in environmental assessment. Figures 6.1 and 6.2 represent stylized DPSI schemes in a space-time grid to illustrate the different scale characteristics of the problem domains of climate change and water resources allocation, respectively. In this representation regional refers to larger political or geographical areas such as Europe, US, China and large trans-boundary basins. For most countries the national scale would be located between regional and local.

With regard to climate change, pressure from the emissions of greenhouse gases is regional and mainly caused by energy consumption and fossil-fuel burning in industrialized countries and increasingly in emerging economies. Drivers related to economic development are global however. Technologies are diffusing globally and lifestyles with similar levels of high energy consumption have been adopted in most countries of the world. Due to the short time scale of mixing processes within the

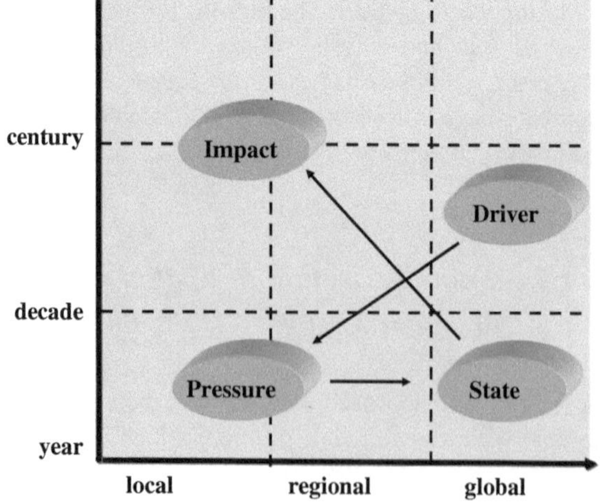

Fig. 6.1 Schematic representation of typical space-time-scale dependence of driver, pressure, state and impact for the climate change problem domain

global atmosphere, regional carbon dioxide emissions are dispersed in the atmosphere at the global scale within a few years. Hence, the decisive change in the state of the environmental system is at the global scale. The impacts of climate change are long-term and will be experienced at regional scales as a result of the regional manifestations of global climate change. The most severe impacts are expected in developing countries that have contributed relatively little to the overall problem of climate change (IPCC 2014). As a consequence of such considerations climate change has been conceptualized as a problem that can only be tackled at the global scale and with a high degree of international cooperation of nation states (IPCC 1992). Climate change mitigation is expected to be associated with additional costs and the efforts of individual countries cannot solve the problem if many other countries do not stop or at least decrease their production of greenhouse gases. Countries that continue to produce greenhouse gases while other countries stop doing so will also profit if climate change and its impacts can be prevented. However, the global United Nations Framework Convention on Climate Change (UNFCC) process has not been particularly effective to date and is even perceived by some as a complete failure (Rogelj et al. 2010; Bodansky 2010). Hence the global discourse has changed and some voices argue for a much stronger role of the regional and even local scales in developing and implementing effective response strategies (Ostrom 2010b, 2012). Proponents of such approaches argue that early movers will have a competitive advantage once all countries join and that reducing the dependence on fossil fuels will mobilize innovative potential and increase autonomy from fossil fuels at a larger scale in the long term. In summary the whole development of the climate change discourse could be described as moving from the dominance of global governance to increasingly engaging regional and local scales and acknowledging that climate change is a multi-level challenge. In the domain of water governance and management the reverse development can be observed.

Figure 6.2 represents a stylized DPSI scheme for the water resources allocation problem domain in a space-time grid. Arrows depicting the causal chain are not included to simplify the representation. Pressures affecting quantity and quality of water resources are inherently local and short term (e.g., high water consumption for irrigation agriculture, fertilizer use in agriculture, industrial pollution). The state of the available water resource in terms of quantity and quality may thus be affected rapidly. The environmental reservoirs of importance are regional aquifers. Large rivers are responsible for directed transport processes over wider spatial areas and longer distances. Pollution incidents may cause impacts downstream within hours and days. Such events are not included in this schematic representation however. Impacts such as the depletion of aquifers or groundwater pollution are experienced in the short to mid-term at regional scales. Natural processes cause the uneven distribution of precipitation and physical water scarcity. Regional water scarcity problems are not counteracted by any physical transport phenomena related to the global hydrological cycle. In general, the whole problem domain is much more localized than climate change. For a long time the importance of the global scale has thus been ignored (Pahl-Wostl et al. 2002). However, in a globalized economy drivers are

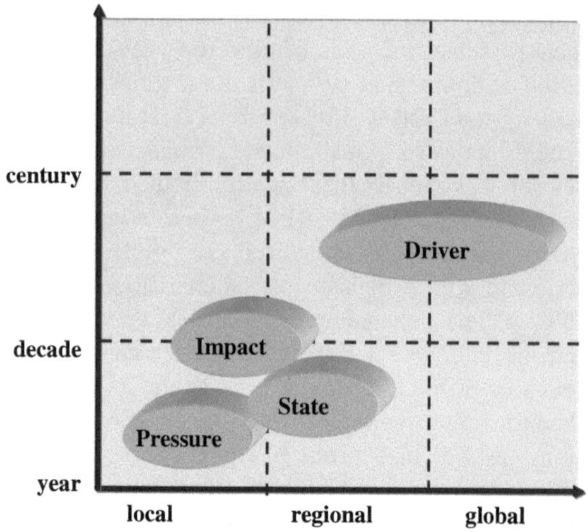

Fig. 6.2 Schematic representation of typical space-time-scale dependence of driver, pressure, state and impact for the problem of water resources allocation

increasingly global (Vörösmarty et al. 2013). Economic development leads to changes in diet and affects global food trade. The interest in renewable energy leads to an increase in the global production of biofuels which affects land-use and food production. Export of agricultural produce has implications for water consumption in producing countries. In summary (vague-adjective needed or another term) development as a whole could be described as moving from the dominance of local governance to increasingly engaging regional and global scales and acknowledging that sustainable water governance and management is a multi-level challenge.

6.1.3 Social Construction of Scale in Water Resource Problems

Irrespective of the considerable uncertainties involved in the DPSIR approach, it provides a structured assessment of the nature of cause-effect relationships in environmental problem domains. However, policy making is not a straightforward process aimed at problem solving where a common, objective understanding of a problem can be developed and communicated and the interest into solving it is shared among decision makers. The discussion of the historical development of the recognition of scales in the climate change and water problem domains has illustrated that "the" scale of a problem and the response to it is strongly influenced or even determined by political processes and interests outside of the problem domain.

In particular literature in critical and political geography has highlighted the importance of the social construction of scales and documented the role of politics in creating and shifting scales (Swyngedouw 1997, 2000; Delaney and Leitner 1997; Sayre 2005; Norman et al. 2012; Marston 2000). Initial policy framing and the selection of problem boundaries depend on who is involved and whose voice is having impact in problem assessment (Lebel 2006; Mostert et al. 2007). Even scientific and technical expertise which should contribute a perspective devoid of vested political interests is influenced by disciplinary practice and experience. The choice of a spatial scale determines system boundaries and which issues and which actors are included and excluded (cf. as well Table 6.1). Actors may have different reasons for up- or down-scaling an environmental problem (Gupta 2008). Change in scale may lead to changes in responsibility and accountability, and in power or access to different kinds of resources. By scaling-up a problem from a national to supra-national level more countries are included which might increase political legitimacy and improve the effectiveness of governance solutions to sustainability problems. At the same time such scaling up may be an attempt by national governments to reduce their own accountability, to postpone decisions and avoid taking measures at the domestic level (ibid, p. 237). Effectiveness of governance may also be improved by scaling down a problem. Scaling down might allow the use of existing institutions. It might help to mobilize local people to design and implement solutions relying on their knowledge. At the same time national governments may use scaling down to detract from their own responsibilities, to divide and control power over and access to resources (ibid, p. 239). Thiel (2015) showed that such scalar dynamics and reorganization within countries depends on institutional settings (viz. constitutional rules) that determine the necessary settings for the winning coalition.

Norman et al. (2012) summarize findings from case studies on water governance which provide evidence for the interdependence between power and social networks and institutional dynamics and scalar constructions. They point out that the recognition of scale as socially constructed and contingent on political struggle might inform analyses of water governance and advance understanding of hydrosocial networks. Based on their analysis of water governance in the Mekong river basin, Lebel et al. (2005, p. 1) argue that "*Acknowledging how actors' interests fit along various spatial, temporal, jurisdictional, and other social scales helps make the case for innovative and more inclusive means for bringing multi-level interests to a common forum.*" Ignoring such relationships and assuming that scales for science, management, and decision making can be unambiguously derived from physical characteristics of water resources might lead to irreconcilable conflicts though.

Overall one can conclude that an 'optimal' spatial or temporal scale on which water should be governed or managed does not exist. Water-related problems are always multifaceted and addressing them requires the inclusion of more than one scale in space and time. Different aspects of water management issues need to be addressed at different scales. In order to assure good governance[3] reliance on

[3]Good governance is characterized by being accountable, transparent, responsive, equitable and inclusive, effective and efficient, follows the rules of law, participatory, consensus oriented (UNESCAP 2009).

transparent and effective cross-scale coordination and negotiation processes is essential. It is important to develop an improved understanding of multi-level water governance and mechanisms of cross-scale coordination, of the role of knowledge and different conceptualizations of scales.

6.2 Conceptualization of Multi-level Water Governance

6.2.1 Multi-level Bureaucracies

Even though the term of "multi-level governance" has become more popular only over the last decade, such phenomena have been the subject of analyses in the political sciences much earlier. Research on federalism has explored delegation of authority from the central state to lower levels and analysed the performance of different kinds of political systems from more centralized (e.g., France) to more federal (e.g. Germany, Switzerland) structures (Ammon et al. 1996). The influence of different levels has always been a theme in international relations studies (Welch and Kennedy-Pipe 2004). Multi-level governance as such has become more prominent with research on the European political system which has become characterized by the complex interaction between the European level, the nation states and regions (Hooghe and Marks 2003; Grande 2000). Hooghe and Marks (2003) reviewed different strands of literature on the changing role of the central state in complex governance arrangements to conclude that there seems to be an overall agreement that governance has become (and should be) multi-level. In particular arguments for increasing flexibility have been put forward to support a shift towards multi-level instead of centralized governance arrangements. However, Hooghe and Marks conclude as well that there is no consensus about how multi-level governance should be organized. This implies questions such as the following need to be addressed: "*Should jurisdictions be designed around particular communities, or should they be designed around particular policy problems? Should jurisdictions bundle competencies, or should they be functionally specific? Should jurisdictions be limited in number, or should they proliferate? Should jurisdictions be designed to last, or should they be fluid?*" (ibid, p. 236). To capture such differences Hooge and Marks introduce a distinction between two types of multi-level settings which they call Type I and Type II governance.

Type I governance is related to the delegation of authority to general-purpose, non-intersecting, and stable jurisdictions of limited number as in typical federal systems. Such jurisdictions are characterized by non-intersecting membership and a limited number of jurisdictional levels. They are the product of purposeful design and are characterized by a systemwide, durable architecture. The European Union with its levels of supra-national, national and provincial governments is a typical example of Type I governance. Type II governance is characterized by task-specific jurisdictions, intersecting membership, many jurisdictional levels and flexible design. These governments are intended to respond flexibly to emerging challenges. Examples at the local level are authorities that provide specific public services such as waste water treatment or water supply. The Emschergenossenschaft, for example is a non-profit, regional government and private sector cooperation responsible for waste water treatment and flood protection in the Emscher basin, a tributary to the Rhine in North-Rhine Westphalia, Germany. Municipalities and business located along the course of the Emscher are members of the association.

The chief benefit of multi-level governance lies in its flexibility with respect to scale, and its ability to deliver public services and address problems at the appropriate scales with appropriate means. However, Tsebelis (2002) argues that an increasing number of powerful actors involved in governance of any type tend to be an impediment to flexibility and change and serve to maintain the status quo. Thus the benefits of multi-level governance do not come for free. They are associated with the transaction costs of coordinating multiple jurisdictions. According to Hooghe and Marks, Type I governance reduces transaction costs by limiting the number of jurisdictions to be coordinated through non-intersecting membership and general-multi-purpose interactions. Type II governance limits coordination costs by constraining the number of interactions across jurisdictions through functional specificity and flexible, policy-specific architecture. Type I and Type II jurisdictions are complementary and in practice one may encounter complex architecture where both types exist alongside and interact with each other. Box 6.1 provides an example of the complex multi-level governance structure that developed in Germany in the context of the implementation of the European Water Framework Directive.

Box 6.1 Multi-level water governance arrangements for the implementation of the European Water Framework Directive (WFD) in Germany illustrated by the Elbe River Basin

The WFD which entered into force in 2000 brought about considerable change in water governance in all European Member states. This is illustrated by the complex multi-level setting that was established in the German Elbe basin in the course of implementation of the WFD (Borowski 2004; Borowski et al. 2008). The WFD prescribes the management of water at the river basin scale. In Germany had not yet introduced basin management prior to the WFD. The dominant level for water governance and management has traditionally been the federal state level (Bundesländer). Germany decided not to introduce another bureaucratic layer of river basin organizations but to establish coordination groups. In the Elbe basin this is the Flussgebietegemeinschaft (FGG) Elbe comprising representatives from all federal states having a share in the German Elbe Basin. One state has the lead on a rotating base. The FGG only has a coordination function—all formal decision-making power remains at the federal state level. The part of the Elbe basin under German jurisdiction is divided in five sub-basins referred to as coordination units. The various committees of the FGG Elbe coordinate the development of management plans for these sub-basins. The FGG Elbe represents Germany in the international river basin coordination group comprising delegates from Germany,

Czech Republic, Poland, Austria and the European Commission. Stakeholder participation is mainly organized at the level of the sub-basins and at the international level. In the meantime coordination groups in most German basins have also taken on the responsibility for implementing the European Floods Directive which entered into force in 2007. This example illustrates that multi-level governance can become quite complex and involved. It is evident that care needs to be exercised to ensure that costs of coordination do not exceed benefits. It is thus not a trivial task to implement effective and efficient institutional multi-level governance settings.

Most work on multi-level governance is restricted to the realm of multi-layered jurisdictions and bureaucratic hierarchies. For the sake of conceptual clarification Rosenau (2004) suggested using the multi-level concept to refer exclusively to levels of government. He then continued to introduce a distinction between spheres of authority deriving from formal and informal rule systems, respectively, to be able to capture the complexity of governance settings where various types of actors participate and different governance modes interact. Furthermore, he introduced a typology of six types of international governance regimes based on the importance of formal and informal structures and the vertical and horizontal flow of authority. We used Rosenau's typology to analyse architectures of global water governance and concluded that the current global water governance regime is a fragmented, or in the terminology used by Rosenau a mobius-web, arrangement with complex formal and informal vertical and horizontal interactions (Pahl-Wostl et al. 2008). I do not follow Rosenau's typology here but focus on an approach which I consider more appropriate for analysing a wider range of governance systems for the local to the global levels: polycentric governance.

6.2.2 *Polycentric Governance*

Work on polycentric governance started with the reorganization of public administration but has developed into a much broader and all-encompassing approach over the past decades. The work of Elinor and Vincent Ostrom laid the foundations for an increased recognition of the importance of polycentricity in the governance of environmental resources. Their original point of departure was to contradict conventional wisdom that public administration and thus the governance of metropolitan areas should be organized in a hierarchical and centralized way (Ostrom et al. 1961; Ostrom 1972). Studies of metropolitan areas provided evidence that smaller units and the wider distribution of responsibilities often outperformed larger centrally controlled structures in delivering services to citizens (Ostrom et al. 1978). Ostrom et al. (1961) stressed from the outset of their research that the distribution of responsibilities must be linked to coordination by a shared set of rules: "*Polycentric connotes* ***many centers of decision making*** *that are formally* *independent of each other... To the extent that they take each other into account in competitive relationships, enter into various contractual and cooperative undertakings or have recourse to central mechanisms to resolve conflicts, the various political jurisdictions in a metropolitan area may* ***function in a coherent manner*** *with consistent and predictable patterns of interacting behavior. To the extent that this is so, they may be said to* ***function as a system***" (ibid, pp. 831–832, own emphasis). At that time this statement was quite revolutionary contradicting conventional theory. More than half a century later, the benefits of wider distribution of authority are generally acknowledged. However, despite numerous governance reforms supporting decentralization systematic analyses of the conditions that affect their performance are lacking. In particular the importance of effective coordination has been largely neglected. Polycentric governance systems must fulfil at least two criteria to function as systems: the presence of multiple centres of decision making and coordination by an overarching system of rules (Ostrom 2010b; McGinnis and Ostrom 2011; Aligica and Tarko 2012). The comprehensive work of Elinor Ostrom highlighted the importance of self-organization as a guiding principle in polycentric systems (Ostrom 2010a). Coordination and rules evolve from negotiations and interactions rather than being imposed by one powerful actor as might be the case in a strictly hierarchical system where coordination is imposed from the top. Elinor Ostrom's work focused on the governance of common-pool resources by local communities. Such local communities must, in general, be nested in a multi-level governance structure where governmental arrangements provide supporting conditions at higher governance levels (Mansbridge 2014). Polycentricity may be described as a kind of systemic logic characterizing many spheres of governance (Aligica and Tarko 2012). Further empirical studies have strengthened the assumption of systemic interdependency in polycentric governance systems. Plummer et al. (2013) confirm from their chronology of management in the Grand River watershed in Canada, a shift in governance towards a more polycentric and cooperative nature. They emphasize the need to understand institutions as being more broadly embedded in the social context. Blomquist and Schlager (2005) observe: "*However complicated they are, polycentric regimes may be seen as incorporating horizontal and vertical elements, with communities of interest, place, or identity acting as principals with respect to organizations or governments that represent them, and those organized communities interacting with each other horizontally to contest and coordinate over watershed governance and management in something more nearly like a system of checks and balances*" (ibid, p. 109).

But developing effective polycentric watershed governance which overcomes both fragmented bureaucracies and rigid central coordination may take considerable time. In her comprehensive account of the several decades of history of river basin management in the Columbia River basin in the US launched already in 1937, Vogel (2012) pointed out that organising management in a river basin territory did not lead to more holistic, balanced management. Management had for several decades been disproportionately focused on hydropower production, while responsibilities, benefits, and influence were distributed in a way to please the most powerful interests rather than to contribute to a more sustainable and equitable sharing of benefits. This can in part be explained by the complex institutional landscape that river basin management was confronted with. The Columbia River basin is situated in two countries, crosses multiple US state lines and one Canadian province, encompasses scores of local jurisdictions, and includes the territories and fishing grounds of fourteen Native American tribes and several Canadian First Nations. However, Vogel (2012) also pointed out that the accountability of river managers to a wide array of constituents from throughout their service area, has contributed over time to the development of institutional capacity. Since the 90s benefits derived from the river basin have diversified (e.g., introduction of fish and wildlife funds) as a response to changes in public opinion and other pressures but also profiting from improved institutional capacity and experience with cooperation. In their analyses of the development of integrated flood management in Europe, Pahl-Wostl et al. (2013a) also demonstrated that change and development of institutional capacity take place on time scales of decades rather than years. This implies that developing an improved understanding of the pathways towards polycentric governance needs to receive more attention in order to facilitate long-term processes of change.

But is polycentricity always a desirable property of governance systems? Will they display increased flexibility and the ability to respond better to complex governance challenges? From a normative point of view it is of significant interest that polycentric systems are assumed to be more flexible, have a higher ability to adapt to a changing environment and their integrity is less affected by sudden changes or failures in parts of the system (Ostrom 2001, 2005; Pahl-Wostl 1995, 2007a, b; Hooghe and Marks 2003). The argument that polycentric systems are more flexible has been a core argument in supporting decentralization of government functions (Hooghe and Marks 2003). I focus more broadly here on the increased adaptive capacity of polycentric systems and arguments supporting this derived from complex adaptive systems theory. A CAS is a complex, nonlinear, interactive system which has the ability to adapt to a changing environment (Pahl-Wostl 1995; Levin 1999). Such systems are characterized by the potential for self-organization, existing in a non-equilibrium environment. In a CAS, many elements interact according to certain rules of interaction. These elements are diverse in both form and capability and they adapt by changing their rules and, hence, behaviour, as they gain experience. Examples include living organisms, the nervous system, the immune system, the economy, corporations, societies, and so on. Often these systems have a modular structure which means that they are composed of connected sub-units. Modular system structure and decentralized control lead to higher degree of adaptiveness and robustness of a system (Miller and Page 2007). Similarly theoretical ecologists have claimed a positive relationship between the increasing diversity of ecological systems and the increasing ability of governments to maintain functional integrity in changing environments (Pahl-Wostl 1995; Ludwig and Walker 1997; Tilman 1999). Polycentric and adaptive governance systems should include a certain degree of redundancy. This claim is in conflict with the short-term maximization of efficiency. An important research question is how to strike an appropriate balance between improving efficiency by reducing redundancy and increasing adaptive capacity.

Which paths will lead tow polycentric systems rather than governmental fragmentation? Essential elements of polycentric governance are a balance between bottom-up and top-down influences and the capacity of actors to self-organize (Andersson and Ostrom 2008; Pahl-Wostl 2009; Huntjens et al. 2011; Marshall et al. 2013; Plummer et al. 2013; Young 2011). To function as a system, polycentric governance requires an overarching set of rules. To realize such a governance system and respect good governance principles is demanding. Good governance is participatory, consensus-oriented, accountable, transparent, responsive, effective and efficient, equitable and inclusive, and follows the rule of law (UNESCAP 2009). Pahl-Wostl et al. (2012) found a strong correlation between the polycentricity of the water governance system and the realization of good governance principles. One may argue about which is the dependent and which is the independent variable. However, such an argument may be futile if polycentricity and good governance are mutually reinforcing characteristics of the governance system. The path towards polycentricity may be essential for building capacity for good governance, in particular when the point of departure is a governance system that does not comply with good governance principles. Polycentric regimes result from emergence and self-organization in combination with purposeful design. Dynamics such as these requires the combination and interactions of different governance modes—networks, bureaucratic hierarchies and markets need to act in concert (Pahl-Wostl 2009). In bureaucratic hierarchies coordination is mainly achieved by top-down control. Within market systems interactions among actors are mainly characterized by competitive relationships. Within networks coordination is mainly based on trust and cooperation. If bureaucratic hierarchies are dysfunctional, because the rule of law is not respected and rent-seeking behaviour of governmental actors prevails, network governance and strengthening the capacity of local communities to exercise their rights and to call for accountability among governmental officials may be essential for improving governmental performance.

In summary, a polycentric governance system can be defined as a complex, modular system where differently-sized governance units with diverse purposes, organization, and spatial locations interact to form together a largely self-organized governance regime. Polycentric governance systems are characterized by high degrees of freedom. Multi-level governance in polycentric systems implies that decision making authority is distributed within a nested hierarchy and does not reside or is not concentrated at one single level, not the top where only the highest level of government is enforcing decisions, nor the intermediate level where only states/provinces enforce decisions beneficial for their region without due consideration to other regions. Polycentric systems balance bottom-up and top-down flow of authority and effective cross-level coordination is essential for effective governance.

6.3 The Nature of Linkages Across Administrative Levels

6.3.1 OECD Studies on Challenges in Multi-level Water Governance

Recognizing that "the current water crisis is largely a governance crisis" (OECD 2011, p. 11), the OECD launched its Water Governance Initiative in 2013,[4] an international stakeholder network to share good practices for improving governance in the water sector. One reason for launching this initiative were findings of the OECD (2011) Report "Water Governance in OECD Countries: a Multi-level Approach" that concluded that despite increasing knowledge of the technical, economic and institutional solutions to water problems, serious obstacles were encountered in their implementation. The report highlighted that "*There is no one-size-fits-all answer, magic blueprint or panacea to respond to governance challenges in the water sector, but rather a plea for home-grown and place-based policies integrating territorial specificities and concerns. The institutions in charge of water management are at different developmental stages in different countries, but common challenges occur—including in the most developed countries—and can be diagnosed ex ante to provide adequate policy responses. To do so, there is a pressing need to take stock of recent experiences, identify good practices and develop pragmatic tools to assist different levels of governments and other stakeholders in engaging effective, fair and sustainable water policies.*" (ibid, p. 11).

With their programme on water governance the OECD has started to tackle a challenge that has not yet been taken up by the scientific community. Pahl-Wostl et al. (2012) deplored the lack of systematic comparative analyses of governance systems. The OECD has started to conduct surveys and comparative assessments on multi-level water governance. The ultimate goal is to identify major governance gaps and to support the exchange of experiences across countries on how to overcome them. Two studies from 17 OECD[5] and 13 Latin American[6] (LTA) countries report on major multi-level policy gaps and coordination instruments that were implemented in the various countries (OECD 2011, 2012). The LTA study group comprises developing and transition countries. The OECD study group comprises mainly industrialized countries. Two countries (Chile, Mexico) are included in both groups. The surveys are based on responses from a limited number of governmental representatives and therefore need to be interpreted with some care. Nevertheless these studies provide an initial overview of the major challenges related to multi-level water governance and differences in challenges encountered based on the state of institutional and economic development. The studies are based on the OECD Multi-level

[4]http://www.oecd.org/gov/regional-policy/water-governance-initiative.htm.

[5]Australia, Belgium, Canada, Chile, France, Greece, Israel, Italy, Japan, Korea, Mexico, Netherlands, New Zealand, Portugal, UK (England), USA (Colorado).

[6]Argentina, Brazil, Chile, Costa Rica, Cuba, Dominican Republic, El Salvador, Guatemala, Honduras, Mexico, Nicaragua, Panama, Peru.

Table 6.2 Multi-level governance gaps in water policy used in the OECD water governance framework (based on Table 3.1 from OECD (2011))

Governance gap	Manifestation in water governance
Policy gap	Overlapping, unclear allocation of roles and responsibilities
Administrative gap	Mismatch between hydrological and administrative boundaries
Information gap	Asymmetries of information between central and sub-national governments
Capacity gap	Lack of technical capacity, staff, time, knowledge and infrastructure
Funding gap	Unstable or insufficient revenues of sub-national governments to effectively implement water policies
Objective gap	Intensive competition between different ministries
Accountability gap	Lack of citizen concern about water policy and low involvement of water users' associations

Governance Framework which is organized around seven governance gaps which are summarized in Table 6.2.

Figure 6.3 shows the multi-level governance challenges for water policy making identified in OECD and in LTA countries, respectively. A higher number of countries report governance gaps in the LTA compared to the OECD country group. The average for all governance gaps is 73 % in the LTA and 52 % in OECD countries. Furthermore the various gaps have different levels of importance in the two groups. All countries in the LTA group reveal the overlapping, unclear allocation of roles and responsibilities as a major problem. This is followed by the lack of citizen concern and low involvement of water user groups. Both of these governance gaps are clearly coordination challenges. Lack of capacity and funding is an obstacle which is seemingly not easily overcome even in industrialized countries. Interestingly, a mismatch between hydrological and administrative boundaries does not appear to be an issue of major concern.

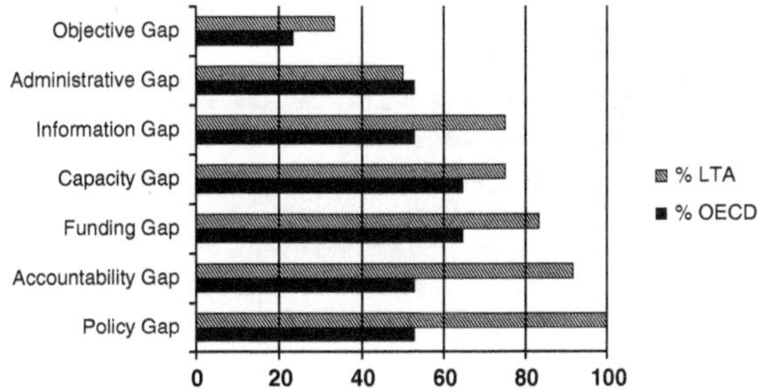

Fig. 6.3 Multi-level governance challenges for water policy making identified in OECD and in Latin American (LTA) countries. The x-axis provides the percentage of countries participating in the surveys that identified the respective gap as important or very important (based on Table 3.2 from OECD (2011) and Table 3.2 from OECD (2012))

Figure 6.4 shows vertical coordination mechanisms identified in OECD and in Latin American (LTA) Countries, respectively. A lower number of countries have vertical coordination mechanisms in place in the LTA compared with the OECD country group. The average for all governance gaps concerned with coordination mechanisms is 38 % in the LTA and 55 % in OECD countries. The most notable differences in the use of instruments between the two country groups are in coordination agencies, performance indicators, financial transfer on incentives and consultation of private stakeholders. One is tempted to conclude that using these instruments in the countries of the OECD group has been the key to successfully addressing the policy and accountability gaps still prevalent in the LTA countries.

Coordination agencies at the sub-national level support coordination between central government and local authorities and may take different forms such as committees, commissions or agencies (OECD 2011, p. 102ff). Such coordination helps in the building of capacity and the sharing of good practices at the sub-national level. Building capacity and facilitating co-ordinated actions across levels of government can also be achieved through performance measurements which comprises the setting of targets and indicators, monitoring and evaluation (ibid, p. 89).

Financial transfers between different levels of government are important for bridging the funding gap. Fiscal transfers from central budgets can be a source of investment capital in some countries and increase the creditworthiness of local municipalities in raising capital for investment from financial markets. Local governments can also play an important role if local authorities are authorised to raise taxes (ibid, p. 92ff).

Consultation of private stakeholders refers primarily to the participation of citizens and organized stakeholder groups in policy implementation (ibid, p. 99ff). Participation can build support for innovative measures and increase the awareness of the public at large. Involving interest groups in policy implementation is essential

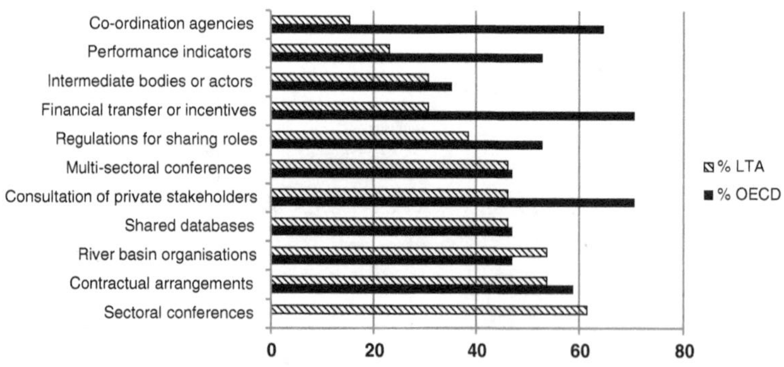

Fig. 6.4 Vertical coordination mechanisms identified in OECD and Latin American (LTA) countries. The x-axis provides the percentage of countries participating in the surveys that identified the respective gap as important or very important (based on Fig. 4.3 from OECD (2011) and Fig. 4.3 from OECD (2012))

for assessing conflictual issues at an early stage and for implementing integrated and innovative approaches.

These instruments are not independent of each other. Applied in appropriate combinations they may lead to improved coordination. Coordination bodies may set up an efficient process of measuring performance of policy implementation and of developing financial transfer schemes to mobilize the resources required for policy implementation.

6.3.2 Representation of Multi-level Governance in the MTF

Few frameworks have been developed that allow systematic and consistent representation and thus structured comparative analyses of multi-level governance systems and processes of vertical and horizontal coordination. The Management and Transition Framework (MTF) is intended to close this gap (Pahl-Wostl et al. 2010).

The MTF represents governance processes as sequences of connected action situations (ASs). Cross-level linkages can result from connections between ASs or from social interactions within ASs. When examining the connection between ASs at different levels one can distinguish between linkages by institutions and knowledge, respectively.

- *Institutions* (formal and informal) created at one level may influence ASs at another level. Typically influence is exercised by formal institutions in a top-down rather than bottom-up way as an expression of a hierarchical governance mode. However, normative claims about institutional change may be developed in a bottom-up process and influence higher institutional levels including legislation.
- *Knowledge* produced at one level may influence ASs at another level. Influence may be exercised both from the top down and from the bottom up. Traditional and local knowledge is typically produced at lower levels and then transferred in a bottom-up process whereas scientific expert knowledge is typically produced at higher levels and transferred in a top-down process.

With regard to social interactions within Ass, one can distinguish between actors:

- *Actors* from one level (e.g., national level) participate in decision processes at another level (e.g., European level or basin level). Actors from lower levels may become actively involved in the production of rules at higher levels that influence them (e.g. design of rules for implementing water user associations). Actors from higher levels may have a leading role in policy implementation at lower levels.
- *Actors* act as boundary spanners and take on the role of bridging organizations by acting at several levels simultaneously.

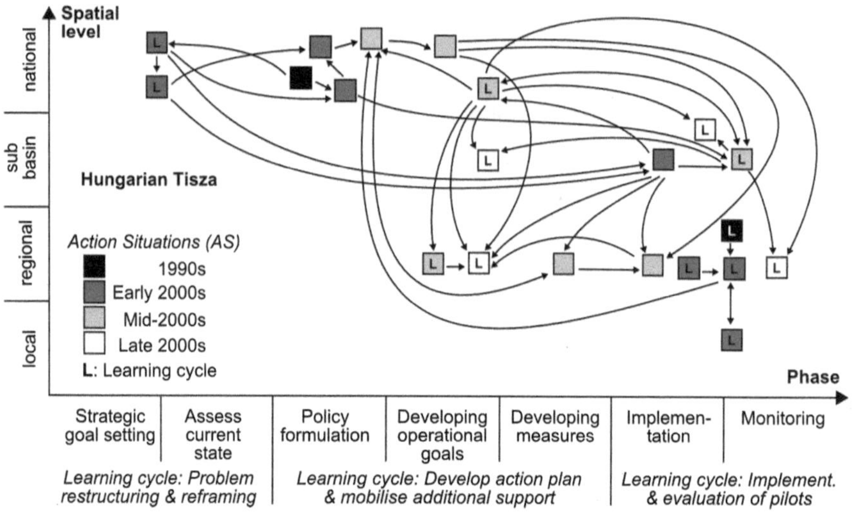

Fig. 6.5 Multi-level governance of floods: formal policy and informal learning and their interdependencies in the Hungarian Tisza. Each *box* denotes an AS and *arrows* denote influence by institutions or knowledge (cf. Fig. 3.8). The different *shadings* denote different time periods. ASs in the learning process are labelled with an L. The phases refer to stylized policy and learning processes, respectively (based on Fig. A1 in Pahl-Wostl et al. 2013a)

Important questions that need and can be addressed include the distribution of authority across levels and institutions and processes in place supporting vertical (across spatial scales and administrative levels) and horizontal (across administrative and sectoral boundaries) coordination. Box 6.2 illustrates the application of the MTF to the example of multi-level flood governance in the Hungarian Tisza basin.

Box 6.2 Using the MTF to analyse multi-level governance in the flood domain (Pahl-Wostl et al. 2013a)

Figure 6.5 shows the development of flood policy in the Hungarian Tisza in a multi-level representation of sequences of connected ASs. In the past decades Hungarian flood policy underwent major reform. Formal flood governance has traditionally been top-down and quite technocratic. A learning process carried by a shadow network succeeded in introducing innovative ideas in the policy reform process (Sendzimir et al. 2007, 2010). One can note that interactions across levels and between policy and learning process are complex. Furthermore policy development does not follow a linear logic from strategic goal setting to implementation but is an iterative process. More in-depth analyses revealed that vertical coordination by institutions (AS1 creates an institution which influences AS2) derives predominantly from the national level with influence on the basin and provincial levels as coordination by hierarchical top-down control in the formal policy process (Pahl-Wostl et al. 2013a). Knowledge is mainly produced and integrated (traditional, scientific, experiential from pilots) in the learning process. The key level of knowledge production is the provincial/federal state level with influence on the higher national and the lower municipal levels. scale-hoppers can be attributed an important role which are actors active at several levels and act thus as boundary-spanners. Scale hoppers can be found both in the group of governmental authorities and in that of civil society organization. Several ministries had a leading role at national levels and were actively involved in implementation processes at lower levels. With regard to NGOs and one research institute scale-hopping was mainly realized by individuals who participated in ASs at several levels. Individuals play a key role in cross-level linkages and in bridging the learning and formal policy processes. However, such linkages proved to be fragile if they hinged on the presence of individuals only and were not stabilized in shared and codified practices.

6.4 Overall Conclusions

An optimal spatial scale upon which to govern water resources does not exist. Water governance problems are complex and require a multi-level approach. The perceived scale of a problem though and its preferred solutions are socially constructed and often influenced by strategic political considerations. Hence transparent negotiation processes are of major importance to address the multi-level and multifaceted nature of water governance problems.

Polycentric governance systems seem to be a promising architecture for dealing with complex governance problems. To move towards polycentric rather than fragmented governance systems cross-scale coordination is essential. Polycentric systems are characterized by the distribution of power and authority combined with effective coordination. They embrace many levels and different modes of governance and balance top-down and bottom-up flow of authority. A lack of effective vertical and horizontal coordination still constitutes one of the major governance challenges both in developed and developing countries. Scientific knowledge and systematic analyses of the requirements for effective and flexible coordination and pathways towards polycentricity are still limited. Regarding instruments for coordination no blue-prints exist. To improve coordination more exchange on successes and failures is important and more analytical capacity is required to support a diagnostic approach.

References

Aligica, P. D., & Tarko, V. (2012). Polycentricity: From Polanyi to Ostrom, and beyond. *Governance: An International Journal of Policy, Administration, and Institutions*, 252237–252262. doi:10.1111/j.1468-0491.2011.01550.x

Ammon, G., Fischer, M., Hickmann, T., & Stemmermann, K. (Eds.). (1996). *Föderalismus und Zentralismus: Europas Zukunft zwischen dem deutschen und dem französischen Modell (Vol. 11, Schriftenreihe des Europäischen Zentrums für Föderalismus-Forschung)*. Baden-Baden, Germany: Nomos Publishers.

Andersson, K. P., & Ostrom, E. (2008). Analyzing decentralized resource regimes from a polycentric perspective. *Policy Science, 41*, 71–93.

Bernauer, T., & Siegfried, S. (2008). Compliance and performance in international water agreements: The case of the Naryn/Syr Darya Basin. *Global Governance, 14*, 479–502.

Blomquist, W., & Schlager, E. (2005). Political pitfalls of integrated watershed management. *Society and Natural Resources, 18*, 101–117. doi:10.1080/08941920590894435

Bodansky, D. (2010). *The International climate change regime: The road from Copenhagen*. Harvard: Harvard Project on International Climate Agreements.

Borja, Á., Galparsoro, I., Solaun, O., Muxika, I., Tello, E. M., Uriarte, A., et al. (2006). The European water framework directive and the DPSIR, a methodological approach to assess the risk of failing to achieve good ecological status. *Estuarine, Coastal and Shelf Science, 66*(1–2), 84–96. doi:10.1016/j.ecss.2005.07.021

Borowski, I. (2004). Public participation in the Elbe Basin. Case study report produced under Work Package 5. Osnabrück: University of Osnabrück.

Borowski, I., Le Bourhis, J., Pahl-Wostl, C., & Barraqué, B. (2008). Spatial misfit in participatory river basin management: Effects on social learning. A comparative analysis of German and French case studies. *Ecology and Society, 13*(1), 7. URL:http://www.ecologyandsociety.org/vol13/iss11/art17/

Cohen, A., & Davidson, S. (2011). An examination of the watershed approach: Challenges, antecedents, and the transition from technical tool to governance unit. *Water Alternatives, 4*(1), 1–14.

Conca, K. (2006). *Governing water: Contentious transnational politics and global institution building*. Cambridge, MA, USA: MIT Press.

Costejà, M., Font, N., Rigol, A., & Subirats, J. (2002). *The evolution of the national water regime in Spain*. Barcelona: Eu.

Delaney, D., & Leitner, H. (1997). The political construction of scale. *Political Geography, 16*(2), 93–97. doi:10.1016/S0962-6298(96)00045-5

Dombrowsky, I. (2008). Integration in the management of international waters: economic perspectives on a global policy discourse. *Global Governance, 14*, 455–478.

EEA. (2010). *The European environment—state and outlook 2010: Synthesis*. Copenhagen: European Environment Agency.

Fischhendler, I. (2015). The securitization of water discourse: Theoretical foundations, research gaps and objectives of the special issue. *International Environmental Agreements: Politics, Law and Economics*, 1–11. doi:10.1007/s10784-015-9277-6

Gleick, P. H., & Lane, J. (2005). Large international water meetings: Time for a reappraisal. *Water International, 30*(3), 410–414. doi:10.1080/02508060508691883

Global Water Partnership (GWP). (2000). Integrated water resources management. In G. W. Partnership (Ed.), *Technical advisory committee background papers* (Vol. 4). Stockholm: Global Water Partnership.

Grande, E. (Ed.). (2000). *Wie problemlösungsfähig ist die European Union? Regieren im europäischen Mehrebenensystem*. Baden-Baden: Nomos Verlag.

Gupta, J. (2008). Global change: Analyzing scale and scaling in environmental governance. In O. Young, L. King, & H. Schröder (Eds.), *Institutions and environmental change: Principal findings, applications, and research frontiers*. Boston, MA: MIT Press.

Gupta, J., & Pahl-Wostl, C. (2013). Global water governance in the context of global and multi-level governance: Its need, form, and challenges. *Ecology and Society, 18*(4), 53. doi:10.5751/ES-05952-180453

Hooghe, L., & Marks, G. (2003). Unraveling the central state but how? Types of multi-level governance. *American Political Science Review, 97*, 233–243.

Hooper, B. P. (2005). *Integrated river basin governance: Learning from international experiences*. London: IWA Publishing.

Huntjens, P., Pahl-Wostl, C., Rihoux, B., Schlüter, M., Flachner, Z., Neto, S., et al. (2011). Adaptive water management and policy learning in a changing climate: A formal comparative analysis of eight water management regimes in Europe, Africa and Asia. *Environmental Policy and Governance, 21*(3), 145–163. doi:10.1002/eet.571

Impress. (2003). Analysis of pressures and impacts. Common implementation strategy for the water framework directive. *Common Implementation Strategy for the Water Framework Directive. Guidance Document* (Vol. Guidance Document Nr. 3.).

IPCC. (1992). Climate change: The IPCC 1990 and 1992 assessments: Overview and policymaker summaries Canada: Intergovernmental panel on climate change.

IPCC. (2014). Summary for policymakers. In C. B. Field, V. R. Barros, D. J. Dokken, K. J. Mach, M. D. Mastrandrea & T. E. Bilir et al. (Eds.), *Climate change 2014: Impacts, adaptation, and vulnerability. Part A: Global and sectoral aspects. Contribution of working group II to the fifth assessment report of the intergovernmental panel on climate change*. Cambridge, United Kingdom: Cambridge University Press.

Jeffrey, P., & Gearey, M. (2006). Integrated water resources management: Lost on the road from ambition to realisation? *Water Science and Technology, 53*(1), 1–8.

Katz, D., & Fischhendler, I. (2011). Spatial and temporal dynamics of linkage strategy: Arab-Israeli water negotiations. *Political Geography, 30*(1), 13–24.

Knüppe, K., & Pahl-Wostl, C. (2013). Requirements for adaptive governance of groundwater ecosystem services: Insights from Sandveld (South Africa), Upper Guadiana (Spain) and Spree (Germany). *Regional Environmental Change, 13*(1), 53–66. doi:10.1007/s10113-012-0312-7

Lebel, L. (2006). The politics of scale in environmental assessment. In W. V. Reid, F. Berkes, T. Wilbanks, & D. Capistrano (Eds.), *Bridging scales and knowledge systems concepts and applications in ecosystem assessment* (pp. 37–58). Washington: Island Press.

Lebel, L., Garden, P., & Imamura, M. (2005). The politics of scale, position, and place in the governance of water resources in the Mekong region. *Ecology and Society, 10*(2), 18.

Lemos, M. C., & Agrawal, A. (2006). Environmental governance. *Annual Review of Environment and Resources, 31*(1), 297–325. doi:10.1146/annurev.energy.31.042605.135621

Levin, S. A. (1999). *Fragile dominion—complexity and the commons*. Cambridge, UK: Perseus Publishers.

Lowi, M. (1993). *Water and power: The politics of a scarce resource in the Jordan River Basin*. Cambridge: Cambridge University Press.

Ludwig, D., Walker, B., & Holling, C. S. (1997). Sustainability, stability, and resilience. *Conservation Ecology, 1*(1), 7. URL:http://www.consecol.org/vol1/iss1/art7/

Mansbridge, J. (2014). The role of the state in governing the commons. *Environmental Science and Policy, 36*, 8–10. doi:10.1016/j.envsci.2013.07.006

Marshall, G. R., Connell, D., & Taylor, B. M. (2013). Australia's Murray-Darling Basin: A century of polycentric experiments in cross-border integration of water resources management. *International Journal of Water Governance, 1*, 197–218. doi:10.7564/13-IJWG17

Marston, S. A. (2000). The social construction of scale. *Progress in Human Geography, 24*(2), 219–242. doi:10.1191/030913200674086272

McGinnis, M. D., & Ostrom, E. (2011). Reflections on Vincent Ostrom, public administration, and polycentricity. *Public Administration Review, 72*(1), 15–25. doi:10.111/j.1540-6210.2011.02488.x

Miller, J. H., & Page, S. E. (2007). *Complex adaptive systems: An introduction to computational models of social life*. Princeton, USA: Princeton University Press.

Molle, F. (2009). River-basin planning and management: The social life of a concept. *Geoforum, 40*(3), 484–494. doi:10.1016/j.geoforum.2009.03.004

Moss, T. (2003). Solving problems of 'fit' at the expense of problems of 'interplay'? The spatial reorganisation of water management following the EU water framework directive. In H. Breit, A. Engels, T. Moss & M. Troja (Eds.), *How institutions change. Perspectives on Social learning in global and local environmental contexts* (pp. 85–121). Opladen: Leske+Budrich.

Mostert, E., Pahl-Wostl, C., Rees, Y., Searle, B., Tàbara, D., & Tippett, J. (2007). Social learning in European river-basin management: Barriers and fostering mechanisms from 10 river basins. *Ecology and Society, 12*(1).

Newson, M. D. (1997). *Land, water and development: Sustainable management of river basin systems*. London: Routledge.

Norman, E. S., Bakker, K., & Cook, C. (2012). Introduction to the themed section: Water governance and the politics of scale. *Water Alternatives, 5*(1), 52–61.

OECD. (2011). Water governance in OECD countries: A multi-level approach, OECD studies on water. In OECD (Ed.), *OECD studies on water*. Paris.

OECD. (2012). Water Governance in Latin America and the Caribbean: A Multi-level Approach. In OECD (Ed.), *OECD studies on water*. Paris.

Ostrom, E. (1972). Metropolitan reform: Propositions derived from two traditions. *Social Science Quarterly, 53*, 474–493.

Ostrom, E. (1990). *Governing the commons. The evolution of institutions for collective action* (Political Economy of Institutions and Decisions). Cambridge: Cambridge University Press.

Ostrom, E. (2001). Vulnerability and polycentric governance systems. *IHDP (International Human Dimensions Programme on Global Environmental Change) Newsletter Update, 3*(1), 3–4.

Ostrom, E. (2005). *Understanding institutional diversity*. Princeton: Princeton University Press.

Ostrom, E. (2010a). Beyond markets and states: Polycentric governance of complex economic systems. *American Economic Review, 100*, 641–672. doi:10.1257/aer.100.3.641

Ostrom, E. (2010b). Polycentric systems for coping with collective action and global environmental change. *Global Environmental Change, 20*, 550–557.

Ostrom, E. (2012). Nested externalities and polycentric institutions: Must we wait for global solutions to climate change before taking actions at other scales? *Economic Theory, 49*, 353–369. doi:10.1007/s00199-010-0558-6

Ostrom, E., Parks, R. B., & Whitaker, G. (1978). *Patterns of metropolitan policing*. Cambridge, MA, USA: Ballinger.

Ostrom, V., Tiebout, C. M., & Warren, R. (1961). The organization of government in metropolitan areas: A theoretical inquiry. *American Political Science Review, 55*, 831–842.

Pahl-Wostl, C. (1995). *The dynamic nature of ecosystems: Chaos and order entwined*. Chichester: Wiley.

Pahl-Wostl, C. (2002). Polycentric integrated assessment [social learning, multi-scale stakeholder processes, participatory integrated assessment, evolutionary change in social systems.]. *Integrated Assessment, 3*(2–3), 220–232.

Pahl-Wostl, C. (2007a). Requirements for adaptive water management. In C. Pahl-Wostl, P. Kabat & J. Möltgen (Eds.), *Adaptive and integrated water management. Coping with complexity and uncertainty* (pp. 1–22). Heidelberg: Springer.

Pahl-Wostl, C. (2007b). Transition towards adaptive management of water facing climate and global change. *Water Resources Management, 21*(1), 49–62.

Pahl-Wostl, C. (2009). A conceptual framework for analysing adaptive capacity and multi-level learning processes in resource governance regimes. *Global Environmental Change, 19*, 354–365.

Pahl-Wostl, C., Becker, G., Sendzimir, J., & Knieper, C. (2013a). How multilevel societal learning processes facilitate transformative change: A comparative case study analysis on flood management. *Ecology and Society, 18*(4), 58. http://dx.doi.org/10.5751/ES-05779-180458

Pahl-Wostl, C., Conca, K., Kramer, A., Maestu, J., & Schmidt, F. (2013b). Missing links in global water governance: A processes-oriented analysis. *Ecology and Society, 18*(2), 33. http://dx.doi.org/10.5751/ES-05554-180233

Pahl-Wostl, C., Gupta, J., & Petry, D. (2008). Governance and the global water system: A theoretical exploration. *Global Governance, 14*, 419–435.

Pahl-Wostl, C., Hoff, H., Meybeck, M., & Sorooshian, S. (2002). The role of global change research for aquatic sciences. *Aquatic Sciences, 64*(4), iv–vi. doi:10.1007/PL00012595

Pahl-Wostl, C., Holtz, G., Kastens, B., & Knieper, C. (2010). Analyzing complex water governance regimes: The management and transition framework. *Environmental Science and Policy, 13*(7), 571–581. doi:10.1016/j.envsci.2010.08.006

Pahl-Wostl, C., Lebel, L., Knieper, C., & Nikitina, E. (2012). From applying panaceas to mastering complexity: Toward adaptive water governance in river basins. *Environmental Science and Policy, 23*, 24–34.

Plummer, R., Spiert, A., FitzGibbon, J., & Imhof, J. (2013). The expanding institutional context for water resources management: The case of the Grand River watershed. *Canadian Water Resources Journal, 30*, 227–244.

Rogelj, J., Nabel, J., Chen, C., Hare, W., Markmann, K., Meinshausen, M., et al. (2010). Copenhagen accord pledges are paltry. *Nature, 464*(7292), 1126–1128. doi:10.1038/4641126a

Rosenau, J. N. (2004). Strong demand, huge supply: Governance in an emerging epoch. In I. Bache & M. Flinders (Eds.), *Multi-level governance* (pp. 31–48). Oxford, UK: Oxford University Press.

Sayre, N. F. (2005). Ecological and geographical scale: Parallels and potential for integration. *Progress in Human Geography, 29*(3), 276–290. doi:10.1191/0309132505ph546oa

Sendzimir, J., Flachner, Z., Pahl-Wostl, C., & Knieper, C. (2010). Stalled regime transition in the upper Tisza River Basin: The dynamics of linked action situations. *Environmental Science and Policy, 13*(7), 604–619. doi:10.1016/j.envsci.2010.09.005

Sendzimir, J., Magnuszewski, P., Flachner, Z., Balogh, P., Molnar, G., Sarvari, A., et al. (2007). Assessing the resilience of a river management regime: Informal learning in a shadow network in the Tisza River Basin. *Ecology and Society, 13*(1), 11.

Smeets, E., & Weterings, R. (1999). *Environmental indicators: Typology and overview. EEA technical reports*. Copenhagen: European Environmental Agency (EEA).

Swyngedouw, E. (1997). Neither global nor local: 'Glocalization' and the politics of scale. In K. Cox (Ed.), *Spaces of globalization: Reasserting the power of the local* (pp. 137–166). New York, NY, USA: Guilford.

Swyngedouw, E. (2000). Authoritarian governance, power, and the politics of rescaling. *Environment and Planning D: Society and Space, 18*(1), 63–76.

Thiel, A. (2015). Constitutional state structure and scalar re-organization of natural resource governance: The transformation of polycentric water governance in Spain, Portugal and Germany. *Land Use Policy, 45*, 176–188. doi:10.1016/j.landusepol.2015.01.012

Tilman, D. (1999). The ecological consequences of changes in biodiversity: A search for general principles. *Ecology, 80*(5), 1455–1474.

Tsebelis, G. (2002). *Veto players: How political institutions work*. Princeton: Princeton University Press.

UNESCAP. (2009). *What is good governance*? Bangkok, Thailand: United Nations Economic and Social Commission for Asia and the Pacific.

Varady, R. G., & Iles-Shih, M. (2011). Global water initiatives: What do the experts think? In A. K. Biswas & C. Tortajada (Eds.), *Impacts of megaconferences on the water sector* (pp. 53–102). Berlin: Springer.

Vogel, E. (2012). Parcelling out the watershed: The recurring consequences of organising Columbia River management within a basin-based territory. *Water Alternatives, 5*(1), 161–190.

Vörösmarty, C. J., Pahl-Wostl, C., Bunn, S., & Lawford, R. (2013). Global water, the anthropocene and the transformation of science. *Current Opinion in Environmental Sustainability, 5*(6), 539–550.

Welch, S., & Kennedy-Pipe, C. (2004). Multi-level governance and international relations. In I. Bache & M. Flinders (Eds.), *Multi-level governance* (pp. 127–144). Oxford, UK: Oxford University Press.

Yang, H., Pfister, S., & Bhaduri, A. (2013). Accounting for a scarce resource: Virtual water and water footprint in the global water system. *Current Opinion in Environmental Sustainability, 5* (6), 599–606. doi:10.1016/j.cosust.2013.10.003

Young, O. R. (2002). *The institutional dimensions of environmental change: Fit, interplay, and scale*. Cambridge: MIT Press.

Young, O. R. (2011). Land use, environmental change, and sustainable development: The role of institutional diagnostics. *International Journal of the Commons, 5*(1), 66.

Zarfl, C., Lumsdon, A. E., Berlekamp, J., Tydecks, L., & Tockner, K. (2015). A global boom in hydropower dam construction. *Aquatic Sciences, 77*(1), 161–170. doi:10.1007/s00027-014-0377-0

Chapter 7
Shaping Human—Environment Interactions

Whereas humanity was still at the stage of hunting and gathering having little impact on the biosphere just a few thousand years ago, humans have become a global force in shaping the planet in the 21st century. This transformation has been accompanied by fundamental changes in the perception of human–environment relationships. Still we suffer from the dichotomy in which societal and ecological systems are conceptualized as separate and at times even antagonistic systems. It is a significant challenge for governance to overcome this dichotomy. To conceptualize human–environment interactions I introduce the concepts of ecosystem services and environmental hazards. Water security is introduced as a risk-based concept integrating the two perspectives. Based on this conceptualization I argue for and highlight the challenges of adaptive governance and management for increased resilience of social-ecological systems.

7.1 Human–Environment Interactions

7.1.1 A Historical Account of a Complex Relationship

The presence of humans on Earth altering their environment has a long history seen from the perspective of a lifespan of an individual human being. Even the presence of humans as a species is short though in the context of the history of the Earth. Most of the history of human-environment interactions was shaped by the desire of humans to impose some control on the environment, to protect themselves from dangers and benefit from environmental goods and services. The first traces of Homo sapiens go back more than 100,000 years. In their initial stage as hunters and gatherers, humans started to use simple tools but did not have much impact on their environment. The Neolithic revolution around 10,000 BC marked the first major transition for humanity and their impact on the environment. The establishment of settlements and the initiation of agricultural activities led to larger human populations for the first time. The first advanced civilisations developed. This transition also witnessed the emergence of significant human-induced environmental impacts through deforestation and overgrazing. But it was only with the Industrial

C. Pahl-Wostl, *Water Governance in the Face of Global Change*,
Water Governance—Concepts, Methods, and Practice,
DOI 10.1007/978-3-319-21855-7_7

Revolution in the 19th century and the transition from agricultural to industrialised societies that human beings became a major force in shaping the Earth at a global scale. The Industrial Revolution has brought unprecedented wealth and improvement in the quality of life to some societies. At the same time global disparities have increased dramatically and humanity is becoming a force that is having an unprecedented influence upon global material cycles. This development has caused a scientific debate on the start of a new geologic epoch: the 'Anthropocene' (Zalasiewicz et al. 2010). This debate has also been associated with discussions about a new responsibility for humanity and a change in human–environment relationships from exploitation to stewardship (Chapin III et al. 2009). It would be hubris to assume that humans have ever been or would ever be able to control the natural environment. The history of human-environment interactions has been characterised by attempts to manage unintended consequences of human activities. But the magnitude of current and potential future impacts has achieved a dimension which threaten the survival of humanity and higher forms of life. This would not be a disaster for the Earth but for humanity. Hence the plea for greater stewardship is first and foremost a claim in the interest of the survival of our own species.

Upon entering the Anthropocene, the sustainable management of global water resources is one of the most pressing environmental challenges of the 21st century. Technological progress has allowed the cultivation of deserts and floodplains. Pushing human activities to the limits of or even beyond the capacities of environmental systems has resulted in many regions in increased vulnerability to environmental extremes. Increasing water security for humans by massive investment in water technology in industrialized countries has been associated with severe degradation of ecosystems (Vörösmarty et al. 2010; Nilsson et al. 2005; Pahl-Wostl et al. 2013a).

So we are now faced with a compelling question: How should we conceptualize human-environment interactions in order to support a transformation towards sustainable water governance that overcomes the dichotomy between humans and nature, and in particular, the trade-off between human and environmental water needs?

7.1.2 Conceptualizing Human–Environment Interactions

The history of human–environment interactions suggests that a distinction between beneficial and harmful interactions, or rather what are perceived by humans as beneficial or harmful interactions, captures the most significant driving forces that are at play in the relationship between humans and nature. Hence the concepts of ecosystem services and environmental hazards are used to characterize the interactions between humans and environment as represented in Fig. 7.1. Services capture the function of an ecological system as provider of different types of services for human activities (e.g., water for irrigation). Hazards are the threats that an ecological system poses to a societal system (e.g., floods).

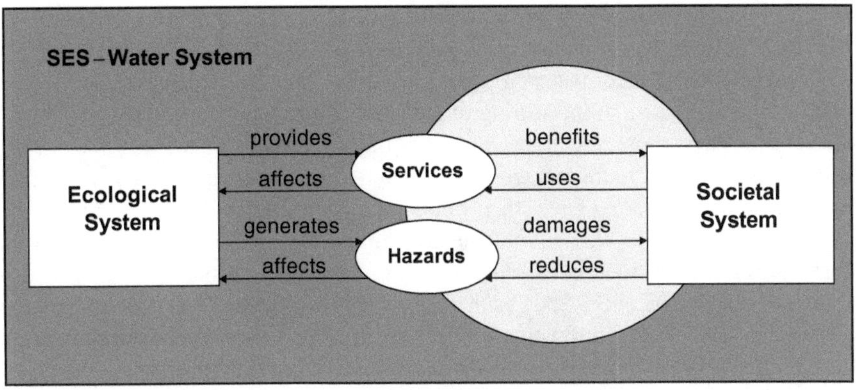

Fig. 7.1 Schematic representation of the interface between ecological and societal systems. In conformance with the terminology of the MTF, the overarching social-ecological system (SES) is called the water system and comprises all environmental and human components (Pahl-Wostl et al. 2010). Ecosystem services and environmental hazards serve as bridging concepts to characterize human-environment interactions. The *light shaded circle* denotes the sphere of influence of governance and management processes

The lightly-shaded circle denotes the potential of governance and management to influence the four kinds of interactions that emanate from the societal system. Services provide benefits to humans. These benefits depend on the societal perception of services and their valuation which can but must not necessarily be expressed in monetary terms. The relative importance attributed to a service may be determined by culture and economic conditions and may vary among different societal groups. The use of services is in general associated with infrastructure such as irrigation technologies which affect quantity and quality of consumption. The quality of a service can be characterized by the externalities that it produces, i.e. damage providing no direct benefit which is caused by the way in which the service is used (e.g., deforestation during the course of building irrigation channels). Hazards impose damages on the societal system. The relative importance attributed to a hazard depends on exposure and economic conditions and may vary among different societal groups. Measures may be undertaken to reduce the occurrence and/or the impacts of hazards. Both the use of services (e.g., groundwater supply) and the management of hazards (e.g., flood control) may lead to changes in the characteristics of an ecosystem (e.g., floodplains, biodiversity, flow variability) which may, in turn, affect the provision of services and the likelihood of hazard events. These feedback loops result in a tight coupling between societal and ecological components and it is more appropriate to talk of a social-ecological system (SES) that has to be analysed as a whole. Conventional flood management provides an example of a positive feedback loop which increases vulnerability to climate change and traps the system in a cycle where it is difficult to break out. In order to reduce flood hazards, rivers are regulated and dikes are built. This leads to the reduction of floodplain biodiversity and a reduction of benefits provided by floodplain services like water retention. Natural buffering capacity is further reduced by an increase in the sealing of soils resulting in an increase in peak flows. At the same time more assets are built into the floodplain. This leads to a higher likelihood of severe flood damage in the event of flooding which imposes further pressure on authorities to strengthen flood protection and build higher dikes. The system as a whole is also vulnerable to climate change. The sensitivity to an increase in extreme flood events is high and adaptation options are reduced as a result of path dependence. Such path dependence is not only manifested in large investments in enduring infrastructure but also in the whole governance structure that has co-evolved with a particular flood management paradigm. Socio-ecological systems thus become trapped in the prevailing pattern of using services and managing risks with little capacity for innovative change.

Such developments can be attributed to the lack of a systemic perspective in governance and management due to limited understanding or to explicit negligence of complex interdependencies driven by strategic considerations.

The increasing concern about the unsustainability of water resources management and the lack of sustainable solutions for water security led Grey and Sadoff (2007, p. 545) to develop a comprehensive approach by defining water security as "*...the availability of an acceptable quantity and quality of water for health, livelihoods, ecosystems and production, coupled with an acceptable level of water-related risks to people, environments and economies*". This definition embraces both beneficial aspects of water, namely ecosystem services and harmful aspects of water, namely hazards. One could rephrase this definition using the notions of services and hazards: **Water security** is the availability of the quantity and quality of water required to deliver an acceptable level of services for health, livelihoods, ecosystems and production, coupled with an acceptable level of risks emanating from water-related hazards to people, environments and economies. Such a framing of water security could guide the adoption and implementation of a more integrated and systemic perspective.

7.2 Ecosystem Services

7.2.1 Definition and Categorization

According to Daily (1997) ecosystem services can be defined as: "*the conditions and processes through which natural ecosystems, and the species that make them up, sustain and fulfil human life*" (ibid p. 4). Ecosystems are recognized for their essential services to human life and the functional aspects of ecosystems that support these services. Many environmental problems arise from negligence or ignorance of the role of vital ecosystem services and the implications of overexploiting some services thereby eroding the functional base of others and potentially generating new sources of environmental hazards. The Millennium Ecosystem Assessment (MA) conducted under the auspices of the United Nations was the first global effort "*to assess the consequences of ecosystem change for human well-being*" (MA 2005a). The MA established "*the scientific basis for actions needed to enhance the conservation and sustainable use of ecosystems and their contributions to human well-being*". The main findings (ibid p. 1) were: "*Over the past 50 years, humans have changed ecosystems more rapidly and extensively than in any comparable period of time in human history, largely to meet rapidly growing demands for food, fresh water, timber, fiber, and fuel. .... The changes that have been made to ecosystems have contributed to substantial net gains in human well-being and economic development, but these gains have been achieved at growing costs in the form of the degradation of many ecosystem services, increased risks of nonlinear changes, and the exacerbation of poverty for some groups of people..... The challenge of reversing the degradation of ecosystems while meeting increasing demands for their services ..... involve significant changes in policies, institutions, and practices that are not currently under way. ....Many options exist to conserve or enhance specific ecosystem services in ways that reduce negative trade-offs or that provide positive synergies with other ecosystem services*". A trade-off between two ecosystem services arises if enhancing the use of one service reduces the provision of another service. The MA provided clear evidence for the need for political will and action to reverse alarming trends.

At the time the MA was conducted an overall agreed framework on how to categorize ecosystem services did not exist. As part of the conceptual framework the MA introduced a now widely used typology for ecosystem services which makes a distinction between provisioning, regulating, supporting and cultural services. Provisioning services comprise, for example, food, drinking water and energy. Regulating services refer to benefits that are not directly perceived and thus often ignored such as climate regulation, waste decomposition, purification of water and air, and pest and disease control. Supporting services such as nutrient dispersal and cycling, seed dispersal and primary production provide the basis for ecosystem functioning. Cultural services comprise, for example, cultural and spiritual inspiration, recreational experiences and processes that contribute to scientific discovery. Table 7.1 lists important freshwater ecosystem services classified according to the MA categories. Supporting services are not explicitly listed since they form the basis for the other three categories rather than being a distinct class of services with direct benefits to humans.

Provisioning services have, in general, received the highest attention since they deliver direct benefits and marketable goods. Regulating services deliver mainly indirect benefits that have no real market value. Ecosystem services may have the nature of public or private goods as well as collective or toll goods, depending on the ease of exclusion and their subtractability (Ostrom 2005, p. 24). Typically

Table 7.1 Important freshwater ecosystem services (Aylward et al. 2005; Russi et al. 2013)

Provisioning services	Regulating services	Cultural services
Water for consumptive use for drinking, domestic use, and agriculture and industrial use Water for non-consumptive use for generating power and for transport/navigation Aquatic organisms for food and medicines	Maintenance of water quality (natural filtration and water treatment) Flood protection through buffering of peak flows, water retention Erosion control through water/land interactions and flood control infrastructure Groundwater recharge and discharge	Recreation—sporting activities Tourism—river and wildlife viewing Existence values—personal satisfaction, preferred housing Spiritual meaning, religious rituals

regulating and cultural services have the nature of public goods with a low possibility of exclusion and subtractability. Recreational activities may also have the nature of private or toll goods if equipment is required and/or access is restricted due to private ownership of land. These distinctions are important since they have significant implications for governance and management of ecosystem services and in particular trade-offs. Table 7.2 lists attributes of ecosystem services that should be taken into consideration in order to develop governance responses appropriate for guiding their sustainable management (Pahl-Wostl et al. 2010). Furthermore it is essential to develop a profound understanding of the complex interdependencies between different services.

7.2.2 Trade-off and Synergies—Interdependencies Between Services

In water governance and management systems overwhelming emphasis has been given to provisioning services, whereas regulating and supporting services and the requirements for maintaining them have been largely ignored for a long time. Provisioning services such as water supply for irrigation provide the most direct socio-economic benefits. Correspondingly, governance and management systems have evolved from the exploitation and guaranteeing of access to these services. Ineffective governance systems and ignorance of complex feedbacks have often led to ineffective use and overexploitation of some services to the detriment of the overall integrity of ecological systems with long-term negative consequences for human well-being.

Trade-offs between provisioning and regulating services are characteristic of intense agricultural production. The production of food, fiber or biofuels depends on the provision of freshwater. Regulating services such as water purification or

Table 7.2 Attributes of ecosystem services (Pahl-Wostl et al. 2010)

Type of service: according to the Millennium Ecosystem Assessment
Excludability: refers to the possibility that users can be excluded from using an ecosystem service. Without excludability it is difficult to charge for using a service and to avoid the presence of free riders who do not contribute to sustaining a service but still enjoy the benefits of using it (a typical problem of public goods). Many regulating services such as climate or disease regulation display the property of public goods
Subtractability: refers to whether an ecosystem service has a limited capacity only, which means that any user is diminishing this capacity and less is available for other uses and users. This is the case for most provisioning services which thus display the characteristics of common pool resources or private goods depending on excludability
Reproducability: refers to whether and how fast an ecosystem service can be reproduced or is irreversibly exploited. The lack of reproducibility asks for cautionary management and considerations of intergenerational equity. Fossil groundwater resources cannot be reproduced. Loss of biodiversity and species extinction is irreversible with potentially serious consequences for current and future human generations
Economic value: an ecosystem service may be priced or be taken into account in some other way in market transactions. This attribute would indicate if a service can be priced in principle and if it is already priced
Degree of variability: refers to the degree of variation in time of the availability of an ecosystem service and the uncertainty associated with this variability

groundwater retention may be severely impeded. A decline in regulating services but also direct pollution may lead to negative impacts on the provision of freshwater not only for agriculture but also for drinking water supply. Interdependencies between services are often complex and are found on different spatial and temporal scales. Negative impacts may be felt only after considerable time lags and spatially dislocated from the activities causing them.

Interdependencies as displayed graphically in Fig. 7.2, may result in different kinds of relationships among the trade-offs between two ecosystem services. The figure is based on elaborations by Elmqvist et al. (2011) on how to manage trade-off between provisioning and regulating ecosystem services. The magnitude and the shape and gradient of the curves representing the trade-offs can be influenced by governance and management strategies. If Service 1 refers, for example, to a provisioning service such as food production and Service 2 to a regulating service such as water retention, the shape is influenced by use of fertilizer and agrochemicals, by deforestation and/or by cropping patterns. However, Fig. 7.2 should not lead to the incorrect conclusion that it would be straightforward or even possible to quantify such relationships. Nevertheless, a qualitative understanding of the factors that influence the shape of the trade-off curves can already guide the development of more sustainable and integrative management strategies.

Adopting an approach which takes into account trade-offs and synergies between freshwater ecosystem services leads to a holistic landscape perspective. The characterization of goods and services and in particular their interdependence requires profound knowledge of ecological functions as well as of governance structures and

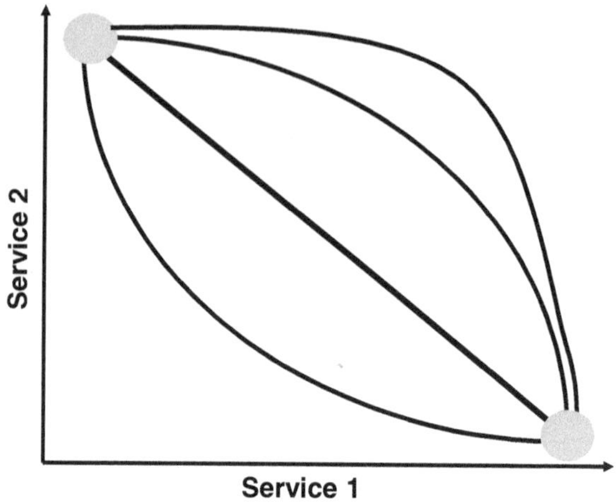

Fig. 7.2 Four potential shapes of the curves representing the trade-offs between two ecosystem services [based on Fig. 1 in Elmqvist et al. (2011)]

decision-making processes. Aquatic and terrestrial ecosystems, sectoral and regional policies cannot be seen in isolation. However, given the complexity of human–environment feedbacks, it is rare to find linear causal patterns of the linkages among ecosystems, ecosystem services, human well-being, human response, and feedbacks to drivers of change (Carpenter et al. 2009; Pahl-Wostl 2007). This is a challenge for developing adequate responses. More knowledge on complex interdependencies does not necessarily imply that uncertainties will be substantially reduced. Governance and management responses must be adequate to deal with uncertainty and surprise. And one must also to be aware that knowledge on complex interdependencies may not be the limiting factor in the politics that impacts the allocation of ecosystem services. Robards et al. (2011) highlighted the importance of power asymmetries and poverty traps for understanding trade-offs and for assessing the sustainability of prevailing patterns of flows of ecosystem services. Overcoming such political barriers is scarcely possible without more fundamental changes in resource governance.

7.2.3 Governance of Ecosystem Services—Beyond Monetary Valuation

The governance of ecosystem services has increasingly been associated with the logic underlying a market mode of governance (cf. Chap. 5) which builds on monetary valuations of ecosystem goods and services (Muradian and Rival 2012; Costanza et al. 2014; Gómez-Baggethun et al. 2010). One may distinguish between accounting schemes for natural capital (Prugh et al. 1999) and market-based instruments guiding exchange transactions such as payments for ecosystem services (PES) schemes. Whereas accounting schemes have not yet found widespread attention in policy making, PES schemes have enjoyed increased popularity over the past decade (Engel et al. 2008; Engel and Schaefer 2013; Wunder et al. 2008; Vatn 2010; Kinzig et al. 2011). Wunder et al. (2008, p. 834) define PES as "*(a) a voluntary transaction where (b) a well-defined environmental service (ES) or a land use likely to secure that service (c) is being 'bought' by a (minimum one) service buyer (d) from a (minimum one) service provider (e) if and only if the service provider secures service provision (conditionality)*". A comprehensive definition like this may have to be considered as an idealized benchmark. In practice, PES schemes do not represent one well-defined policy instrument but may vary substantially. Numerous examples for PES schemes in developed and developing countries have been documented over the past decade (Wunder et al. 2008; Russi et al. 2013; Pagiola 2008; Wunder and Albán 2008; Turpie et al. 2008; Bennett 2008; Corbera et al. 2007). They differ among other things in the design of the schemes such as government- or user-financed, involvement of government and degree of formal regulation, or level (local, regional, national). A prominent and often cited example of the effectiveness of PES in reconciling conflicting economic interests is the Vittel case in France (see Box 7.1 for more details). In this case the economic interest of a mineral water producer coincided with the preservation of high quality groundwater and he was willing to pay farmers to change their practices in order to protect this regulating service.

Box 7.1 PES in voluntary cooperation—The Vittel case [based on Perrot-Maître (2006)]: In the 1980s the continued production of the famous mineral water, Vittel, was threatened by potential pollution of the mineral spring by nitrate due to an intensification of agriculture in the catchment. Since regulations did not allow the treatment of mineral water, the potential pollution posed an existential threat to mineral water production. After elaborating on a range of measures Vittel (part of Nestlé Waters) decided to provide incentives to farmers to voluntarily change their practices. In collaboration with a leading national research organization (Institut National de la Recherche Agronomique—INRA) they launched an action-research programme providing financial incentives to farmers to experiment with innovative management practices and analysing the impact of farming practices on nitrate levels in the aquifer. All farms in the region engaged in contractual relationships which implied financial support and technical assistance. Payments were not conditional on a quantitative nitrate reduction target but on the willingness of farmers to adopt new practices. Nitrate levels and farming practices are now regularly monitored and recommendations are made if adjustments in management practices appear to be warranted. Overall the programme has proven to be highly effective and is always portrayed as one of the success stories of PES schemes. But it should also be highlighted that one factor in the success has been the trust that developed between farmers and the company over the years. This can be seen as example for a successful combination of market and network governance.

Success stories as the Vittel case suggest that monetary valuation of ecosystem services may be a strong driving force for their preservation. TEEB (The Economics of Ecosystems and Biodiversity-www.teebweb.org) is a wide-scale international initiative to draw attention to the economic benefits of biodiversity and ecosystem services. It is hosted by UNEP (United Nations Environmental Program) and is financed by a number of national and international donors. The main activities focus on knowledge production (natural accounting, assessments of market-based policy instruments) and knowledge transfer to policy-makers at various levels as well as business (TEEB 2009, 2010, 2012a, b). A number of countries have commissioned studies to assess the state and trend of natural capital and ecosystem services. TEEB has the firm intention of supporting and even leading a path-breaking change in the conservation of natural assets by providing a quantitative, factual basis for decision making and by assessing the effectiveness of market-based instruments.

However, the emphasis on market based approaches and the trends towards monetization and commodification of ecosystem services has increasingly brought critics of such an interpretation of the ecosystem services concept on the scene (Gomez-Baggethun and Ruiz Perez 2011; Peterson et al. 2010; Wegner and Pascual 2011; Kosoy and Corbera 2010; Norgaard 2010). In their account of the history of the ecosystem services concept, Gómez-Baggethun et al. (2010) showed how the meaning of ecosystem services has changed over the past decades. Initially it was introduced by ecologists to draw attention to the dependence of society on ecological life support systems (Westman 1977; Ehrlich and Mooney 1983). The seminal paper by Costanza et al. (1997) can be seen as the start of a shift in the focus towards monetary valuation. Muradian and Rival (2012) identify the limitations of market-based policy tools for enhancing the provision of ecosystem services. In particular they highlight limitations in dealing with complexity and uncertainties, with trade-offs between services and with the character of common pool resources. They argue that hybrid regimes are more suitable (compared to pure markets or hierarchies) for addressing the governance challenges resulting from the characteristics of ecosystem services.

Many existing PES schemes can already be identified as hybrid regimes either by design or even more likely by the dynamics unfolding during the implementation process. An example is the payment scheme for hydrological services implemented by the Mexican government described in Box 7.2 (Muñoz-Piña et al. 2008). Vatn (2010) concludes from a review of a large number of case studies that PES in practice depends rather fundamentally on state and/or community engagement. However, Vatn argues for more attention to unexpected and undesirable developments and more systematic analyses of PES schemes. While payments may, for example, strengthen community relations and promote action for environmental care, they may also introduce a purely instrumental rationality and in some cases worsen the environmental status by crowding out environmental virtues.

Box 7.2 PES implemented by government—hydrological services of Mexico's forests [based on Muñoz-Piña et al. (2008)]

To combat increasing deforestation and exacerbation of water scarcity problems Mexico's government launched The Payment for Hydrological Environmental

Services (PSAH) Programme in 2003. Its goal was to provide economic incentives to avoiding deforestation where other policies had largely failed. These were, in particular, regions where water problems were caused by deforestation but where commercial forestry was less profitable than agriculture or ranching. Direct payments between beneficiaries and users were unrealistic since hydrological services provided by forests are public goods on a regional scale. The programme has thus been financed by a national water fee which is intended to also contribute to poverty alleviation and a balancing of regional income disparities. From the perspective of economic efficiency and effectiveness, PSAH should just compensate opportunity costs incurred by forest owners who abstain from deforestation and thus other land use options. Interestingly, public knowledge and a wide-spread belief that forests are important for water supply were instrumental for a largely unobstructed policy design process. Scientific evidence supporting the effectiveness of the proposed measures was, however, uncertain and influenced by context-specific variables. The government set a fixed price since auctions were judged to be too complicated and too unfamiliar for local communities. Within areas that were designed to be eligible for the programme forest owners could sign up voluntarily. Overall the programme has proven to be a success. However, analyses showed that many payments were made in areas with low deforestation risk since opportunity costs where close to zero. This is symptomatic of a national programme which cannot take into account local needs and conditions. In a second phase, local governments and communities are encouraged to set up comparable programmes at the regional scale. This example illustrates that PES schemes evolve over time and combine market-based instruments, hierarchical governmental steering and community governance.

I argue here as well that the combination of different governance modes—namely, markets, regulatory mechanisms, bureaucratic hierarchies and learning networks—is essential for the sustainable implementation of the ecosystem services concept. Regulation is needed to set boundary conditions—target states and negotiation rules. Networks support deliberation and learning—which are essential for addressing complexities and uncertainties. Even when in practice many PES schemes can already be described as hybrid regimes more systematic considerations and purposeful design seem to be warranted. The EU project, ESAWADI, for example, analysed the potential of the ecosystem services approach to support deliberative processes and help to identify and negotiate complex trade-offs during the implementation of the European Water Framework Directive (ESAWADI 2013). Insights gained supported the value of the ecosystem services approach as a communication and education tool to capture the complexity of ecosystem functions and how they affect the benefits provided to society. The ecosystem services approach has thus the potential to support an integrative and holistic rather than a utilitarian perspective. Such a holistic perspective is essential to assess complex interdependencies and to develop appropriate governance and management schemes able to embrace uncertainties. Overall, there appears to be agreement that market-based instruments in general and PES in particular are no panaceas and that systematic evaluation of experiences in different contexts and the development of more pluralistic frameworks and tool boxes are needed (Kinzig et al. 2011; Wegner and Pascual 2011; Vatn 2010; Matzdorf et al. 2013).

7.3 Environmental Hazards

7.3.1 Hazards to Humans Emanating from Natural Events

An environmental hazard to society may be defined as an environmental event that has the potential to cause societal harm and damage. Human activities have always been threatened by environmental hazards ranging from earthquakes, volcanic eruptions to hurricanes, floods, droughts and disease. With increasing development societies have greater means to protect themselves from the impacts of natural hazards. The number of people killed has decreased over the past century. At the same time economic damage has increased due to an increased density of assets and human population.

The hazards most prevalent in the water domain are floods and droughts. Despite tremendous efforts to reduce impacts these hazards continue to pose major risks to economic activities and human life. This is evident in a comparison of the development in the number of natural disasters, the number of people killed and the number of people affected since 1900 as represented in Fig. 7.3. The statistics have been obtained from EM-DAT, the International Disaster Data Base (Guha-Sapir et al. 2014). For a disaster to be entered in the database the following criteria have to be met: ten (10) or more people reported killed; one hundred or more people reported affected; a declaration of a state of emergency; and a call for international assistance.

EM-DAT divides natural disasters into five categories (Guha-Sapir et al. 2014, http://www.emdat.be/new-classification). Geophysical disasters refer to events originating from solid earth such as earthquakes or volcanic eruptions. Meteorological disasters refer to hazards "*caused by short-lived/small to meso-scale atmospheric processes (in the spectrum from minutes to days*". Hydrological disasters encompass different kinds of hazards "*caused by the occurrence, movement, and distribution of surface and subsurface freshwater and saltwater*". Climatological disasters encompass extreme temperature, drought and wildfire events caused by "*long-lived/meso to macro scale processes ranging from intra-seasonal to multi-decadal climate variability*". Finally biological disasters are "*caused by the exposure of living organisms and toxic substances*".

It is evident that the number of disasters worldwide has increased exponentially over the past century, in particular over the past several decades (Fig. 7.3a). One reason for the overall trend may be a change in reporting procedures in particular in developing and threshold countries. However, there are highly significant differences in the magnitude of the trends. In particular the increase in the number of hydrological disasters over the past few decades is remarkable. Looking at the number of people killed (Fig. 7.3b) in comparison with the number of people affected (Fig. 7.3c), the former has decreased significantly but the latter has increased enormously. Better protection measures have reduced the number of casualties. However, increasing population density has increased the number of people that are affected by an event.

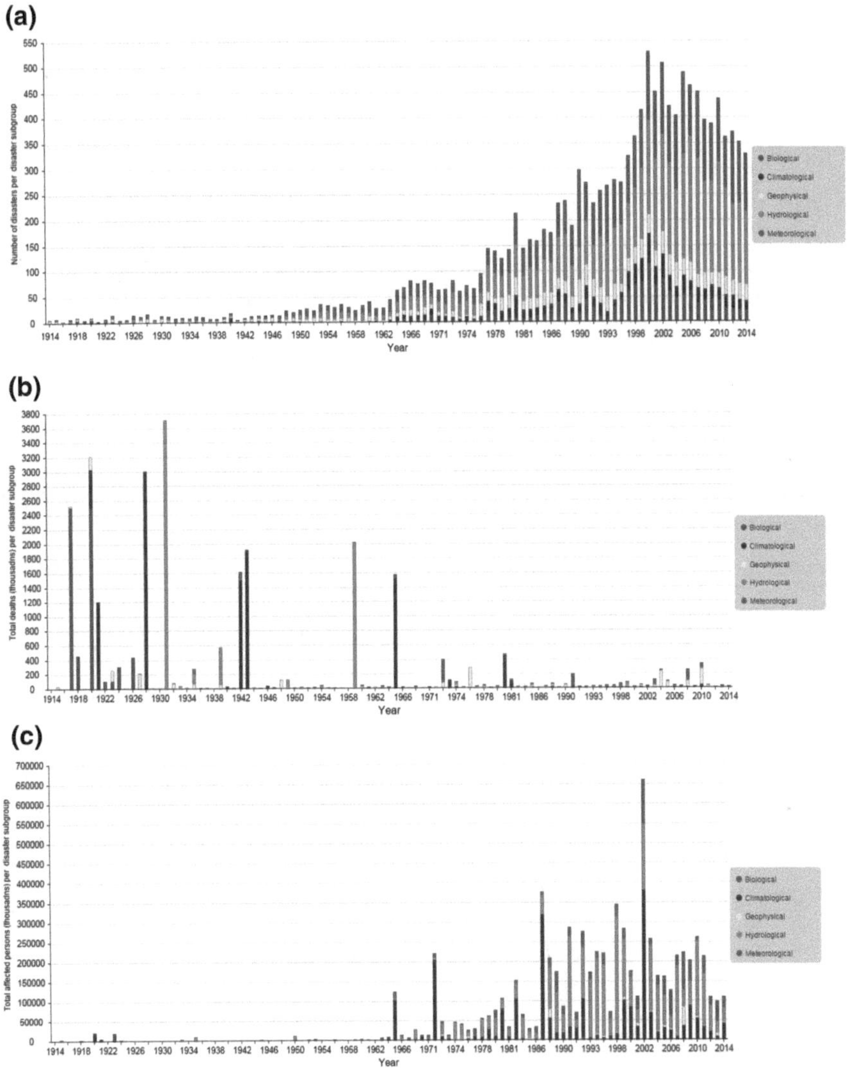

Fig. 7.3 EM-DAT statistics for the time period 1914–2014. **a** Number of natural disasters, **b** number of people killed, **c** number of people affected. *Source* Guha-Sapir et al. (2014)

This development is even more pronounced if one has a closer look at the trend floods and droughts. Figure 7.4 presents the number of floods and droughts worldwide over the past 50 years between 1964 and 2014. The number of floods (including riverine and flash floods) shows a remarkable increase in the past few decades. Changes in the number of people killed and those affected are documented in Table 7.3 for the ten most important natural disasters in the China P.R. since 1900. In terms of casualties, the most severe disasters (including four floods) occurred during the first half of the last century. In terms of people affected

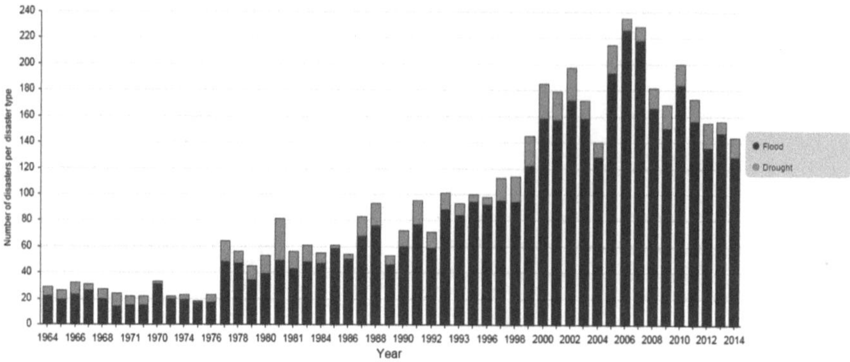

Fig. 7.4 EM-DAT statistics for the time period 1964–2014 for the number of floods (including riverine and flash floods) and droughts. *Source* Guha-Sapir et al. (2014)

Table 7.3 Most significant natural disasters in China P.R. listed in order of number of people killed and affected

Disaster	Date	No killed (in Mio)	Disaster	Date	No Affected (in Mio)
Flood	Jul 31	3.70	Flood	01.07.1998	238.973
Drought	1928	3.00	Flood	01.06.1991	210.232
Flood	Jul 59	2.00	Flood	30.06.1996	154.634
Epidemic	1909	0.50	Flood	23.06.2003	150.146
Drought	1920	0.50	Flood	29.05.2010	134.000
Flood	Jul 39	0.50	Flood	15.05.1995	114.470
Earthquake (seismic activity)	27.07.1976	0.24	Flood	15.06.2007	105.004
Earthquake (seismic activity)	22.05.1927	0.20	Flood	23.06.1999	101.024
Earthquake (seismic activity)	16.12.1920	0.18	Flood	14.07.1989	100.010
Flood	1935	0.14	Storm	14.03.2002	100.000

Figures refer to millions of people. *Source* Guha-Sapir et al. (2014)

disasters, the most sever disaster (only floods) occurred all over the past 25 years. Despite uncertainties in the data the overall trends are clear.

In terms of economic damage it is far more difficult to find any reliable and comparable data over longer time periods due to diverse assessment procedures and the difficulty of comparing absolute damage figures from different years. Hence it is largely impossible to compare damage figures over a whole century. Over shorter time scales one may adjust damage figures by the rate of inflation to improve comparability. An excellent database on natural disaster statistics over the past decades is provided by Munich RE, one of the world's leading reinsurance companies.[1] Figure 7.5 shows the development of weather-related loss events since

[1]Munich RE NatCatSERVICE, Downloadcenter for statistics on natural catastrophes, www.munichre.com/natcatservice.

1980. The number of events has a pronounced positive linear trend. In particular, the number of hydrological events (Fig. 7.5a) shows a major increase which confirms the overall trend observed already for the development of the number of events over one century (Fig. 7.3a). The magnitude of financial losses is increasing even faster than the number of events (Fig. 7.5b). The extrapolation suggests a non-linear increase which is a reason for concern.

The pattern of impacts shows large inter-continental disparities. Table 7.4 provides the percentage distribution of loss events, fatalities, and overall and insured losses for the different continents and for the three disaster categories. The occurrence of events is not evenly distributed. Asia, North America and Europe together account for 76 % of all events. Asia clearly has the largest death toll with 69 % of all fatalities compared to a share of 30 % of all loss events. North America has the largest economic damage with 44 % compared to a share of 25 % of all loss events. Such differences among these regions cannot be explained by differences in the shares of the various disaster categories. One can deduce from Table 7.4 that the three disaster categories differ in the magnitude of damage. Storm events tend to have a slightly higher proportion of fatalities and economic losses than one would expect by simply extrapolating from the share of events. The reasons for the differences among the continents must rather be sought in the state of economic development and population densities. As a result of the high level of economic development and wealth in North America more economic assets are endangered by hazard events, whereas human lives are well protected. Higher population densities and presumably less effective disaster management strategies lead to more fatalities in Asia. One can also note that North America is the continent with the highest level of precautionary measures in the form of insurance. The overall economic losses in Africa are negligible from a global perspective and insurance is non-existent. However, these losses may be detrimental for local populations and national economies. Overall one can conclude that impacts of natural disasters depend on the likelihood of the occurrence of a natural hazard event, the density of vulnerable populations, the exposure and value of economic assets and the effectiveness of disaster protection. The interest in an improved understanding of these relationships has increased tremendously with the prospect of a changing climate and an expected increase in disasters resulting from extreme weather events, viz. floods and droughts.

7.3.2 Assessing and Managing Risks Emanating from Natural Hazards

Analysing the impacts of natural hazards and assessing potential response strategies is in the centre of a broad range of research activities on vulnerability and disaster management. It is beyond the scope of this book to review this thriving and at times

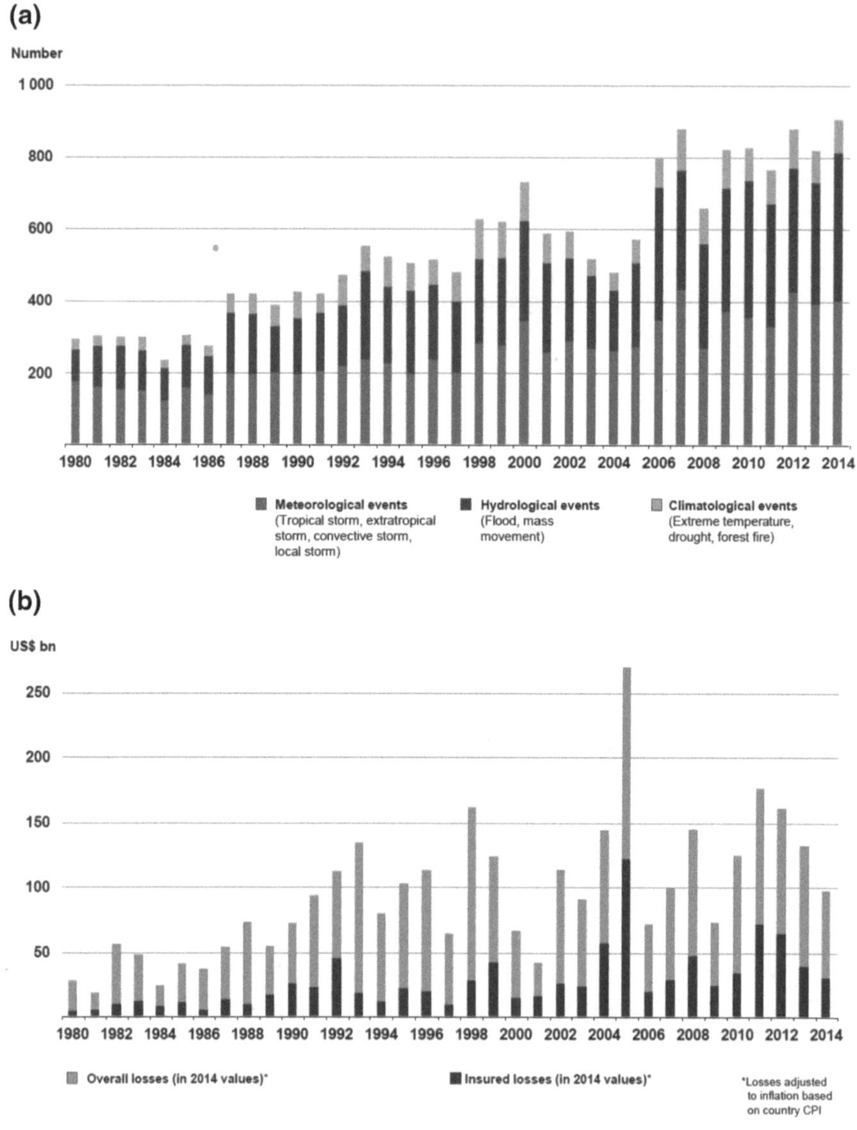

Fig. 7.5 Development of weather related loss events since 1980. **a** Number of loss events, **b** magnitude of overall and insured losses in billion US$. *Source* MunichRe

controversial field of research and practice (for reviews see for example Adger 2006; Füssel 2007; Smit and Wandel 2006). The diversity of approaches can be attributed to the diversity of application fields and scales addressed. Differences exist, for example, in what is considered as being vulnerable—a society, a social-ecological system, specific social groups, individuals or organizations. One

Table 7.4 Weather related loss events worldwide during the period 1980–2014 (*Source* MunichRe)

	Loss events (19,400 total)	Fatalities[a] (850,000)	Overall losses[b] (US$3300bn)	Insured losses[c] (US$940bn)
Percentage distribution by continent				
North America	25	7	44	68
South America	6	3	3	<1
Europe	21	17	17	19
Africa	10	4	1	<1
Asia	30	69	32	9
Australia/Oceania	8	<1	4	3
Percentage distribution by type of event[c]				
Meteorological	46	51	51	79
Hydrological	41	27	32	12
Climatological	13	21	17	9

[a]Number of fatalities without famine
[b]Losses in 2014 values adjusted to inflation based on country CPI
[c]Meteorological events include tropical, extratropical, convective and local storms; hydrological events include flood, mass movement, climatological events include extreme temperature, drought, forest fire

stream of research emanating from human geography and political ecology falls largely under the label of social vulnerability (Bohle et al. 1994; Adger 1999, 2006; Cutter et al. 2003). The concept of social vulnerability emphasizes social factors such as poverty or race as determinants of the vulnerability of diverse social groups. Another stream of work which has its roots in natural sciences and engineering focuses on natural hazards and their attributes and employs more of a technical risk assessment approach (Burton et al. 1993). However, both the factors that are intrinsic to the hazard and the factors that are related to the social community need to be taken into consideration when vulnerability is assessed (Adger 2006; Füssel 2007). The widely-cited definition of vulnerability provided by the IPCC (2007, p. 883) states that "*Vulnerability is the degree to which a system is susceptible to, and unable to cope with, adverse effects of climate change, including climate variability and extremes. Vulnerability is a function of the character, magnitude, and rate of climate change and variation to which a system is exposed, its sensitivity, and its adaptive capacity*". This definition embraces the elements underlying most conceptualizations of vulnerability: exposure, sensitivity and capacity to adapt. Exposure refers to likelihood that and extent to which a system experiences a natural hazard arising from climate change. Sensitivity refers to the degree to which a system is expected be affected by climate change. Adaptive capacity refers to "*the ability of a system to adjust to climate change (including climate variability and extremes), to moderate damages or cope with consequences*" (IPCC 2007, p. 869). The definition of vulnerability by the IPCC is hazard-specific:

$$V(H) = f(E(H), S(H), C(H)).$$

where V refers to vulnerability, E to exposure, S to sensitivity and C to adaptive capacity. (H) denotes that the corresponding variable depends on a hazard H. Focusing on hazard-specific components of sensitivity and capacity may detract from more general characteristics of systems that affect the impact they experience and being able to cope with a wide range of disturbances affecting, for example, financial resources or institutional capacity. Therefore Gallopín (2006) argues in favour of a definition that treats both sensitivity and adaptive capacity as general attributes of the system independent of a specific hazard. I agree to these broader definitions. Furthermore, proponents of the concept of social vulnerability dismiss a systemic focus and argue in favour of group specific-vulnerability assessment that emphasises the influence of societal and economic factors and distributional issues. Hence I have adapted the IPCC definition and define hazard-specific **vulnerability** as the degree to which a system or a social group is susceptible to, and unable to cope with, the adverse effects of a natural hazard. **Vulnerability** (V) is a function of a system's or a social group's exposure (E) to a hazard (H), of a system's or a social group's sensitivity (S_{tot}) and its capacity (C_{tot}) to deal with disturbance where both sensitivity and capacity have hazard-independent and hazard-specific components.

$$V(H) = f(E(H), S_{tot}(S(H) + S), C_{tot}(C(H) + C)).$$

Total capacity embraces both more hazard specific components but also a general capacity to deal with disturbance which depends on what is now generally referred to as a system's resilience. Folke et al. (2010) defined resilience as "*the capacity of a system to absorb disturbance and reorganize while undergoing change so as to still retain essentially the same function, structure and feedbacks, and therefore identity*", that is, the capacity to change in order to maintain the same identity. However, I do not want to enter here the debate about the different interpretations of the concepts vulnerability, adaptive capacity and resilience and the relationships among them (Gallopín 2006). Hence I am reluctant to use these concepts together in one definition. Albeit somewhat cumbersome, this broad definition of vulnerability has the advantage of making a range of different responses for reducing vulnerability explicit.

As pointed out earlier in this chapter concerns about the impacts of climate change in the context of general global change have triggered a surge of interest in research on vulnerability, impacts of natural disasters and risk management strategies. In particular the management of risks related to floods and droughts may have to undergo fundamental changes since key assumptions on which risk management is based do not hold anymore. Figure 7.6 shows a stylized representation of strategies guiding the management of risks that derive from variability and associated extremes in precipitation. It is assumed that variability follows approximately a normal distribution and that the shape of the distribution and the probability of extreme events can be derived by analysing historical time series (Mays 2011). In the past, risk management strategies were primarily aimed at reducing

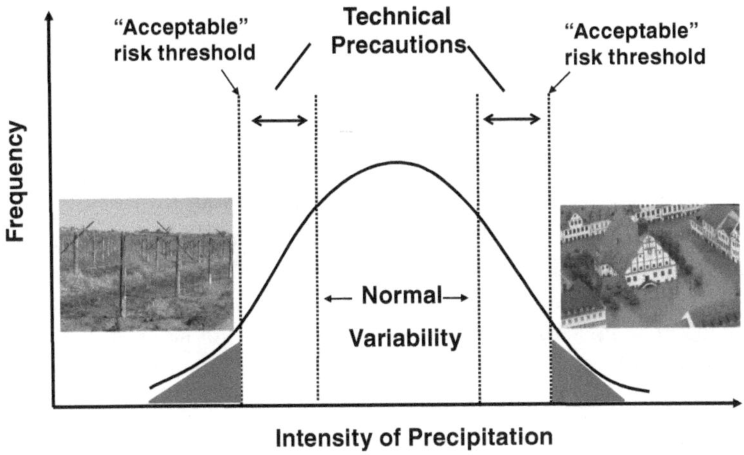

Fig. 7.6 Stylized representation of the management of risks caused by variability and extremes in precipitation (based on Kabat and Bates 2003, p. 25)

exposure, namely the likelihood of a hazardous event occurring. The regulation of rivers and the building of dikes have allowed an increase in the threshold for what is considered to be a damaging flood, thus to larger events. Acceptable risk denotes the threshold at which the costs for further protection measures exceed the expected benefits of risk reduction. Protection has contributed to an increase in assets in areas that were previously exposed to flooding. Once a flood disaster occurs economic damage in built-up areas is normally high (cf. also Fig. 7.5, Table 7.4 with statistics)—flood-specific sensitivity has increased. If risk is quantified by the probability of a flood event times the potential damage, the introduction of (further) flood protection measures shift the acceptable risk threshold to higher water levels. Flood protection measures are often designed to provide a certain degree of protection from a '1-in-X-year-flood'.[2] Engineering solutions have always been able to adapt to the considerable degree of uncertainty associated with such calculations. However, climate change may, in the near future, completely undermine basic premises upon which the engineering design criteria are based (Appleton 2003; Bates et al. 2008; Milly et al. 2008). The largest changes are expected in the tails of probability distributions and the assumptions that time series are stationary do not hold anymore. Such prospects support arguments in favour of transformation towards a more integrated approach to flood management. An integrated ecosystem-based approach in the management of floods will also reconcile the trade-off between flood protection and preserving vital regulatory ecosystem services (Opperman et al. 2009; WMO 2009) and thus increase the overall adaptive capacity of the social-ecological systems. I elaborate on this below Sect. 7.3.3.

[2]For example, a 1-in-1000-year-flood (X = 1000) is a flood event that has a 0.1 % probability of occurring in any given year. The statistical recurrence interval is thus 1000 years.

Similar kinds of developments can be observed for the left tail of the precipitation distribution. Irrigation and the construction of water storage facilities has allowed agriculture to take place in dry regions and the development of large urban centres in areas that were hostile to human settlement. Again, exposure has been reduced but both susceptibility and sensitivity to droughts has increased. The extreme drought in California over the past few years has led to major increases in food prices across the US (United States Department of Agriculture: Economic Research Service 2014) and has triggered for the first time severe measures to reduce water consumption. Despite its location in an arid region the average per capita residential water consumption in Los Angeles is approximately four to five times higher than in Germany, in inland regions of California even about one order of magnitude.[3] Water consumption in Los Angeles has, until recently, been driven only by demand (e.g., for high consumption uses such as watering gardens) rather than being restricted by limited supply.

Increased sensitivity to droughts and the threat of extreme economic losses has increased the pressure to implement further measures in order to assure water security. An example is the controversial South-North Transfer from the Yangtze to the Yellow River in China (The Economist 2013). 300,000 people had to be relocated to build the channel of a length of 1200 km. The new channel is supposed to deliver 14.8 billion cubic metres of water per year to the Yellow River. The entire transfer scheme comprising three channels is supposed to transfer 44.8 billion cubic metres per year from the South to the thirsty North of China (Moore 2014). This corresponds approximately to the annual discharge of the Danube in Germany. It is only about 5 % of the annual discharge of the Yangtze but more than 50 % of the annual discharge of the Yellow River. The risk of unanticipated environmental consequences is considerable but this did not prevent the Chinese government from embarking on this enormous endeavour.

What is considered to be an appropriate response to environmental hazards seems to vary considerably over times and among different countries and cultures. One should refrain from premature, normative judgements on the appropriateness of responses. More systematic analyses on the effectiveness of the management of hazards are needed to provide a more substantive knowledge base for making normative judgements. Table 7.5 lists attributes of disasters caused by environmental hazards that should be taken into consideration in order to develop and compare governance responses that are appropriate for guiding the effective management of environmental hazards.

In summary, management interventions to reduce risks by floods and droughts has mainly focused on exposure, a reduction in the probability that a harmful event will occur despite highly variable precipitation levels. In most parts of the world, we have been pushing the limits of human activities in such a way that they can

[3]Average per capita water consumption in liter/day is 575 Los Angeles. 1054 in Sacramento, 1183 in Fresno and 2792 in Palm Springs (Rogers and St. Fleur 2014) compared to about 127 in Germany.

Table 7.5 Attributes of disasters arising from environmental hazards

Frequency-intensity-distribution: magnitude versus frequency distribution with associated information on uncertainties
Potential damage: refers to the severity of damage
Affected target groups: most vulnerable target groups in terms of both exposure and their ability to adapt/cope with a hazard
Time scale: duration, response time after early warning signals

tolerate greater extremes in the distribution of precipitation. This has resulted in increased sensitivity to extreme events that exceed the limits of the available protection measures. Increasing population densities and economic development expose greater numbers of people and assets to natural disasters, in particular in developing and threshold countries. Reducing the risks of floods and droughts has often contributed to the degradation of ecosystem functions and associated ecosystem services. For example, wetlands have been destroyed by flood protection measures as much as by overutilization of water of irrigation agriculture (Russi et al. 2013). This degradation in ecosystems may, in turn, lead to the loss of adaptive capacity of the social-ecological system as a whole and impact in particular livelihoods of marginalized groups. The mismatch between meeting societal needs and environmental requirements that characterize the prevailing approaches to managing natural resources must be overcome if we are to sustain the base for survival of humankind on this planet.

7.3.3 Hazards for the Environment Emanating from Human Activities

Rivers are in crisis! The findings of the first global-scale analysis which quantified the impact of human-induced stressors on human water security and riverine biodiversity attracted widespread attention in the scientific, policy and even business communities (Vörösmarty et al. 2010). The study revealed the extent to which freshwater resources are faced with acute human-induced threats in both developed and developing countries. Human water security in developed economies has been achieved with the help of highly-engineered solutions that emphasize the treatment of symptoms rather than the protection of resources with measures that often prove too costly to be adopted by poorer nations. The results of this study are not unexpected. They confirm the continuation of a worrying trend which has been highlighted by several scholars since the mid-90s (e.g., Dynesius and Nilsson 1994; Nilsson et al. 2005; Palmer et al. 2008; Gleick 2003).

The Millennium Ecosystem Assessment identified the main direct drivers of impacts on biodiversity and ecosystems over the past several decades (MA 2005a). For inland waters it identified habitat change and pollution as the most important drivers with a rapid increase in the extent of these impacts, followed by invasive species, overexploitation and climate change. Pollution levels have improved in industrialized countries whereas habitat change remains a serious problem. This was clearly reflected in the classification of the status of all European freshwater bodies prescribed by the European Water Framework Directive (WFD, European Parliament (2000)). The WFD classification revealed significant ecological deficits. For example, while 88 % of the surface water bodies in Germany have reached good chemical status, only 10 % of these bodies have good ecological status, up to 34 % are classified as poor, and 23 % are even considered to have bad ecological status (BMU 2010). Water quality improvements, albeit costly, can be achieved by investing in technical measures. Improvements in the ecological status require changes that are potentially more controversial such as changes in patterns of land and water use (Pahl-Wostl 2006).

Wetlands are one of the most valuable and diverse but also the most threatened of all ecosystems (Russi et al. 2013; MA 2005b). About 42 % of all wetlands areas were lost in the period between 1997 and 2011 (Costanza et al. 2014). The percentage is even higher, namely 63 %, if only freshwater wetlands (swamps, floodplains) are taken into consideration. These insights are not new, however. About a decade ago the MA already concluded that degradation and loss of wetlands were more rapid than that of other ecosystems (MA 2005b). Table 7.6 summarizes the most significant of all direct drivers of impacts on wetlands affecting both ecosystem function and associated ecosystem services.

The increased withdrawal of freshwater for use in agriculture, industry, and households has changed flow regimes and has reduced the amount of water available for maintaining the ecological character of many inland water systems (Nilsson et al. 2005; Poff and Zimmerman 2010). The operation of reservoirs and

Table 7.6 Direct drivers of change in freshwater wetlands affecting ecosystem function and associated services [based on Fig. 4.1 MA (2005b)]

Driver	Effects on ecosystem functions
Dam	Interruption of connectivity of river systems, disruption of fish spawning and migration. Alteration of seasonal flood regimes and retention of sediments required to maintain productivity of floodplain agriculture
River channelization	Reduction of habitat. Alteration of flood patterns
Large-scale irrigation and river diversions	Alteration of natural flow regimes. Reduction of downstream water availability and drying up of wetlands
Overharvesting of wild resources	Overharvesting of fish, in particular, driven by subsistence needs and/or commercial exploitation with negative consequences for long-term security of food and livelihoods
Agricultural expansion	Conversion of wetlands. Increase in soil salinity through evaporation. Off-site pollution
Roads and flood control infrastructure	Interruption of wetland connectivity. Disruption of aquatic habitat
Urban and industrial pollution	Reduction of water quality. Negative impacts on diversity and abundance of aquatic organisms

dams is optimized for hydropower production, irrigation or flood control not taking into account the requirements of ecosystems. Ecological integrity does not only depend upon the availability of a base flow but also on diurnal and seasonal flow dynamics (Poff et al. 1997; Bunn and Arthington 2002). Agricultural expansion, urban development, road construction and flood protection infrastructure reduce the space available for wetlands, interrupt wetlands connectivity and disrupt aquatic habitats. Increased pollution by chemicals and excessive nutrient loading from diffuse sources has severe impacts on aquatic biodiversity.

Table 7.7 indicates the impacts (positive or negative) of various types of interventions on ecosystem services. Costanza et al. (2014) estimated that this shrinking of freshwater wetlands corresponded to a loss in value of some 2.7 trillion US$ per year (in 2007 dollars). Russi et al. (2013) provided evidence from numerous regional and local studies on the economic value of wetlands and the losses associated with their degradation. One may argue about the usefulness of attributing an economic value to ecosystems in general. However, given the fact that economic considerations have often led to the destruction of wetlands it makes sense to highlight and try to quantify losses of natural capital. Furthermore losses associated with the degradation of wetlands are often incurred in public goods, whereas the benefits translate into private profits. Wetlands provide buffering capacity in times of floods, recharge groundwater, increase water quality by filtering out pollutants, and provide essential habitats for birds and wildlife (Barbier 2011; Russi et al. 2013). The MA pointed out that taking into account both the marketed and non-marketed economic benefits of wetlands the total economic value of unconverted wetlands was often greater than that of converted wetlands (MA 2005b). Increasingly quantitative evidence is available to substantiate these claims (Russi et al. 2013). Despite such compelling, widely publicized arguments the overall trend of increasing degradation has not changed over the past decade. More knowledge on the economic valuation of ecosystems does not appear to be solving the problem. Obstacles may stem from distributional issues. The economic value of wetlands refers mainly to public goods, in contrast to private benefits from an overexploitation of services that leads to the degradation of wetlands.

7.3.4 Trade-off and Synergies—Complex Interdependencies Between Services and Hazards

Management interventions are intended to reduce the risks caused by environmental hazards and, in doing so, may negatively affect ecosystem services and may even indirectly and unintentionally increase the hazard potential. A case in point is flood management. At least at the level of discourse the need for changes in the way natural resources are managed is now widely recognized (Opperman et al. 2009; Pahl-Wostl et al. 2011; Shabman and Scodari 2012). Flood-control infrastructure (e.g., dikes, dams, levees) prevents water during high-flow periods from entering

Table 7.7 Impacts of selected interventions to increase one service in the range of ecosystem services (based on Table 2.2 MA 2005a, p. 48)

Intervention	Provisioning services		Regulating services			Recreation
	Food production	Water consumption	Flood protection	Groundwater recharge	Water purification	Ecotourism
Agriculture intensification to increase food production	P	N	–	–	N	–
Expansion of agriculture to increase food production	P	N	N	N	N	N
Damming rivers to increase water availability	P	P	P/N	N		P/N

P refers to positive, *N* to negative

floodplains in order to make it available for other uses—agriculture, urban settlements or industry, as well as protecting human uses in the floodplain itself. As a result, wetlands disappear and with them important ecosystem services. Natural flood control capacity is diminished. As highlighted by Opperman et al. (2009) floodplain reconnection would accomplish several objectives beneficial to humans and nature: flood-risk reduction, an increase in floodplain goods and services, and resiliency to potential climate change impacts. Flood-risk reduction would result, on one hand, from increased natural buffering capacity because of the reconnection of floodplains. On the other hand, periodic inundations would encourage a shift to flood-tolerant land uses that are less susceptible to being negatively impacted by floods. According to Opperman et al. (2009) such an approach would only be economically and politically viable if reconnected floodplain areas could remain largely under private ownership, generating revenue through, for example, agriculture. Financial transfers raised by actors who enjoy benefits from ecosystem services could compensate those actors who suffer financial losses. Opperman and colleagues rightly emphasize the importance of economic considerations and compensation payments. However, the proposed change in water management is much more profound and entails more than the economic dimension. It requires

Table 7.8 Comparison of several characteristics of the current state of regulated and controlled rivers with features of a potential future state to illustrate the magnitude of transformative change required for different dimensions [modified from Pahl-Wostl (2006)]

Dimensions	Current state with regulated and controlled rivers	Future state with a multi-functional dynamic landscape
Stakeholder groups and their role	Authorities as regulators in a highly regulated environment Engineers construct and operate dams, reservoirs and levees Environmental protection groups fight for floodplain restoration Insurance companies selling insurance against flood damage Homeowners live in floodplains largely ignorant of the flood risk Agriculture using land in the vicinity of rivers Shipping industry interested in well-functioning water-ways	Authorities as facilitators of an adaptive management process with shared responsibilities Landscape architects are engaged in the design of multi-functional landscapes Engineers who have skills in systems design and cooperate with ecologists And environmental protection groups Insurance companies offer new types of insurances to house owners Homeowners with property in a floodplain facing higher risk of being flooded due to lower protection levels need to manage this risk Tourism industry and tourists using the floodplains for recreation

(continued)

Table 7.8 (continued)

Dimensions	Current state with regulated and controlled rivers	Future state with a multi-functional dynamic landscape
Stakeholder participation	Low level of stakeholder participation—sometimes consultation where various stakeholder groups and the public at large are asked to provide their opinion on a management plan or scenario prepared by experts	Stakeholders and the public are actively involved in river basin management. Active involvement may include contributing to policy development (co-designing), influencing decisions (co-decision-making), or even full responsibility for (parts of) river basin management
Flood management paradigm	Management as control. Technology driven. Risk can be quantified and optimal strategies can be chosen. Zero-sum-games in closed decision space Implementation of controllable and predictable technical infrastructure (reservoirs, dams) based on fixed regulations for acceptable risk-thresholds	Adaptive and integrated water management. "Living with water". Acceptable decisions are negotiated Implementation of a multi-functional landscape and increased adaptive capacity of the system. Moderated risk dialogues and cascade of adaptation measures for living with extremes
Adaptive capacity	"Hard" approach to systems design with the goal of implementing optimal solutions. Adaptive capacity is in general quite low due to high levels of sunk costs in infrastructure and often inflexible legal regulations (e.g., water use rights allocated for decades, technological norms that prescribe good practice and prevent innovation)	"Soft" approach to systems design which allows the taking into account of new insights and responding to changing environmental and socio-economic boundary conditions. This is more in line with the new paradigm of adaptive water management

structural transformation along various dimensions. Table 7.8 compares several dimensions of the current and the desirable future state characterized as 'current state with regulated and controlled rivers' and 'future state with a multi-functional and dynamic landscape', respectively (Pahl-Wostl 2006).

This transformation will require substantial changes in the role and power of the different stakeholder groups involved. Engineers have to extend their skills and share the responsibility for the work with ecologists and landscape architects. However, water engineering is a profession with established rules of good practice that engineers have to follow in order to be recognized in their community. Such rules are not easy to change even when convincing alternatives are available. Some homeowners will face a loss in the value of their property if certain areas now protected from flooding are rezoned for temporal flooding. Responsibility for managing risks will be transferred to some extent from government to individual homeowners who will have to make their houses flood-proof. Residents may also be concerned about rising groundwater tables and an increase in mosquito populations as a consequence of periodic inundations. The influence of governmental authorities will decline to some extent if the management scheme as a whole becomes more participatory. Such fundamental changes cannot be designed and implemented without some delegation of authority and active participation of citizens and organized stakeholder groups (Pahl-Wostl et al. 2013b). Box 7.3 illustrates the complexity of feedback processes and the various influences of competing paradigms for the Tisza basin as an example. Given the complexity of interdependent processes implementation will always be associated with considerable uncertainties and encounter obstacles. Climate change adds another dimension of uncertainty. Adaptive governance and management are thus required for both supporting change and sustaining a new integrated floodplain management paradigm.

Box 7.3 Feedback processes and competing paradigms in flood management illustrated for the Hungarian Tisza basin

Flood management in the Hungarian Tisza River basin has experienced challenges as a consequence of the reigning management paradigm over the past few decades (Sendzimir et al. 2010 and Box 6.2). Pahl-Wostl et al. (2011) defined a management paradigm as a set of basic assumptions about the nature of the system to be managed, the goals associated with managing the system and the ways in which these goals can be achieved. Based on this definition and using elements from the MTF, Halbe et al. (2013) developed a method to identify paradigms underlying system dynamic models of flood management. Figure 7.7 summarizes the results from this type of analysis for the Hungarian Tisza basin. Different and in part competing, paradigms were identified at the level of sub-system components. The paradigms reflect different perspectives of the overall system and strategies for dealing with the flood management problem. The presence of different paradigms can be interpreted as an indicator of an ongoing transformation process.

7.4 Drivers of Change—Integrated Governance of SESs

Given the alarming developments in and continuing trends towards the degradation of ecosystems, one is inclined to be pessimistic about the future. However, I would like to highlight several, to some degree interdependent, discourses that may support change towards a more integrated and sustainable governance of SESs: climate change, green infrastructure and water security.

The climate change debate has increased awareness of the influence of ecosystem services on adaptive capacity and resilience of social-ecological systems. Investments in nature have been promoted as an efficient and effective approach in climate change adaptation to increase the capacity of social-ecological systems to deal with uncertainty and surprise. Scientific scholars and practitioners have increasingly argued that more emphasis should be devoted to how investment in natural capital and ecosystem services can reduce vulnerability to climate change by

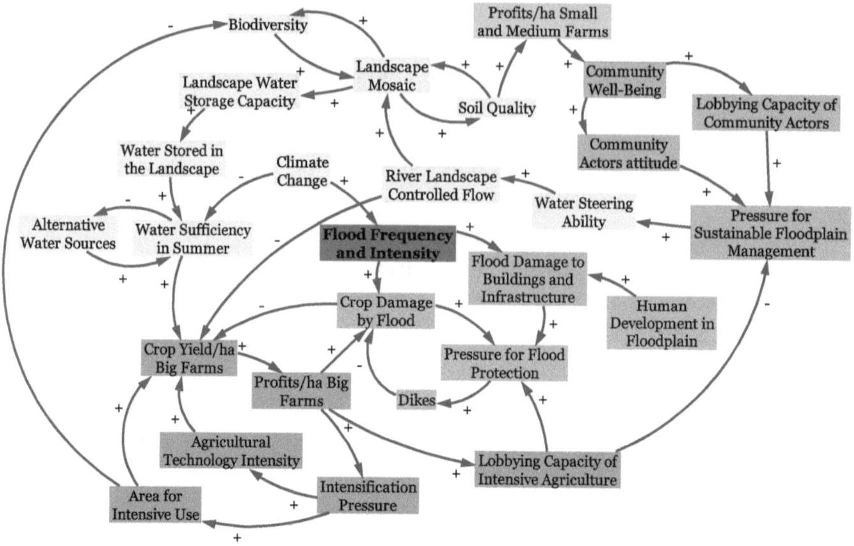

	"Economics" Paradigm	**"Predict and Control" Paradigm**	**"Adaptive Management" Paradigm**	**"Community Involvement" Paradigm**	**"Tradition" Paradigm**
System Perspective	Large farms (intensive agriculture)	River and protected values	Floodplain landscape	Flood prone communities	Small farms
Solution Strategies	Economies of scale; rationalization	Dike construction	River-Landscape controlled flows	Community involvement	Traditional farming methods
Risk and Uncertainty Management	Reduce flooding risk and uncertainties	Reduction of uncertainty	Accept flood risk; Adaptive Management (through experimentation)	Uncertainty dialogue	Build upon experience from the past

Fig. 7.7 Illustration of the presence of different and in part competing paradigms in a system dynamics representation of complex interdependencies in flood management in the Hungarian Tisza (Reproduced from Fig. 1 Halbe et al. 2015, with permission)

building adaptive capacity and by increasing resilience of social-ecological systems in general (Opperman et al. 2009; Smith and Barchiesi 2009). This conceptualization takes limitations in predictability into account and emphasizes the need to build capacity in dealing with uncertainties and unexpected surprises. It promotes a systemic and integrated approach to addressing ecosystem services in contrast to the still prevailing fragmented and one-dimensional perspectives. It deviates thus from the more narrow approach of assessing vulnerability and developing responses to individual environmental hazards. Furthermore, building resilience rather than reducing vulnerability puts more emphasis on synergies and beneficial aspects of management strategies.

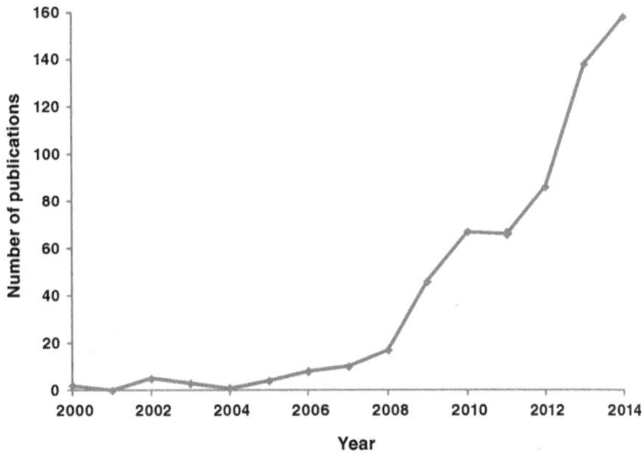

Fig. 7.8 Number of publications which refer to green infrastructure in title, abstract or keywords based on a SCOPUS analysis (search conducted 25 May 2015)

The idea of making intelligent use of ecosystem services instead of technical infrastructure is not new. It is a core principle of ecological engineering (Mitsch 1993; Barrett 1999). However, such ideas were marginalized for a long time by the dominant technology-focused engineering paradigm. Currently, this dominance seems to be weakening. The idea of green infrastructure has enjoyed sudden popularity in recent years. This is illustrated in Fig. 7.8 with the sharp increase in the use of the term "green infrastructure" in the scientific and technical literature. The European Commission adopted a Green Infrastructure Strategy in 2013 (European Commission 2013). One argument put forward is that making use of natural or green infrastructure builds resilience against climate change and thus enhances water security in general (Smith and Barchiesi 2009; Opperman et al. 2009; Pahl-Wostl et al. 2013c).

The concept of water security has encountered a remarkable increase in popularity over the past decade (Cook and Bakker 2012). One of the most widely used definitions of water security is the one by Grey and Sadoff (2007) introduced in Sect. 7.1. This definition identifies four dimensions of water security and highlights economic, social and environmental trade-offs as a matter of concern. By using the attribute, 'acceptable', Grey and Sadoff highlight that water security is a social construct which must be negotiated in a societal discourse. Hence, governance and, in particular, respect of good governance principles are central to defining and implementing a sustainable approach to water security. However, prevailing governance systems and approaches to assess the various dimensions of water security are not particularly supportive of an integrated approach (Pahl-Wostl et al. 2013c). Sectoral fragmentation and a different system of logic for determining what is acceptable render a comprehensive analysis of trade-offs difficult. The ecosystem services concept and green infrastructure could be central notions and boundary objects that help to overcome fragmentation if embedded in the appropriate enabling governance structures and processes of societal learning.

7.5 Concluding Remarks

Given the pervasive nature of water, human-environment interactions are particularly complex. Water is a central agent for transmitting global change effects. It influences and is influenced by decisions in different policy domains such as agriculture or energy. It plays a critical role in economic growth and human development. Achieving water security for the wide range of, at times, competing human activities is a governance domain characterized by debates and conflicts. As a result, the needs of the environment are often neglected with detrimental consequences for ecosystems and in the long-term for socio-ecological systems as a whole. The concepts of ecosystem services and environmental hazards are promising bridging concepts to bring about more sustainable patterns of interactions between humans and nature. There are hopeful signs that change is occurring—at least at the level of discourse. The challenge is to translate discourse into transformative change that allows us to overcome the limitations and weaknesses of prevailing governance and management systems.

References

Adger, W. N. (1999). Social vulnerability to climate change and extremes in coastal Vietnam. *World Development, 27*(2), 249–269. doi:10.1016/S0305-750X(98)00136-3

Adger, W. N. (2006). Vulnerability. *Global Environmental Change, 16*(3), 268–281. doi:10.1016/j.gloenvcha.2006.02.006

Appleton, B. (Ed.). (2003). *Climate changes the water rules: How water managers can cope with today's climate variability and tomorrow's climate change*. Delft, The Netherlands: Dialogue on Water and Climate.

Aylward, B., Bandyopadhyay, J., & Belausteguigotia, J. (2005). Freshwater ecosystem services. In K. Chopra, R. Leemans, P. Kumar, & H. Simons (Eds.), *Ecosystems and human well-being: Policy responses* (Vol. 3, pp. 213–254). Washington DC: Island Press.

Barbier, E. B. (2011). Wetlands as natural assets. *Hydrological Sciences Journal, 56*(8), 1360–1373. doi:10.1080/02626667.2011.629787

Barrett, K. R. (1999). Ecological engineering in water resources. *Water International, 24*(3), 182–188. doi:10.1080/02508069908692160

Bates, B. C., Kundzewicz, Z. W., Wu, S., & Palutikof, J. (Eds.). (2008). *Climate change and water (technical paper of the intergovernmental panel on climate change)*. Geneva: Switzerland IPCC Secretariat.

Bennett, M. T. (2008). China's sloping land conversion program: Institutional innovation or business as usual? *Ecological Economics, 65*(4), 699–711. doi:10.1016/j.ecolecon.2007.09.017

BMU (2010). *Water framework directive: The way towards healthy waters. Results of the German river basin management plans 2009*. Berlin: Federal Ministry for the Environment, Nature Conservation and Nuclear Safety, BMU

Bohle, H. G., Downing, T. E., & Watts, M. J. (1994). Climate change and social vulnerability: Toward a sociology and geography of food insecurity. *Global Environmental Change, 4*(1), 37–48. doi:10.1016/0959-3780(94)90020-5

Bunn, S., & Arthington, A. (2002). Basic principles and ecological consequences of altered flow regimes for aquatic biodiversity. *Environmental Management, 30*(4), 492–507.

Burton, I., Kates, R. W., & White, G. F. (1993). *The Environment as hazard* (2nd ed.). New York: Guilford Pubn

Carpenter, S. R., Mooney, H. A., Agard, J., Capistrano, D., DeFries, R. S., Díaz, S., et al. (2009). Science for managing ecosystem services: Beyond the millennium ecosystem assessment. *Proceedings of the National Academy of Sciences, 106*(5), 1305–1312. doi:10.1073/pnas.0808772106

Chapin III, F. S., Kofinas, G. P., & Folke, C. (Eds.). (2009). *Principles of ecosystem stewardship: Resilience-based natural resource management in a changing world.* Berlin: Springer.

Cook, C., & Bakker, K. (2012). Water security: Debating an emerging paradigm. *Global Environmental Change, 22*, 94–102.

Corbera, E., Kosoy, N., & Martínez Tuna, M. (2007). Equity implications of marketing ecosystem services in protected areas and rural communities: Case studies from Meso-America. *Global Environmental Change, 17*(3–4), 365–380. doi:10.1016/j.gloenvcha.2006.12.005

Costanza, R., d'Arge, R., de Groot, R., Farber, S., Grasso, M., Hannon, B., et al. (1997). The value of the world's ecosystem services and natural capital. *Nature, 387*, 253–260.

Costanza, R., de Groot, R., Sutton, P., van der Ploeg, S., Anderson, S. J., Kubiszewski, I., et al. (2014). Changes in the global value of ecosystem services. *Global Environmental Change, 26*, 152–158. doi:10.1016/j.gloenvcha.2014.04.002

Cutter, S. L., Boruff, B. J., & Lynn, W. S. (2003). Social vulnerability to environmental hazards*. *Social Science Quarterly, 84*(2), 242–261. doi:10.1111/1540-6237.8402002

Daily, G. C. (1997). *Nature's services—societal dependence on natural ecosystems.* Washington: Island Press.

Dynesius, M., & Nilsson, C. (1994). Fragmentation and flow regulation of river systems in the northern third of the world. *Science, 266*(5186), 753–762. doi:10.1126/science.266.5186.753

Ehrlich, P. R., & Mooney, H. A. (1983). Extinction, substitution, and ecosystem services. *BioScience, 33*, 248–254.

Elmqvist, T., Tuvendal, M., Krishnaswamy, J., & Hylander, K. (2011). *Managing Trade-offs in Ecosystem Services. Ecosystem Services Economics (ESE), Working Paper Series* (Vol. 4). Nairobi.

Engel, S., Pagiola, S., & Wunder, S. (2008). Designing payments for environmental services in theory and practice: An overview of the issues. *Ecological Economics, 65*(4), 663–674. doi:10.1016/j.ecolecon.2008.03.011

Engel, S., & Schaefer, M. (2013). Ecosystem services—a useful concept for addressing water challenges? *Current Opinion in Environmental Sustainability, 5*(6), 696–707. doi:10.1016/j.cosust.2013.11.010

ESAWADI. (2013). *Synthesis report.* Utilizing the Ecosystem Services Approach for Water Framework Directive Implementation ESAWADI Project.

European Commission. (2013). Communication from the commission to the European parliament, the council, the European economic and social committee and the committee of the regions. In E. Commission (Ed.), *COM/2013/0249 final.* Brussels.

European Parliament. (2000). Council Directive 2000/60/EC of 23 October 2000 on establishing a framework for Community action in the field of water policy.

Folke, C., Carpenter, S. R., Walker, B., Scheffer, M., Chapin, T., & Rockström, J. (2010). Resilience thinking: Integrating resilience. *Adaptability and Transformability. Ecology and Society, 115*(4), 20.

Füssel, H.-M. (2007). Vulnerability: A generally applicable conceptual framework for climate change research. *Global Environmental Change, 17*(2), 155–167. doi:10.1016/j.gloenvcha.2006.05.002

Gallopín, G. C. (2006). Linkages between vulnerability, resilience, and adaptive capacity. *Global Environmental Change, 16*(3), 293–303. doi:10.1016/j.gloenvcha.2006.02.004

Gleick, P. (2003). Global freshwater resources: Soft-path solutions for the 21st century. *Science, 302*, 1524–1528.

Gómez-Baggethun, E., de Groot, R., Lomas, P. L., & Montes, C. (2010). The history of ecosystem services in economic theory and practice: From early notions to markets and payment schemes. *Ecological Economics, 69*(6), 1209–1218. doi:10.1016/j.ecolecon.2009.11.007

Gomez-Baggethun, E., & Ruiz Perez, M. (2011). Economc valuation and the commodification of ecosystem services. *Progress in Physical Geography, 35*, 613–628.

Grey, D., & Sadoff, C. W. (2007). Sink or swim? Water security for growth and development. *Water Policy, 9*, 545–571.

Guha-Sapir, G., Below, D.R., & Hoyois, Ph. (2014). EM-DAT: International Disaster Database. Université Catholique de Louvain—Brussels. www.emdat.be. Accessed August 19, 2014.

Halbe, J., Adamowski, J., & Pahl-Wostl, C. (2015). The role of paradigms in engineering practice and education for sustainable development. *Journal of Cleaner Production, in press.*

Halbe, J., Pahl-Wostl, C., Sendzimir, J., & Adamowski, J. (2013). Towards adaptive and integrated management paradigms to meet the challenges of water governance. *Water Science and Technology, 67*(11), 2651–2660. doi:10.2166/wst.2013.146

IPCC (Ed.). (2007). Climate change 2007: Impacts, adaptation and vulnerability. In M. L. Parry, O. F. Canziani, J. P. Palutikof, P. J. van der Linden & C. E. Hanson (Eds.), *Contribution of working group II to the fourth assessment report of the intergovernmental panel on climate change.* Cambridge: Cambridge University Press.

Kabat, P., & Bates, B. C. (2003). The Evidence. In B. Appleton (Ed.), *Climate changes the water rules: How water managers can cope with today's climate variability and tomorrow's climate change* (pp. 1–25). Delft, The Netherlands: Dialogue on Water and Climate.

Kinzig, A. P., Perrings, C., Chapin, F. S, I. I. I., Polasky, S., Smith, V. K., Tilman, D., et al. (2011). Paying for ecosystem services—promise and peril. *Science, 334*, 603–604.

Kosoy, N., & Corbera, E. (2010). Payments for ecosystem services as commodity fetishism. *Ecological Economics, 69*(6), 1228–1236. doi:10.1016/j.ecolecon.2009.11.002

MA (2005a). *Ecosystems and human well-being: Synthesis. Millennium ecosystem assessment.* Washington DC: Island Press.

MA (2005b). *Ecosystems and human well-being: Wetlands and water synthesis. Millennium ecosystem assessment.* Washington DC: World Resources Institute.

Matzdorf, B., Sattler, C., & Engel, S. (2013). Institutional frameworks and governance structures of PES schemes. *Forest Policy and Economics, 37*, 57–64. doi:10.1016/j.forpol.2013.10.002

Mays, L. W. (2011). Probability, risk, and uncertainty analysis for hydrological and hydraulic design. In L. W. Mays (Ed.), *Water resources engineering* (2nd ed., pp. 361–399). Hoboken, USA: Wiley.

Milly, P. C. D., Betancourt, J., Falkenmark, M., Hirsch, R. M., Kundzewicz, Z. W., Lettenmaier, D. P., et al. (2008). Stationarity is dead: Whither water management? *Science, 319*(5863), 573–574.

Mitsch, W. (1993). Ecological engineering a cooperative role with the planetary life-support system. *Environmental Science and Technology, 27*(3), 438–445. doi:10.1021/es00040a600

Moore, S. M. (2014). Modernisation, authoritarianism, and the environment: the politics of China's south-north water transfer project. *Environmental Politics, 23*(6), 947–964. doi:10.1080/09644016.2014.943544

Muñoz-Piña, C., Guevara, A., Torres, J. M., & Braña, J. (2008). Paying for the hydrological services of Mexico's forests: Analysis, negotiations and results. *Ecological Economics, 65*(4), 725–736. doi:10.1016/j.ecolecon.2007.07.031

Muradian, R., & Rival, L. (2012). Between markets and hierarchies: The challenge of governing ecosystem services. *Ecosystem Services, 1*(1), 93–100. doi:10.1016/j.ecoser.2012.07.009

Nilsson, C., Reidy, C. A., Dynesius, M., & Revenga, C. (2005). Fragmentation and flow regulation of the world's large river systems. *Science, 308*, 405–408.

Norgaard, R. B. (2010). Ecosystem services: From eye-opening metaphor to complexity blinder. *Ecological Economics, 69*(6), 1219–1227. doi:10.1016/j.ecolecon.2009.11.009

Opperman, J. J., Galloway, G. E., Fargione, J., Mount, J. F., Richter, B. D., & Secchi, S. (2009). Sustainable floodplain through large-scale reconnection to rivers. *Science, 326*, 1487–1488.

Ostrom, E. (2005). *Understanding institutional diversity.* Princeton: Princeton University Press.

Pagiola, S. (2008). Payments for environmental services in Costa Rica. *Ecological Economics, 65*(4), 712–724. doi:10.1016/j.ecolecon.2007.07.033

Pahl-Wostl, C. (2006). The importance of social learning in restoring the multifunctionality of rivers and floodplains. *Ecology and Society, 11*(1), 10. [online] URL:http://www.ecologyandsociety.org/vol11/iss11/art10/

Pahl-Wostl, C. (2007). The implications of complexity for integrated resources management. *Environ. Modelling & Software, 22*, 561–569.

Pahl-Wostl, C., Arthington, A., Bogardi, J., Bunn, S., Hoff, H., Lebel, L., et al. (2013a). Environmental flows and water governance: Managing sustainable water uses. *Current Opinion in Environmental Sustainability, 5*, 341–351.

Pahl-Wostl, C., Becker, G., Sendzimir, J., & Knieper, C. (2013b). How multilevel societal learning processes facilitate transformative change: A comparative case study analysis on flood management. *Ecology and Society, 18*(4), 58. doi:10.5751/ES-05779-180458

Pahl-Wostl, C., Holtz, G., Kastens, B., & Knieper, C. (2010). Analyzing complex water governance regimes: The management and transition framework. *Environmental Science & Policy, 13*(7), 571–581. doi:10.1016/j.envsci.2010.08.006

Pahl-Wostl, C., Jeffrey, P., Isendahl, N., & Brugnach, M. (2011). Maturing the new water management paradigm: Progressing from aspiration to practice. *Water Resources Management, 25*(3), 837–856. doi:10.1007/s11269-010-9729-2

Pahl-Wostl, C., Palmer, M. A., & Richards, K. (2013c). Enhancing water security for the benefits of humans and nature—the role of governance. *Current Opinion in Environmental Sustainability, 5*, 676–684.

Palmer, M. A., Reidy-Liermann, C., Nilsson, C., Florke, M., Alcamo, J., Lake, P. S. et al. (2008). Climate change and world's river basins: Anticipating management options. *Frontiers in Ecology and the Environment, 6*, 81–89.

Perrot-Maître, D. (2006). *The Vittel payments for ecosystem services: A "perfect" PES case?* Project Paper (Vol. 3). London: International Institute for Environment and Development (IIED) and Department for International Development (DFID).

Peterson, M. J., Hall, D. M., Feldpausch-Parker, A. M., & Peterson, T. R. (2010). Obscuring ecosystem function with application of the ecosystem services concept. *Conservation Biology, 24*(1), 113–119.

Poff, N., Allan, J., Bain, M., Karr, J., Prestegaard, K., Richter, B., et al. (1997). The natural flow regime: A paradigm for river conservation and restoration. *BioScience, 47*(11), 769–784.

Poff, N. L., & Zimmerman, J. K. H. (2010). Ecological responses to altered flow regimes: A literature review to inform the science and management of environmental flows. *Freshwater Biology, 55*(1), 194–205. doi:10.1111/j.1365-2427.2009.02272.x

Prugh, T., Costanza, R., Cumberland, J. H., Daly, H. E., Goodland, R., & Norgaard, R. B. (1999). *Natural capital and human economic survival*. Boca Raton, FL: Lewis Publishers.

Robards, M. D., Schoon, M. L., Meek, C. L., & Engle, N. L. (2011). The importance of social drivers in the resilient provision of ecosystem services. *Global Environmental Change, 21*, 522–529. doi:10.1016/j.gloenvcha.2010.12.004

Rogers, P., & St. Fleur, N. (2014). California Drought: Database shows big differences between water guzzlers and sippers. (2014, July 02). *San Jose Mercury News*.

Russi, D., ten Brink, P., Farmer, A., Badura, T., Coates, D., Förster, J., et al. (2013). *The economics of ecosystems and biodiversity for water and wetlands*. Ramsar Secretariat, Gland: Institute for European Environmental Policy—IEEP, London and Brussels; Ramsar Secretariat, Gland.

Sendzimir, J., Flachner, Z., Pahl-Wostl, C., & Knieper, C. (2010). Stalled regime transition in the upper Tisza River Basin: The dynamics of linked action situations. *Environmental Science & Policy, 13*(7), 604–619. doi:10.1016/j.envsci.2010.09.005

Shabman, L., & Scodari, P. (2012). *Towards integrated water resources management: A conceptual framework for U.S. Army Corps of Engineers water and related land resources implementation studies*. Working Paper. *Institute for Water Resources*: U.S. Army Corps of Engineers.

Smit, B., & Wandel, J. (2006). Adaptation, adaptive capacity and vulnerability. *Global environmental change, 16*(3), 282–292. doi:10.1016/j.gloenvcha.2006.03.008

Smith, D. M., & Barchiesi, S. (2009). Environment as infrastructure: Resilience to climate change impacts on water through investments in nature. In IUCN (Ed.), *Perspectives on water and climate change adaptation*. Gland, Switzerland: International Union for Conservation of Nature (IUCN).

TEEB (2009). The economics of ecosystems and biodiversity for national and international policy makers—Summary: Responding to the value of nature.

TEEB (2010). The economics of ecosystems and biodiversity: Mainstreaming the economics of nature. A synthesis of the approach, conclusions and recommendations of TEEB.

TEEB (Ed.). (2012a). *The Economics of Ecosystems and Biodiversity in Business and Enterprise*. London and Washington: Earthscan. (Edited by Joshua Bishop).

TEEB (Ed.). (2012b). *The economics of ecosystems and biodiversity in local and regional policy and management*. London and Washington: Earthscan. (*Edited by Heidi Wittmer and Haripriya Gundimeda*).

The Economist. (2013, October 12). All dried up: Northern China is running out of water, but the government's remedies are potentially disastrous. http://www.economist.com/news/china/21587813-northern-china-running-out-water-governments-remedies-are-potentially-disastrous-all

Turpie, J. K., Marais, C., & Blignaut, J. N. (2008). The working for water programme: Evolution of a payments for ecosystem services mechanism that addresses both poverty and ecosystem service delivery in South Africa. *Ecological Economics, 65*, 788–798.

United States Department of Agriculture: Economic Research Service. (2014). *California drought 2014: Food prices and consumers*. Accessed August 20, 2014, from http://ers.usda.gov/topics/in-the-news/california-drought-2014-farm-and-food-impacts/california-drought-2014-food-prices-and-consumers.aspx#.U_R1sPl_ugY

Vatn, A. (2010). An institutional analysis of payments for environmental services. *Ecological Economics, 69*(6), 1245–1252. doi:10.1016/j.ecolecon.2009.11.018

Vörösmarty, C. J., McIntyre, P. B., Gessner, M. O., Dudgeon, D., Prusevich, A., Green, P., et al. (2010). Global threats to human water security and river biodiversity. *Nature, 467*, 555–561.

Wegner, G., & Pascual, U. (2011). Cost-benefit analysis in the context of ecosystem services for human well-being: A multidisciplinary critique. *Global Environmental Change, 21*, 492–504.

Westman, W. E. (1977). How much are nature's services worth? *Science, 197*(4307), 960–964. doi:10.1126/science.197.4307.960

WMO (2009). Integrated Flood Management Concept Paper. (URL:http://www.apfm.info/pdf/concept_paper_e.pdf). Geneva, Switzerland World Meteorological Organization (WMO).

Wunder, S., & Albán, M. (2008). Decentralized payments for environmental services: The cases of Pimampiro and PROFAFOR in Ecuador. *Ecological Economics, 65*(4), 685–698. doi:10.1016/j.ecolecon.2007.11.004

Wunder, S., Engel, S., & Pagiola, S. (2008). Taking stock: A comparative analysis of payments for environmental services programs in developed and developing countries. *Ecological Economics, 65*(4), 834–852. doi:10.1016/j.ecolecon.2008.03.010

Zalasiewicz, J., Williams, M., Steffen, W., & Crutzen, P. (2010). The new world of the anthropocene. *Environmental Science and Technology, 44*(7), 2228–2231. doi:10.1021/es903118j

Chapter 8
A Theory on Water Governance Dynamics

The theoretical foundations for describing and explaining the dynamics and transformative change of water governance are weak. This deficit is hardly astonishing if one takes into consideration that the concept of governance itself is still subject to controversy (Mayntz 2004; Grande 2012). Is governance largely a descriptive concept which takes developments observed in real world policy into account? Or does it provide the base for a new normative theory of political steering and the role of the nation state? This chapter contributes to the building of the missing foundations by developing a conceptual and theoretical framework of water governance systems. It is largely a framework of analysis but also entails a normative dimension by identifying characteristics which are considered here to be essential for dealing with complex governance challenges. Even when the focus is on water governance the scope of theoretical considerations is not limited to this domain of governance.

Initially, I develop the essential elements of a theoretical framework that captures the dynamics of governance systems. Then I address processes of social learning and adaptation before moving to the theme of transformative change. Finally, I elaborate on the potential for and limitations to governing transformative change for sustainability.

8.1 Elements of a Theoretical Framework

Figure 8.1 depicts the elements of a theoretical framework aimed at capturing the dynamics and the process of transformative change within governance systems.

First, it is necessary to develop a representation of the state of a governance system. I refer explicitly to a dynamic state in order to emphasize the fact that governance systems are not static. In the language of dynamic systems theory a system can be described by a state space which comprises all possible states of the system. A specific state of the system corresponds to a unique point in the state space. Stable systems tend to remain within a bounded basin of attraction in the state space and move towards an equilibrium state, the so-called attractor of the system. A point attractor implies that the equilibrium state is static and can only

C. Pahl-Wostl, *Water Governance in the Face of Global Change*,
Water Governance—Concepts, Methods, and Practice,
DOI 10.1007/978-3-319-21855-7_8

Fig. 8.1 Elements of a theoretical framework to capture dynamics and transformative change of governance systems

adopt one single configuration. A swinging pendulum is a system with a point attractor. Unless in a vacuum without friction, the pendulum comes to a standstill if no additional energy is supplied. Rain forests are characterized by spatial replacement of individual species whereas spatially averaged biomass is quite stable. In democracies governments change whereas constitutional rules and the form of government remain the same. A gradual or abrupt shift in long-term averages of variables characterizing the state of the system indicates a transition to a new state.

A dynamic equilibrium implies that the system may adopt different states within its basin of attraction. Societal systems are never static. One may also argue the value of using the notion of an equilibrium state at all since societal system continually change and develop. Nevertheless, it makes sense to distinguish between adaptation of a governance system where a certain dynamic state and its structural configuration are retained and transformation where the structural configuration and thus the basin of attraction are entirely changed.

At this point it is worth bearing in mind that water governance has been defined as a societal function and that a water governance system refers to the ensemble of interdependent elements that perform this function (Chap. 3).

A **Transformation** in a governance system is now defined as structural change in several interdependent system components and change in the overall system logic which is determined by the underlying governance and management paradigm.

Transformative capacity has been defined as the ability of a governance system to first adapt and if required transform structural elements as a response to current or anticipated changes in the social or natural environment (Chap. 3). When introducing this definition I have pointed out that it is analytically quite difficult to make a distinction between adaptive and transformative capacity when developing operational measures for empirical analysis. Transformations may be smooth and there may be considerable overlap since adaptation may be a first step towards transformation. Furthermore it is also not straightforward to develop operational indicators in order to identify what may be called transformative change—not in hindsight but as the transformation occurs. How can one identify a transformation of a system to a new state is really taking place? One can more easily identify radical transformations in political systems often associated with the breakdown of the old regime such as the French revolution, the fall of the Berlin Wall or the end of Apartheid in South Africa. The final tipping point though may be virtually impossible to predict. Transformations in resource governance systems are assumed to be longer-term and smoother and thus more difficult to capture analytically.

The understanding of system dynamics developed in this book is based on complex adaptive systems theory (Levin 1999; Pahl-Wostl 1995). Complex adaptive systems (CASs) comprise many interacting elements that are organized in a modular structure. System structure is largely an emergent product of processes of self-organization, of feedback between emergent macro-level control variables and micro-level interactions. Such a broad definition applies to all kind of CASs from collective intelligence in ant colonies to the human brain or societal communities. Human beings, however, are characterized by intentionality and anticipatory action. Understanding governance as a societal function of regulation requires taking into account instances of purposeful design that are based on processes of social interaction. Armed with this conceptualization I move towards explicating a diagnostic approach linking the state of a governance system with its performance.

8.2 Understanding the State of Water Governance Systems

A diagnostic approach as introduced in Chap. 3 makes a distinction between a water governance system (WGS), its functional performance and the context (economic, social, political, cultural, environmental) in which the system is embedded. Context reflects variables that are external to and/or whose boundaries exceed the water governance system. Societal norms are, for example, considered as context, whereas water legislation and technical norms with respect to water infrastructure would be within the water governance system.

8.2.1 *Performance*

The ability of a WGS to fulfil its societal function can be evaluated by its functional performance which is defined at the level of the system as a whole. As a WGS is to some extent or wholly the product of purposeful design (e.g., water legislation with clearly stated goals), one measure of performance is the extent to which a WGS fulfils its intended purpose. Further performance measures can be identified that capture different aspects of evaluation:

- *Fulfilment of intended purpose*: delivery of public goods and services based on comparing stated with realized intentions. This is largely an internal performance measure defined by the society in which the WGS is embedded.
- *Sustainable enhancement of water security for humans and nature*: this is to some extent a normative measure of performance based on universally accepted

principles. Details of what is deemed sustainable and the kinds of trade-offs among economic, social and environmental aspects of water security that are acceptable are subject to debate and negotiation (Pahl-Wostl et al. 2013b).

- *Respect for good governance principles*: good governance is participatory, consensus oriented, accountable, transparent, responsive, equitable and inclusive, effective and efficient and follows the rule of law (UNESCAP 2009). This is largely a normative measure of performance based on universally accepted principles even when some principles may be differently interpreted and further context specific principles are added.
- *Adaptive capacity and resilience*: ability to deal with internal and external changes, shocks and surprise. This is a normative measure that largely complements general sustainability principles. Details of what this might imply in a specific context and for different societal groups is subject to negotiation (Folke et al. 2010; Pahl-Wostl 2009).

These measures of performance are motivated by a realistic and pragmatic approach which links performance closely to function, in our case the sustainable governance and management of water resources. Performance is not informed by idealistic principles of any societal utopia. Normative principles of sustainability and good governance can be interpreted from a value-laden perspective. They can also be interpreted from a more instrumental point of view that only adhering to such normative principles guarantees water security for a wide range of human activities and societal groups in the long term.

Measures of performance can be used for the internal (by those within and directly affected) and external (by those outside and not directly affected such international NGOs or advocacy groups) validation of success. This does not imply automatically that systemic performance guides individual action. Nor does it imply that all actors in a WGS come to the same conclusions with respect to performance along different dimensions. Actors may judge performance based on individual rather than collective outcomes and collective outcomes may be evaluated differently by different actor groups. The relationships between collective and individual outcomes and collective and individual action depend on the characteristics of the WGS. A key guiding question that needs to be addressed in the theory is: What are the requirements for a water governance system to perform well and thus fulfil the social function of regulating development and management of water resources and provisions of water services (cf. Chap. 3)?

8.2.2 Characteristics of a Water Governance System

I have argued that governance systems must be integrated and adaptive in order to perform well. Integrative implies coordination across levels and sectors as required to realize a systemic approach which enhances synergies and reduces trade-off between different aspects of water security. Adaptive implies the ability to respond

Table 8.1 Assumed characteristics of integrated and adaptive water governance and management systems (further developed from Pahl-Wostl 2007a, b)

Variable	Characteristic properties
Guiding paradigm	Governance and management as learning in complex adaptive systems
Regulatory frameworks	Flexible, allowing some degree of interpretation and tailoring to context, regulation of process rather than outcomes
Governance architecture	Polycentric, distribution of power and authority combined with effective and efficient coordination; balance between bottom-up and top-down flows of influence
Governance modes	Diverse combinations and hybrid forms of governance modes: hierarchies, markets and networks
Integration of learning cycles	Effective links between informal learning and innovation and formal policy processes
Sectoral integration	Cross-sectoral assessments and policy coordination, adoption of a nexus perspective
Information and knowledge management	Open, shared information sources that fill gaps and facilitate integration; acknowledgement of various kinds of knowledge and co-production of new knowledge
Uncertainties and risk governance	Acknowledgement of different perspectives and world views; conscious decision-taking under (irreducible) uncertainty; innovative approaches to manage uncertainty and risk
Infrastructure	Appropriate scale, diverse sources of design—decentralized—centralized; combination of traditional technology and green infrastructure
Finances	Financial resources diversified using a broad set of private and public financial instruments
Human-environment interface	Explicit acknowledgement of a wide range of ecosystem services; Instruments in place to assess and govern trade-offs

to new insights gained during policy implementation, to developments in societal and environmental conditions and to changes in external factors such as climate change. Pahl-Wostl (2007a, b) introduced an initial set of characteristics of such Integrated and Adaptive Governance and Management Systems (IAGMS) derived mainly from concepts and empirical evidence of individual system elements. Further conceptual advancements and more systemic empirical analyses have proposed revisions to this set of characteristics (Pahl-Wostl 2009; Pahl-Wostl et al. 2011, 2012, 2013a; Pahl-Wostl and Knieper 2014). The resulting set of revised characteristic properties is summarized in Table 8.1 and explained below. Justifications provided are concise since they largely build on arguments already elaborated in preceding chapters.

Guiding Paradigm: The guiding paradigm refers to the dominant cultural-cognitive institution that shapes water governance and management (c.f. Chap. 4, Sect. 4.1.2). This paradigm influences system understanding, how boundaries are delineated, and how problems and their solutions are determined. The guiding system metaphor of

IAGMSs is that of complex adaptive systems where governance and management are more appropriately portrayed as invoking many instances of learning rather than control. Instead of trying to reduce degrees of freedom by attempting hierarchical and centralized control (e.g., large-scale technologies, highly regulated top-down governance), retaining freedom and adaptive capacity builds on the strengths of complex adaptive systems to perform well in uncertain environments. This does not preclude the regulation and sanctioning of certain unsustainable or socially unacceptable actions.

Regulatory Frameworks: To enable adaptive governance regulatory frameworks need to be flexible without compromising on provision of guidance and stability (Garmestani and Benson 2013; Green et al. 2013). This may be achieved by providing more regulations for processes rather than outcomes. Transparency must be ensured so that it is clear who decides, why a decision is taken, and which type of evidence provides the basis for a decision to respond to new insights, new developments and changes in external factors. Transparency is required to avoid the abuse of flexible regulatory frameworks by powerful groups who impose their vested interests. Transparency is also essential for supporting effective and flexible coordination between policy fields.

Governance Architecture: Conceptual consideration and empirical evidence strongly suggest that polycentricity is an essential characteristic of IAGMSs (Ostrom 2001, 2010; Pahl-Wostl 2009; Pahl-Wostl et al. 2012; Pahl-Wostl and Knieper 2014; Silveira and Richards 2013). These findings support the more general understanding that complex adaptive systems are both more effective and efficient than centralized systems in the allocation of scarce resources in dynamic and uncertain environments. Polycentric systems combine the distribution of power and authority with effective and efficient coordination and balance bottom-up and top-down governance (Huntjens et al. 2010, 2011; Pahl-Wostl et al. 2012; Pahl-Wostl and Knieper 2014).

Governance Modes: IAGMSs combine processes of emergence and self-organization with purposeful design. Such types of dynamics require the combination and interactions of different governance modes—networks, bureaucratic hierarchies and markets need to act in concert (Pahl-Wostl 2009; Pahl-Wostl and Knieper 2014). Rather than advocating the dominance of a single governance approach, IAGMSs integrate bureaucratic hierarchies, markets and network governance in flexible and diverse ways.

Integration of Learning Cycles: Informal spaces and diverse actor networks are important for supporting the integration of knowledge and experimentation with innovative approaches. Effective links between informal learning and innovation and formal policy processes are required. Connections between learning and policy processes (e.g., hinge on individual actors) are fragile if innovative approaches are not codified in formal institutions and widely shared practices (Pahl-Wostl et al. 2013a).

Sectoral Integration: Cross-sectoral analysis and integration is required to identify problems that may emerge, deal with persistent problems and coordinate sectoral policy implementation. The concept of the Water-Energy-Food nexus, for example, provides a means of holistically integrating different policy fields from the outset (Benson et al. 2015). Flexible and effective instruments for horizontal coordination still need to be developed to achieve the desired combination of sectoral integration and flexibility and learning in management (Jordan and Lenschow 2010; Pahl-Wostl et al. 2012).

Information and Knowledge Management: A comprehensive understanding of complex water problems and their solutions can only be achieved by open, shared information sources that fill gaps and facilitate integration. Inclusive and transparent processes are required to translate information into knowledge that is validated and legitimized and has a shared meaning for different actor groups (Folke et al. 2005, 2007b).

Uncertainty and Risk Governance: Enhancing water security sustainably in an increasingly uncertain and complex world requires that water governance and management systems perform under conditions of irreducible uncertainty and surprise. Sustainability can only be achieved if a systemic and long-term perspective on risk is adopted which embraces the full complexity and interdependencies of the social-ecological systems to be managed (Renn 2005; Pahl-Wostl 2015). Risks have often been managed by prescribing technical standards such as regulations for the required size of flood protection systems. However, due to increased uncertainties, resulting from climate change for example, the conditions under which such regulations were passed may no longer be fulfilled. An acceptable risk needs to be negotiated in participatory processes.

Infrastructure: Large-scale infrastructure (e.g., with a life-span of decades) provides few opportunities for learning and may easily lead to lock-in situations (Pahl-Wostl 2002; Tillman et al. 2005). Adaptive governance and management is mainly limited to the operational level. Careful consideration at the appropriate scale, an increased use of decentralized technologies, making use of green infrastructure, diverse sources of design adapted to the regional context are all promising strategies for achieving sustainable and integrated water management (Gleick 2003; Pahl-Wostl 2007b; Opperman et al. 2009; Smith and Barchiesi 2009; Hallegatte 2009).

Finances: Large-scale infrastructure projects can produce enormous sunk costs which reduce the flexibility of economic instruments. For example, once infrastructure for drinking water provision and wastewater treatment is in place, the price of water is largely determined by the fixed costs and not by the amount of water used by consumers. Hence, water pricing becomes an ineffective instrument for controlling water consumption. Furthermore large-scale investment may be more prone to abuse in weak governance settings with a lack of accountability to and among governmental officials. Diversification of infrastructure (e.g., combination small and large scale waste-water treatment facilities) and innovative approaches for handling uncertainty and risk also require the diversification of financial resources using a broad set of private and public financial instruments.

Human-Environment Interface: Recognition of a wide range of ecosystem services is required to raise awareness of the importance of ecosystem functions for the resilience of social-ecological systems (Folke 2006; MA 2005) and to support negotiations on the trade-offs among economic, social and environmental dimensions of water security (Pahl-Wostl and Knüppe 2015). Payments for ecosystems services can be effective instruments to overcome trade-offs between different ecosystem services (Engel et al. 2008). However, instruments for the implementation of the ecosystem services concept need to go beyond monetary valuation and embrace a combination of governance modes (Gómez-Baggethun et al. 2010; Gomez-Baggethun and Ruiz Perez 2011; Wegner and Pascual 2011).

8.2.3 Context

Environmental and societal context may have a strong influence on the relationship between the characteristics of a WGMS and its performance. In a system science context this refers to those factors that cannot be directly influenced by the system under consideration over the temporal and spatial scales of interest. This does not preclude that, in the long-term, innovation in the water sector has repercussions for other domains. Context may, however, prevent the realization of certain governance configurations. It is, for example, not likely that good governance principles will be broadly adopted in the water sector if they are ignored or violated in all other governance domains. The overall state of a country's economic and institutional development and the nature of the political system thus have an influence on a WGMS. Here I make a distinction between economic development which refers mainly to economic performance measured by indicators (such as GDP per person) and institutional development. The latter captures institutional capacity, the effectiveness of formal institutions and level of corruption, the presence of a civil society, and freedom of speech and means of expressing public opinion.

Weak economic development constrains the governmental capacity to implement policies due to a lack of financial resources and skilled labour. A weak state of institutional development constrains governmental capacity to develop and implement policies because of a lack of sound governance processes and the inability to manage financial resources effectively even if they become available. This implies that in such countries institutional development needs to precede or advance in parallel with economic development in order to improve the performance of a WGMS. This includes investment in water infrastructure as well as water resources development and management.

Another factor influencing the performance of a WGMS is what can be summarized as hydro-complexity (Grey et al. 2013; Grey and Sadoff 2007). Hydro-complexity captures the characteristics of the biophysical environment. High variability combined with low predictability of hydrological conditions and physical water scarcity are all conditions that pose significant challenges for the sustainable governance and management of water resources.

8.3 Understanding Processes of Transformative Change

8.3.1 Sustainability Transformations—An Evolutionary Perspective

Understanding change implies comprehension of two kinds of processes: the ability to adapt to changing circumstances and new insights within the structural constraints provided by a WGMS and the ability to transform the structure of a WGMS if structural constraints do not allow effective adaptation to new challenges.

The theoretical understanding of structural transformations in societal systems is limited (c.f. Chap. 4). Transformation may result from the breakdown of whole political systems such as the communist regimes in Eastern Europe. Transformation may also result from technical innovation. The advent of the internet and digital communication has led to a transformation in many parts of society and aspects of daily life. Neither of these types of transformative change seems to be appropriate for understanding the transformation towards sustainability of a WGMS. A breakdown of a political system is usually the result of complete failure of the prevailing governance system and/or irreconcilable conflicts between supporters of different political paradigms. A breakdown may also result in a greatly reduced or non-functioning governance system—at least during a transition period. In countries with reasonably well-functioning governance systems this is clearly a development within an essential societal function such as water governance that should be avoided. The situation may present itself differently in countries characterized by dysfunctional governance in all domains of public live. In such cases transformation in water governance systems is closely connected with societal development as a whole. Water governance is a social function and hence its transformation happens within and in parallel to changes in the societal and political systems. The market governance mode does not dominate. In contrast, technical innovation is largely driven by the market. I develop here a different model for transformative change in governance systems. Transformative change may be incremental at times, combining self-organization with purposeful design and building on the reflexive capacity of society and not only of individual actors.

Pahl-Wostl (2009) introduced an evolutionary approach to transformative change based on the concept of triple-loop learning. Societal learning is assumed to be an exploratory, stepwise process where actors experiment with innovation until they meet constraints and new boundaries. Single-loop learning refers to an incremental improvement in actions without questioning the underlying assumptions guiding actions. Double-loop learning refers to a revisiting of these assumptions (e.g., about cause-effect relationships) within a value-normative framework. In triple-loop learning underlying values and beliefs are reconsidered as well as world views if assumptions within a world view do not hold anymore. Essential assumptions and elements of the concept of triple-loop learning were introduced in Chap. 4. As pointed out, the interplay and interdependence between structure and agency are essential for understanding change. Agency, the capacity

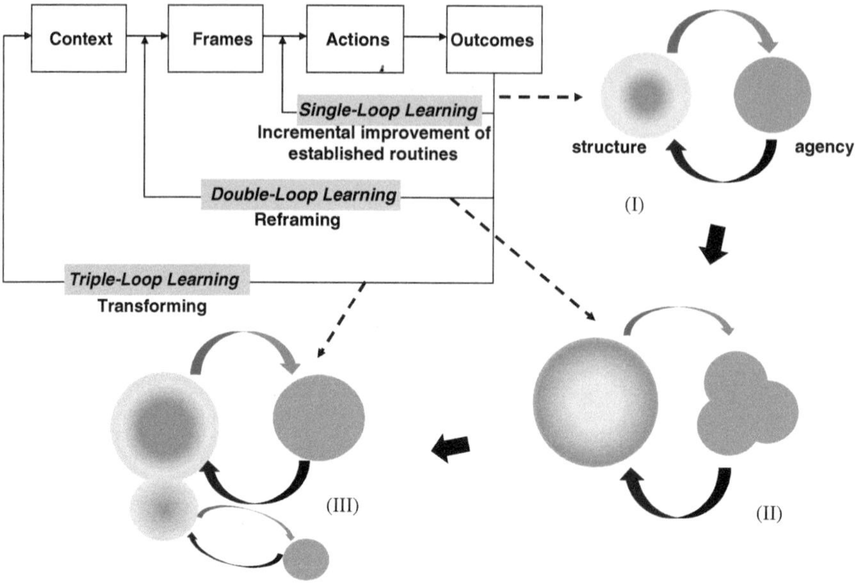

Fig. 8.2 Different stages of triple-loop learning and schematic representation of corresponding changes in the interplay between structure (*blue spheres—left*) and agency (*red spheres—right*)

of actors to act independently and make free choices, is limited by structural constraints which provide stability and predictability. Actors are heterogeneous and some actors are more constrained than others. They have different values, interests and influences. In their actions, actors interpret and may strengthen or weaken structural constraints. I argue that such processes apply more generally to transformative change in a governance system even when the focus here is on the transformation of water governance and management systems. Furthermore, in Sect. 8.2 I derived a normative model of desirable characteristics of a WGMS. In the remainder of this section, I focus on transformative change towards a more adaptive and integrated WGMS.

Figure 8.2 shows the stages of the triple-loop learning concept and a schematic representation of corresponding changes in the interplay between structure and agency. This simplified and stylized representation of complex learning processes highlights what I consider important stages of a structural transformation of a WGMS.

The blue sphere on the left in stages I, II, and III denotes the interpretation space within the current governance structure. The governance structure reflects a constellation of actor networks including power distribution, multi-level interactions and the relative importance of governance modes and combinations thereof (c.f. Chap. 4, Table 4.1). The notion of interpretation space captures the extent to which the current governance structure is challenged by reinterpretation and experimentation with what? The meaning and interpretation of prevailing institutions may for example be quite uniform or start to broaden out or even diverge from each other. The red sphere denotes agency. Agency seen from a systems perspective comprises the diversity and breadth of interpretations voiced by actors, and the diversity of modes of interaction with and enacting/reproduction of governance structures.

I. In single-loop learning the interpretation space is narrow and focused on a central paradigm and/or on a prevailing system logic. Agency is quite uniform with little diversity. Prevailing governance structure and agency reinforce each other as denoted by the feedback loops.
II. In double-loop learning where the first instances of triple-loop learning occur, the interpretation space broadens out. Prevailing institutions are reinterpreted by many parties. Established norms and routines are called into question. Innovative groups experiment with new approaches using loopholes to overcome potential constraints imposed by regulatory frameworks. Societal discourses and experimentation extend the prevailing structural space towards its very boundaries and the core meaning is fading. New institutions may already be introduced in some water governance domains. Agency becomes more diverse and the feedback from structure to agency weakens. One could argue that structural constraints lose their grip on agency. This phase is characterized by expansion and diversification.
III. Triple-loop learning leads to the establishment of a new structural regime. Remnants of the old regime may still be present and may co-exist with the now dominant new regime. The new regime differs compared to the initial state since transformation is assumed to lead to an increase in adaptive capacity. Flexible and adaptive governance systems would and should always include some elements of double-loop learning which implies challenging and reinterpreting prevailing structural constraints. By contrast, the traditional command-and-control water governance and management systems are characterized by less freedom in and little space for interpretation with a correspondingly limited capacity to adapt, innovate and change. This difference is indicated by the larger sizes of the spheres in III compared to I.

8.3.2 Interplay Between Formal and Informal Processes

Figure 8.3 provides a stylized representation of a formal policy process on the left and an informal learning process on the right. I argue that transformative change requires informal learning cycles that are connected to formal policy processes. In order to qualify as a learning cycle a process must fulfil the following requirements:

- It must involve a (partially) informal network of actors who meet regularly. Informal implies that rules for the group (boundaries—who is involved and what is included in the analysis is open; leadership is allowed to emerge; rules guiding how the group operates are negotiable etc.) are not formally prescribed,

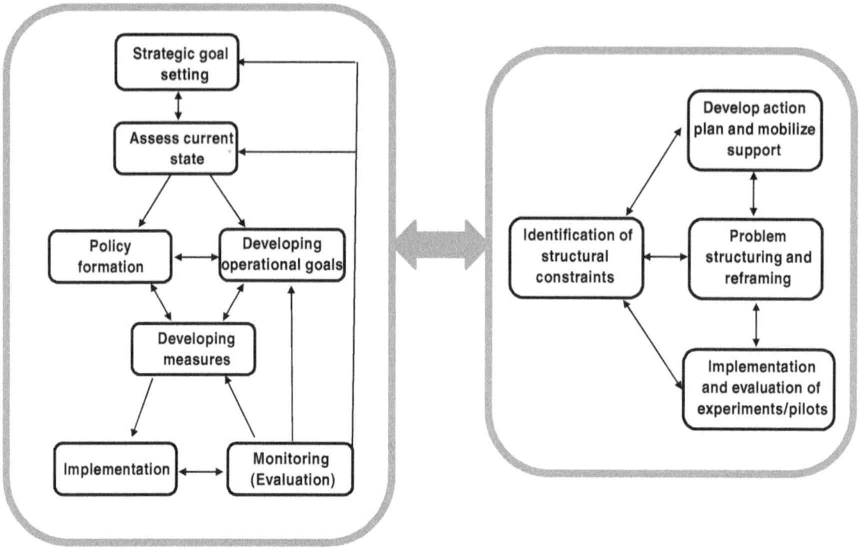

Fig. 8.3 Stylized representations of formal policy (*left*) and informal learning (*right*) processes

that their mandate is open and the results of the policy process are not immediately formally binding. A purely informal network would be a bottom-up process, a kind of shadow network that has no formal link at all to the formal policy and management cycle. The opposite of an informal network would, for example, be a formalized negotiation process in a transboundary river basin commission with the goal of agreeing upon a formal treaty to be signed by all countries. Such a process would not qualify as a learning network.

- It involves an issue-specific network formed to address a specific problem or problematic domain.
- It must comprise a sequence of connected Action Situations (ASs). This might also imply a sequence of individual activities at the outset that are connected at a later stage.
- It has a clear objective to find solutions to a problem and is open and willing to explicitly experiment with a range of innovative approaches.
- It engages in activities that enable double- or even triple-loop learning.
- The actor network qualifies as a community of practice sensu Wenger (1998) with joint and shared practices and tangible products.

To understand transformative change it is of particular interest to analyse the different forms of linkages of learning cycles to formal policy processes in order to understand their influence on the learning process and the effectiveness of the outcomes of the process. Linkages between the two kinds of processes are indicated by the broad arrow. Diverse types of influence are possible. I have refrained from including all possible linkages though to avoid overcrowding of the figure with arrows. Some important pathways are addressed in the explanation of the various phases of the policy process below.

The policy cycle is used in the sense of an analytical devise and not as a strictly normative model. Obviously it is assumed that the different phases in the cycle are required in order to develop and implement any policy or management plan. Management without measurable goals or an evaluation if they are achieved is quite meaningless. However, the phases may overlap, run in parallel, be less clearly separated than the graphical representation suggests. The various phases comprise the whole policy process from early stages of policy development to the implementation of operational measures through to evaluation (Pahl-Wostl et al. 2010).

Strategic goal setting: The strategic goals for the policy/management process are set to determine a desirable state of the water system—e.g., "good ecological status"—as described by the European Water Framework Directive; "water saving society—China"; "Living with water" which guides flood management in the Netherlands; implementation of IWRM, etc. The strategic goals are determined for the water system as a whole and are formally binding for all actors. However, they may not be agreed upon by all actors in the water system and they may be interpreted differently by the various actors.

Assess current state: The current state of the water system is assessed to determine the degree to which it satisfies the predefined criteria and, accordingly, the need for change. In the analysis of some policy processes such a phase could also be explicitly taken into account prior the setting of any strategic goals—as a form of problem identification. After the setting of strategic goals an assessment of the current state provides initial insight into the degree of deviation of the current state from desired goals.

Policy formation: Policies are developed that provide coherent approaches on how to improve the current state of the water system. This phase may include an interpretation and refinement of strategic goals, the development of a policy framework at a lower administrative level—e.g., national, regional or basin, and the selection of preferred types of policy instruments—e.g., market-based. This phase can be highly political with actors pursuing their goals and trying to influence the policy process.

Developing operational goals: Operational goals means measurable goals that allow the assessment of efficiency and effectiveness of measures and that provide the basis for monitoring programs.

Developing measures: A plan with specific measures including an assessment of their costs and their anticipated effectiveness is developed. This phase is likely to also include an assessment of the degree of uncertainty associated with the expected effects and costs of measures. In this phase potential winners and losers may become apparent.

Implementation: In this phase the various measures are implemented at the appropriate level, e.g., constructing new wastewater treatment plants, introduction of new pricing policies.

Monitoring: Monitoring serves to assess if the measures implemented have led to the achievement of the established goals and to detect potential unexpected and undesired consequences. This phase may—but often does not—include an evaluation process. Evaluation is essential, however, for adaptive management which implies the potential need to reconsider the programme of measures and the entire implementation process. This revision process may even go back to the initial step placing strategic goals under scrutiny.

The *learning cycle* may represent processes of reframing and the broader interpretation of existing structural constraints as well as introducing a novel paradigm and promoting radical innovation. Such learning processes may also be triggered by and receive attention and support from more established circles if incremental improvement of current practices does not lead to improved performance and/or the achievement of the envisaged goals. For example, during the iterative process of defining operational goals and developing sets of measures to achieve them, significant uncertainties in the effectiveness of a measure and potential responses for addressing these uncertainties may be identified. A learning cycle may be initiated at this stage. A typical example is the implementation of new non-structural measures where no prior experience exists (e.g., water pricing, water markets, water user associations).

In contrast to purely informal bottom-up processes taking place in shadow networks largely external to the formal policy and management process, I suggest here a normative model for the design of governance processes. These processes include learning cycles as a core element of adaptive governance (cf. Table 8.1) with an explicit link to the formal governance policy process. Learning cycles may be initiated by those responsible for water management, e.g., competent local/regional authorities during a policy implementation process. Such a process would be linked to the formal management process. However, to qualify as informal the results of these processes are not formally binding but policy informing.

The stylized model of a learning cycle is thus to be used in both an analytical and normative sense. First it serves to structure empirical data from case studies and make them comparable. Based on such comparative analyses the normative model can be further developed. However, similar to the policy cycle, the normative model does not imply that the phases depicted in the learning cycle proceed in linear and sequential mode. It does imply though that the different phases outlined are important and that the three aggregated phases—problem structuring and reframing, development of action plan and mobilizing additional support, implementation and evaluation of pilots/experiments—are needed to support effective learning processes (Pahl-Wostl et al. 2010). In the following I will explain these phases in more detail.

Problem structuring and reframing: The dissatisfaction of actors with the dominant governance and management approaches may cross a threshold when many actors engage in double-loop learning and start questioning the prevailing approach. The trigger for such reframing may come from outside the formal policy/management process i.e. shadow networks. It could also stem from within the formal policy process arising, for example, from the need to implement new kinds of measures, from significant uncertainties in the process and/or the identification of obstacles. It may also derive from (repeated) failures or new insights about future developments (e.g., climate change).

One or several groups of actors normally engage in processes of social learning. Even when the process is convened by a formal management board, learning requires that these actors have a certain degree of freedom in the process design in order to adapt the process to the needs arising from the problem to be addressed (e.g., agree on ground rules) or that new members be included in the network to connect to other stakeholder groups and/or to include different types of knowledge. This phase is characterized by "learning to understand different perspectives and to deal with them constructively". The group may succeed in reframing and restructuring the problem which might imply a shift in priorities, boundaries around the needs to be included or excluded in the analysis, or the types of solutions to be addressed.

To embrace different perspectives such groups should not only be comprised of the like-minded; otherwise potential obstacles might be ignored and the learning network may become closed and disconnected from other networks.

Develop action plan and mobilize additional support: Actors identify potential barriers and bridges as well as significant uncertainties surrounding the implementation of strategies in order to achieve the defined goals. Actors may also identify structural constraints that prevent them from realizing their goals. At this stage further stakeholders may be brought into the process to gain access to additional knowledge and build a critical mass of support.

Actors may identify viable paths including the need and possibility for experimentation, and the potential for structured learning processes. They will then assess the resources required for these additional needs (e.g., financial, knowledge, human resources) and develop a plan for securing these resources. The participants in the learning platform then analyse specific possibilities for action along the paths that have been identified. Finally, more intensive interactions with the formal management process are important at the end of this phase. Systemic innovations such as new approaches to flood management may often be linked to larger-scale experiments.

The learning process may stop if the establishment of structured learning and experimentation, and thus a continuation of the learning platform, is judged to be a low priority or infeasible. Another learning cycle aimed at transformation of structural constraints may be initiated if insurmountable constraints (e.g., legal constraints, severe opposition of powerful groups) are identified that prevent any kind of implementation of pilot studies and experiments on the ground (e.g., stringent regulation may prevent the introduction of innovative approaches to deal with flood risks).

Implementation and evaluation of pilots/experiments: Further stakeholder groups may be involved throughout the process of running of the pilots or experiments. Pilots can serve as demonstration projects with prototype experiments at smaller scales as well as sustain momentum of the learning process and develop capacity by accumulating further resources (e.g., access to knowledge, financial resources). This activity requires new methods for structuring participation and public information campaigns.

Learning processes may stop at this stage if pilots remain isolated activities. Innovation may gain momentum though if experiences can be shared and made visible in wider networks. For the latter to happen requires an evaluation of results, the development of new insights by comparing results from different regions and/or experimental pilots and thinking about up-scaling and linking up to further actor networks.

Identification of structural constraints: During the course of experimentation with innovative practices actors may encounter structural barriers—this could result in the initiation of another learning cycle in order to overcome or change structural barriers. However, it might also be that actors decide to compromise and try to scale down their ambitions in order to realize the possible within the constraints provided by the current regime.

Constraints may already have been identified in the first stage during reframing. It is more likely, however, that they are recognized when actions plans and scenarios are developed or during the phase of implementing experiments.

This description should not convey the impression that such learning processes can and should be well structured and tightly managed processes along the lines of a cook book approach. The process will always entail considerable degrees of self-organization. Learning networks may also develop from connecting various currently isolated innovation activities.

8.3.3 A Multi-level Perspective on Transformative Change

The framework presented here resonates with some assumptions of the multi-level perspective on transitions of socio-technical regimes (MLP). Transition research uses a broad regime concept. In general socio-technical regimes are seen as relatively stable configurations of institutions, techniques and artefacts, as well as rules, practices and networks and fulfil socially-valued functions (Smith et al. 2005). The MLP builds on complex systems theory which makes a distinction between the macro, meso and micro level to understand system dynamics (Liljenstrom and Svedin 2005; Geels 2002; Pahl-Wostl 1995). In order to represent socio-technical transitions as multi-level processes the MLP operationalized the three levels as niche, regime and landscape, respectively (Smith et al. 2010). The socio-technical regime constitutes the dominant way of realizing a societal function. Niches are protected spaces where innovative socio-technical approaches can be tested without being exposed to selection pressure exerted by the regime. The MLP has a focus on socio-technical innovation and systemic transformations where markets play a major role. Following a Darwinian understanding of evolutionary change as survival of the fittest, evolution is portrayed as the sequence of the processes of variation, selection, and retention. Market forces in the regime constitute the selection environment which determines which kind of innovation survives. The landscape embraces the context in which a regime is embedded (e.g., regulatory frameworks, culture). The landscape may stabilize but also exert pressure on a regime, e.g., change in public perception, new actors entering the market. The emphasis on markets as the dominant governance mode makes the MLP less suitable for understanding transformative change in water governance and management where markets do not play a major role. As argued previously, I assume that transformative change is triggered by innovative ideas challenging the prevailing paradigm. Selection pressure is exerted by the reigning paradigm, and by strong cultural-cognitive institutions, viz. paradigms, on new ideas and artefacts which embody a new paradigm. However, a purely Darwinian approach to understanding evolutionary change focusing on competition and survival of the fittest does not capturing the complexity and dynamics of such systemic transformations. One needs to take into account competitive and collaborative relationships, interdependencies, co-evolutionary change and path dependence.

Innovative ideas and knowledge are drivers of systemic transformations. They may enter societal discourse and become reified in artefacts such as shared practices that make innovative use of natural infrastructure in flood management. New practices may initially be tested in pilot projects. A perspective that makes a distinction between micro, meso and macro levels to analyse system innovations is useful for capturing such multi-level interactions. Its applicability does not depend on the dominant governance mode be it markets, hierarchies or networks. I refer to the different levels of analysis as *niche* (micro), *governance and management system* (meso) and the *socio-ecological system* (SES) as a whole (macro), respectively.

The SES, in general, provides a stabilizing context for a governance regime but it may also impose pressure on it. The latter has occurred in recent years in water governance and management as a result of global and climate change and the overall increase in the pace of socio-economic developments. If the SES context is favourable, innovative ideas find fertile ground and make it to the level of societal discourse. Catastrophic events such as extreme flooding increase media coverage and public awareness of the need for innovation for a limited amount of time. They may also provide windows of opportunity where innovative practices developed in the context of innovation platforms are taken up by mainstream policy and practice (Pahl-Wostl et al. 2013a).

As I have emphasized repeatedly, understanding the interplay and interdependence between structure and agency is essential for understanding transformative change. Agency, the capacity of actors to act independently and make free choices, is limited by structural constraints which provide stability and predictability. In their actions, actors interpret and may strengthen or weaken structural constraints. Figure 8.4 portrays schematically how innovation and learning platforms in multi-party processes are embedded in a multi-level structure.

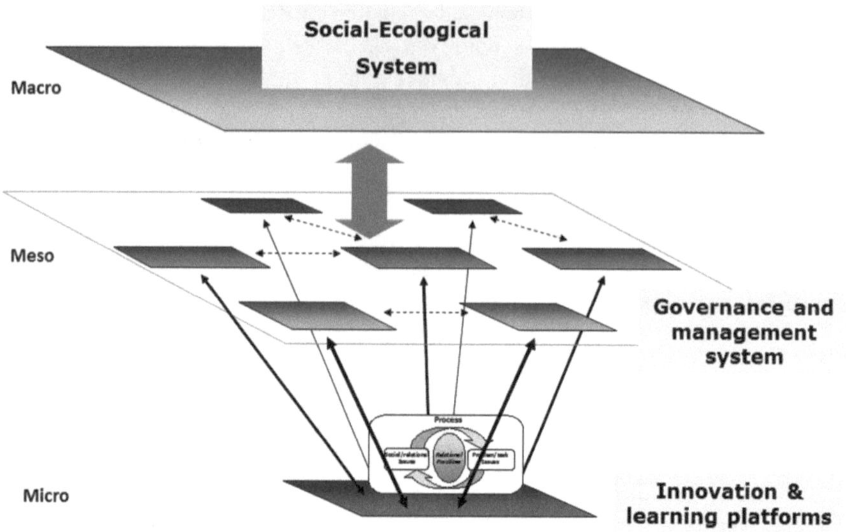

Fig. 8.4 Multi-level representation and cross-level interactions of social and societal learning processes (modified from Fig. 2, Pahl-Wostl et al. 2007a)

The distinction of the three levels resonates to some extent with distinctions among niche, regime and landscape in the MLP of socio-technical transitions. At the micro-level, innovation and learning platforms constitute the locus of innovative agency. At the meso-level, governance and management systems embrace structural configurations that keep the system largely in its current state (institutions, networks of collective actors, power constellations). The social-ecological system at macro-level mainly provides the context—discourse, informal institutions, norms and social practices. The sequence of levels does not have to be identical with increasing spatial scales. Governance and management systems/networks can embrace global scales. The levels refer to different kinds of social arenas rather than nested spatial scales.

Innovation- and learning-platforms exchange intensively with the governance and management system. Individual actors who participate in innovation platforms may often represent their constituencies and interact with other actors within their own organizations thus conveying new ideas, and broader interpretations of current institutions and practices. Innovation platforms are those informal spaces where social learning takes place which is symbolized by the small process circle at the bottom of Fig. 8.4 is derived from Fig. 4.2.

Table 8.2 summarizes the factors which are assumed to support or impede transformative change towards integrative and adaptive water governance and management at the micro, meso and macro levels, respectively.

Even when change is a desirable property of sustainability transformations, a fine balance between stabilizing and change-supporting elements within a governance regime is required. Regulatory frameworks and cultural values provide

Table 8.2 Factors at the three levels introduced in Fig. 8.4 that are assumed to support or impede, respectively, transformative change towards more sustainable water governance and management

Factors that	Micro—innovation and learning platforms	Meso—water governance and management system	Macro—SES water system
Support transformative change	Presence of shadow network Pilot studies—experimentation Diversity of perspectives	Effective links between formal and informal processes Balance between bottom-up and top-down processes Hybrid forms of governance linking governance modes with (predominantly) network governance	Catastrophic events Presence of avenues for public dissent
Impede transformative change	Closed groups Absence of links to wider networks Lack of resources	Rigid water bureaucracies Opportunities for participation in governance functions confined to selected groups	Patronage networks based on strong informal institutions Rent seeking elites

long-term stability whereas flexibility and change are provided by learning and negotiation processes in dynamic actor networks, where the interpretation of rules may be substantially renegotiated or where rules may even be changed. A certain degree of stability is needed for actors to develop their expectations regarding future developments that will influence their own decision making. Processing information, negotiating and changing rules are resource-intensive activities that should be limited to what is perceived by stakeholders themselves and by the policy analyst, respectively, to be necessary for coping with the emerging challenges of water resources management in a rapidly-changing socio-economic and environmental context.

Collaborative platforms may become a de facto part of the governance structure and play a key role in cross-scale linkages, both in terms of geographic and organizational scales. They may improve horizontal and vertical interplay in water governance regimes. This does not imply that such platforms should be fully formalized in terms of membership, procedural rules, roles and the distribution of decision making power. Formalization may destroy the characteristics of open platforms embedded in dynamic networks which render them so valuable in adaptive governance in the first place.

8.4 Conclusions

Even when it is analytically useful to make a distinction, similar factors influence adaptive and transformative capacity of water governance systems. Both adaptive and transformative capacity are strengthened by reflection on prevailing practices, recognition of different types of knowledge, flexible regulatory frameworks, collaboration among actors from different sectors and experimentation in dealing with uncertainties and unexpected developments. It appears to be more a question of degree rather than fundamental differences in types of activities as to when certain processes and experimentation with innovative ideas cross a threshold and lead to transformative change.

Empirically-validated insights into the dynamics of water governance systems which could be used to test and refine theories are scarce, and this remains a major research task. The theoretical framework introduced in this chapter is intended to make a significant contribution in this respect by guiding the synthesis of available empirical results and the design of further empirical studies and their analysis. Chapter 9 introduces a methodological framework to support such systematic empirical analyses. The insights that we have been obtained from empirical analyses so far are summarized in Chap. 10.

References

Benson, D., Gain, A. K., & Rouillard, J. J. (2015). Water governance in a comparative perspective: From IWRM to a 'Nexus' approach? *Water Alternatives, 8*(1), 756–773.

Engel, S., Pagiola, S., & Wunder, S. (2008). Designing payments for environmental services in theory and practice: An overview of the issues. *Ecological Economics, 65*(4), 663–674. doi:10.1016/j.ecolecon.2008.03.011

Folke, C. (2006). Resilience: The emergence of a perspective for social–ecological systems analyses. *Global Environmental Change, 16*(3), 253–267. doi:10.1016/j.gloenvcha.2006.04.002

Folke, C., Carpenter, S. R., Walker, B., Scheffer, M., Chapin, T., & Rockström, J. (2010). Resilience thinking: Integrating resilience, adaptability and transformability. *Ecology and Society, 115*(4), 20.

Folke, C., Hahn, T., Olsson, P., & Norberg, J. (2005). Adaptive governance of social-ecological systems. *Annual Review of Environment and Resources, 30*, 441–473.

Garmestani, A. S., & Benson, M. H. (2013). A framework for resilience-based governance of social-ecological systems. *Ecology and Society, 18*(1), doi:10.5751/ES-05180-180109

Geels, F. W. (2002). Technological transitions as evolutionary reconfiguration processes: A multi-level perspective and a case-study. *Research Policy, 31*(8–9), 1257–1274. doi:10.1016/S0048-7333(02)00062-8

Gleick, P. (2003). Global freshwater resources: Soft-path solutions for the 21st century. *Science, 302*, 1524–1528.

Gómez-Baggethun, E., de Groot, R., Lomas, P. L., & Montes, C. (2010). The history of ecosystem services in economic theory and practice: From early notions to markets and payment schemes. *Ecological Economics, 69*(6), 1209–1218. doi:10.1016/j.ecolecon.2009.11.007

Gomez-Baggethun, E., & Ruiz Perez, M. (2011). Economic valuation and the commodification of ecosystem services. *Progress in Physical Geography, 35*, 613–628.

Grande, E. (2012). Governance Forschung in der Governance Falle. *Politische Vierteljahresschrift, 4*, 565–592.

Green, O. O., Garmestani, A. S., van Rijswick, H. F., & Keessen, A. M. (2013). EU water governance: Striking the right balance between regulatory flexibility and enforcement? *Ecology and Society, 18*(2), doi:10.5751/ES-05357-180210

Grey, D., Garrick, D., Blackmore, D., Kelmen, J., Muller, M., & Sadoff, C. W. (2013). Water security in one blue planet: Twenty-first century policy challenges for science. *Phil. Trans. R. Soc. A, 371*, 20120406. doi:10.1098/rsta.2012.0406

Grey, D., & Sadoff, C. W. (2007). Sink or swim? Water security for growth and development. *Water Policy, 9*, 545–571.

Hallegatte, S. (2009). Strategies to adapt to an uncertain climate change. *Global Environmental Change, 19*(2), 240–247. doi:10.1016/j.gloenvcha.2008.12.003

Huntjens, P., Pahl-Wostl, C., & Grin, J. (2010). Climate change adaptation in European river basins. *Regional Environmental Change, 10*(4), 263–284. doi:10.1007/s10113-009-0108-6

Huntjens, P., Pahl-Wostl, C., Rihoux, B., Schlüter, M., Flachner, Z., Neto, S., et al. (2011). Adaptive water management and policy learning in a changing climate: A formal comparative analysis of eight water management regimes in Europe, Africa and Asia. *Environmental Policy and Governance, 21*(3), 145–163. doi:10.1002/eet.571

Jordan, A., & Lenschow, A. (2010). Environmental policy integration: A state of the art review. *Environmental Policy and Governance, 20*(3), 147–158. doi:10.1002/eet.539

Levin, S. A. (1999). *Fragile dominion—Complexity and the Commons*. Cambridge, UK: Perseus Publishers.

Liljenstrom, H., & Svedin, U. (Eds.). (2005). *Micro-meso-macro: Addressing complex systems coupling*. Singapore: World Scientific Publishing Company.

MA (2005). *Ecosystems and human well-being: Synthesis. Millennium ecosystem assessment*. Washington DC: Island Press.

Mayntz, R. (2004). Governance Theory als fortentwickelte Steuerungstheorie? *MPIfG Working Papers*. Cologne, Germany Max-Planck-Institute for the Study of Societies.

Opperman, J. J., Galloway, G. E., Fargione, J., Mount, J. F., Richter, B. D., & Secchi, S. (2009). Sustainable floodplain through large-scale reconnection to rivers. *Science, 326*, 1487–1488.

Ostrom, E. (2001). Vulnerability and polycentric governance systems. *IHDP (International Human Dimensions Programme on Global Environmental Change) Newsletter UPDATE, 3*(1), 3–4.

Ostrom, E. (2010). Polycentric systems for coping with collective action and global environmental change. *Global Environmental Change, 20*, 550–557.

Pahl-Wostl, C. (1995). *The dynamic nature of ecosystems: Chaos and order entwined*. Chichester: Wiley.

Pahl-Wostl, C. (2002). Towards sustainability in the water sector—The importance of human actors and processes of social learning. *Aquatic Sciences, 64*, 394–411.

Pahl-Wostl, C. (2007a). Requirements for adaptive water management. In C. Pahl-Wostl, P. Kabat, & J. Möltgen (Eds.), *Adaptive and integrated water management. Coping with complexity and uncertainty* (pp. 1–22). Heidelberg: Springer.

Pahl-Wostl, C. (2007b). Transition towards adaptive management of water facing climate and global change. *Water Resources Management, 21*(1), 49–62.

Pahl-Wostl, C. (2009). A conceptual framework for analysing adaptive capacity and multi-level learning processes in resource governance regimes. *Global Environmental Change, 19*, 354–365.

Pahl-Wostl, C. (2015). Water security, systemic risks and adaptive water governance and management. In C. Pahl-Wostl, J. Gupta, & A. Bhaduri (Eds.), *Handbook on water security*. Cheltenham, UK: Edward Elgar.

Pahl-Wostl, C., Becker, G., Sendzimir, J., & Knieper, C. (2013a). How multilevel societal learning processes facilitate transformative change: A comparative case study analysis on flood management. *Ecology and Society, 18*(4), 58. doi:10.5751/ES-05779-180458

Pahl-Wostl, C., Craps, M., Dewulf, A., Mostert, E., Tabara, D., & Taillieu, T. (2007a). Social learning and water resources management. *Ecology and Society, 12*(2), 5. URL: http://www.ecologyandsociety.org/vol12/iss12/art15/

Pahl-Wostl, C., Holtz, G., Kastens, B., & Knieper, C. (2010). Analyzing complex water governance regimes: The management and transition framework. *Environmental Science & Policy, 13*(7), 571–581. doi:10.1016/j.envsci.2010.08.006

Pahl-Wostl, C., Jeffrey, P., Isendahl, N., & Brugnach, M. (2011). Maturing the new water management paradigm: Progressing from aspiration to practice. *Water Resources Management, 25*(3), 837–856. doi:10.1007/s11269-010-9729-2

Pahl-Wostl, C., & Knieper, C. (2014). The capacity of water governance to deal with the climate change adaptation challenge: Using fuzzy set qualitative comparative analysis to distinguish between polycentric, fragmented and centralized regimes. *Global Environmental Change, 29*, 139–154.

Pahl-Wostl, C., & Knüppe, K. (2015). Water security and environmental water needs: The role of the ecosystem services concept and transformation of governance systems. In C. Pahl-Wostl, J. Gupta, & A. Bhaduri (Eds.), *Handbook on water security*. Cheltenham, UK: Edward Elgar.

Pahl-Wostl, C., Lebel, L., Knieper, C., & Nikitina, E. (2012). From applying panaceas to mastering complexity: Toward adaptive water governance in river basins. *Environmental Science & Policy, 23*, 24–34.

Pahl-Wostl, C., Palmer, M. A., & Richards, K. (2013b). Enhancing water security for the benefits of humans and nature—The role of governance. *Current Opinion in Environmental Sustainability, 5*, 676–684.

Pahl-Wostl, C., Sendzimir, J., Jeffrey, P., Aerts, J., Berkamp, G., & Cross, K. (2007b). Managing change toward adaptive water management through social learning. *Ecology and Society, 12*(2), 30. URL:http://www.ecologyandsociety.org/vol12/iss32/art30/

Renn, O. (2005). Risk governance: Towards an integrative approach. In I. R. G. Council (Ed.), *White paper* (Vol. 1). Geneva: International Risk Governance Council.

Silveira, A., & Richards, K. S. (2013). The link between polycentrism and adaptive capacity in river basin governance systems: Insights from the River Rhine and the Zhujiang (Pearl River) Basin. *Annals of the Association of American Geographers, 103*(2), 319–329.

Smith, D. M., & Barchiesi, S. (2009). Environment as infrastructure: Resilience to climate change impacts on water through investments in nature. In IUCN (Ed.), *Perspectives on water and climate change adaptation*. Gland, Switzerland: International Union for Conservation of Nature (IUCN).

Smith, A., Stirling, A., & Berkhout, F. (2005). The governance of sustainable socio-technical transitions. *Research Policy, 34*(10), 1491–1510. doi:10.1016/j.respol.2005.07.005

Smith, A., Voß, J.-P., & Grin, J. (2010). Innovation studies and sustainability transitions: The allure of the multi-level perspective and its challenges. *Research Policy, 39*(4), 435–448. doi:10.1016/j.respol.2010.01.023

Tillman, D. E., Larsen, T. A., Pahl-Wostl, C., & Gujer, W. (2005). Simulating development strategies for water supply system. *Journal of Hydroinformatic, 7*, 41–51.

UNESCAP. (2009). *What is good governance?*. Bangkok, Thailand: United Nations Economic and Social Commission for Asia and the Pacific.

Wenger, E. (1998). *Communities of practice: Learning, meaning, and identity*. Cambridge, UK: Cambridge: University Press.

Wegner, G., & Pascual, U. (2011). Cost-benefit analysis in the context of ecosystem services for human well-being: A multidisciplinary critique. *Global Environmental Change, 21*, 492–504.

Chapter 9
A Methodological Framework for Empirical Analysis

The theoretical framework introduced in Chap. 8 is a necessary but not sufficient condition for developing a profound understanding of transformation towards sustainability. Theoretical propositions should be supported by sound empirical evidence. An appropriate research design needs to capture as much of the complexity of processes in governance systems as possible. This suggests a comparative case-study approach and methodological pluralism. One problem arises: the processes of transformation and change occur over long time scales. Ideally longitudinal studies should be conducted that compare trajectories of change over decades. Alternatively, cases can be compared at different stages of development and transformative change. In order to compare insights from a range of cases and case study designs it is important to have a sound methodology and standardized data collection approaches.

This chapter first elaborates on different methods for comparative analysis. Then operational approaches for the standardization of case study data and various approaches for data collection are introduced.

9.1 Methods for Comparative Analysis

A diverse range of methods have been used in comparative case study research. I refrain from presenting a narrow definition or providing an overview of the various definitions of a case study. In essence, a case study is an in-depth study of a defined topic which also embraces relevant contextual conditions. Such a broad definition rules out approaches where, for example, only one or two variables are collected per sample and compared with methods of linear regression. It is compatible, however, with a wide range of quantitative and qualitative approaches in case study analyses.

Pahl-Wostl and Lebel (2011b) developed a typology of methods for comparative analysis with a focus on water governance studies which is represented in Table 9.1. The first level categorization makes a distinction between more systemic analyses of governance and methods which focus on comparing only one or a few variables or relationships such as the effect of water pricing on irrigation efficiency (e.g.,

C. Pahl-Wostl, *Water Governance in the Face of Global Change*,
Water Governance—Concepts, Methods, and Practice,
DOI 10.1007/978-3-319-21855-7_9

Table 9.1 Typology of methods for comparative case study research in water governance based on Pahl-Wostl and Lebel (2011b)

Typology of comparisons			What is compared
Specific variables	Quantitative	Meta-analysis	Values of a particular governance attribute or describing an attribute's relationship with performance for well-defined set of cases
		Survey	As above but non-systematic selection of cases (i.e. weak inclusion/exclusion or search criteria)
	Qualitative	Systematic review	Descriptions of a particular governance attribute for well-defined set of cases
		Unsystematic review	As above but non-systematic selection of cases
Systemic	Structure	Social networks	Diagrams of actor relationships (e.g., power, influence, authority, communication)
		Organizational	Diagrams of responsibilities and accountability relationships compared
		Systems	Diagrams of governance and other water system components
	Dynamics	Transitions	Set of variables about same location at different times (e.g., reform process)
		Pathways	Pathways of change in different locations
	Contextual	Questions	Responses to a common set of analytical questions
		Narrative	Integrated descriptions of a governance regime
	Indicator based	Checklist	Presence/absence of governance attributes
		Scoring	Ordinal scale measure of governance attributes

Scheierling et al. 2006). Since only a systemic research design can capture the complexity of water governance systems only those methods for systemic comparisons are presented here.

Systemic approaches can be divided into four groups (Table 9.1). Structural approaches typically compare structural features of governance systems and the core analytical products are diagrams of relationships. These may describe relationships among actors, organizations, institutions or, in more complex versions (systems method), several components of the water governance system. The class diagram of the Management and Transition Framework (MTF), for example, falls in the subcategory of 'systemic-structural-systems' since it allows the comparison of the various components of a water system including the management and governance regime (Knieper et al. 2010; Pahl-Wostl et al. 2010). Structural approaches may make use of quantitative methods as in social network analysis where a range of metrics have been developed to characterize network properties such as the centrality of individual actors or the centralization of the network as a whole (Borgatti et al. 2013; Scott 2000).

In approaches focusing on dynamics features, timelines or storylines of change play a special role. Transition approaches focus on changes in a place, typically institutional reforms, such as the introduction of river basin or similar organizations (van der Brugge et al. 2005), new infrastructure (Foran and Manorom 2007) or a new flood plain policy (Sendzimir et al. 2007; Pahl-Wostl et al. 2013). Pathway approaches are more suitable for comparing changes in process in different locations. A good example of the latter is a recent book about policy entrepreneurs which analyses and compares lessons from 13 transitions in river basins (Huitema and Meijerink 2010).

Contextual approaches are probably the most common method. They compare a few cases in different locations by applying the same framework, and usually a set of questions, to each water governance system. The answers to these questions are often context-sensitive and complex, so qualitative methods of analysis are essential. Examples are the comparison of the Murray-Darling basin in Australia and the Mekong basin in Southeast Asia by Chenoweth et al. (2002) or the institutional arrangements in eight river basins around the world by Blomquist et al. (2005). In some instances analysis is carried out on prior narrative work which is not organized in a form conducive to question-and-answer analysis. In this case less emphasis may be given to a set of specific questions and more to broad set of questions about how regime characteristics interact and co-evolve (e.g., Molle et al. 2009; Lebel et al. 2007). Pahl-Wostl and Lebel (2011b) referred to this as the *narrative method.*

Indicator approaches are usually semi-quantitative. At a minimum they focus on the presence/absence of a large number of governance attributes sometimes on the basis of aggregate indices of these attributes. Pahl-Wostl and Lebel (2011b) called this the *checklist method.* Such an approach was used by Borowski et al. (2008) to compare the influence of social learning and spatial misfits for the implementation of the European Water Framework Directive in Germany and France. If more nuanced levels for many of the regime attributes are observed then this is called the *scoring method* (e.g., Mostert et al. 2007; Huntjens et al. 2011; Pahl-Wostl et al. 2012). Indicator approaches may be applied using a more descriptive mode by listing an inventory of differences up by more analytical and explanatory approaches which seek to identify patterns and causal explanations of differences between cases.

Methods used to compare water governance systems differ in dimensions other than those highlighted in the typology in Table 9.1. For example, they also differ in terms of who makes the analysis, or whose knowledge is used for an assessment. Some approaches are entirely driven by the judgements and understanding of the researcher(s), whereas other methods are explicitly designed to promote understanding among several experts or a wider body of stakeholders. Some use primary data while others base comparisons on meta-analyses of available data.

Another important aspect which applies to all comparative methods is the choice of an appropriate set of cases to be compared. Comparing very different water governance systems (e.g., Germany and Uzbekistan) or similar ones (e.g., Germany and Netherlands) has different implications for analytical power on the one hand,

and generalizability, on the other. Internal validity is increased by choosing similar cases. But this increase in internal validity comes often at the expense of external validity, the relevance of the findings for a wider range of cases.

Irrespective of such considerations, all comparative analyses have to rely on a minimum level of standardized data collection. Otherwise comparisons may become quite arbitrary and suffer from non-transparent interpretation. In a small number of cases this may be achieved by an individual researcher conducting the analysis for all the cases. But such internal coherence does not imply that comparison is possible with studies undertaken by other researchers on a related theme for different or even for similar cases. The absence of a minimum set of shared standards for conceptualizing variables and conducting data analysis may be one reason why relatively few analyses exist in resource governance research in general, and water governance research, in particular, that adopt a systemic approach and compare a large number of case studies.

9.2 Harmonized Databases and Governance Indicators

9.2.1 Standardization in the Social Sciences

The knowledge base in water governance and in governance of social-ecological systems in general is quite fragmented. Individual case studies abound. The use of a diversity of concepts, terminology and methods render more systematic comparisons of different studies difficult if not impossible. Such a fragmented research landscape seems to be quite typical for the social sciences. One notable exception is neo-classical economics—if one accepts that economics is a social science as I do. However, the dominance or even hegemony of neo-classical economics, despite all its disputable assumptions and weaknesses in dealing with real-world complexity, also reveals the danger of theoretical orthodoxy and suppression of alternative views. Only in recent years when the weaknesses of dominant economic theory have become all too obvious, have more critical voices start to be heard, also from within the discipline (Krugman 2009). Diversity in approaches that does not lead to fragmentation though seems to be desirable for fostering progress and preventing stagnation in scientific scholarship.

In her comparison of natural and social sciences (Mayntz 2005) pointed out that scientific progress in the natural sciences is often cumulative. By contrast this is hardly the case in the social sciences where the object of research cannot be detached from theory and methods used for analysis. Theories are often quite narrow in scope and in their ability to provide explanations and replace rather than build on each other. This is a striking difference from the natural sciences. I recall from my own experience the lack of understanding from natural science colleagues when social scientists use most of their presentation time elaborating on the conceptual and theoretical underpinnings of their research rather than presenting results. Admittedly the natural sciences might profit from more reflection on the basic assumptions that underpin their work and the social sciences would almost certainly profit from more agreement on shared concepts and methodologies. We might witness such developments in particular in inter-disciplinary research on social-ecological systems (SES). A pioneer in this respect was Elinor Ostrom with her later work on the SES framework, a framework for analysing the sustainability of SESs (Ostrom 2007, 2009). The SES framework builds on the IAD framework and introduces a more in depth representation of the ecological system. Figure 9.1 shows the first-tier components of the revised SES framework. Elinor Ostrom rightly pointed out that the analysis of complex systems such as SES and developing diagnostic capabilities requires a shared terminology.

Elinor Ostrom argued that frameworks are largely "theory-free". She further argued in favour of making a clear distinction among the terms "framework," "theory," and "model" that are used by many analysts almost interchangeably (Ostrom 2005, 2010; McGinnis and Ostrom 2014). "*A framework provides the basic vocabulary of concepts and terms that may be used to construct the kinds of causal explanations expected of a theory. Frameworks organize diagnostic, descriptive, and prescriptive inquiry. A theory posits specific causal relationships*

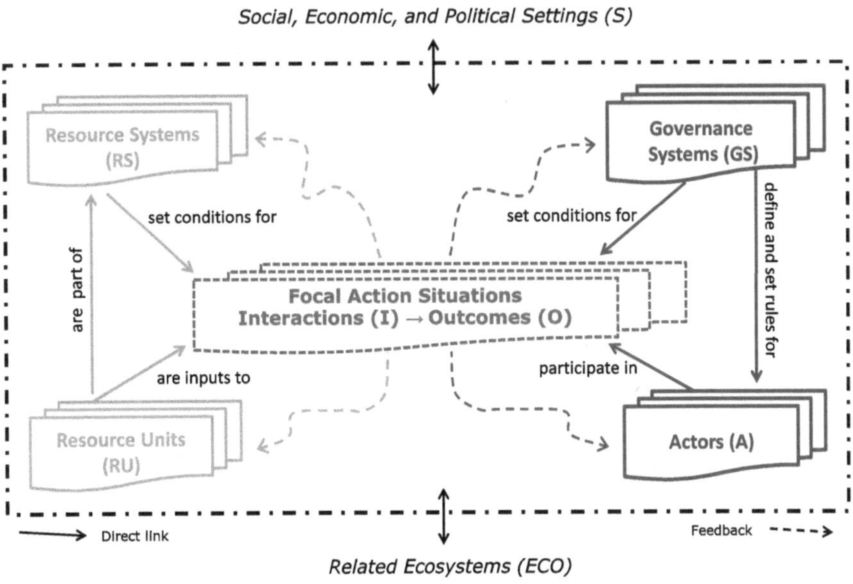

Fig. 9.1 Multiple first-tier components of the revised SES framework (reproduced from Fig. 2 in McGinnis and Ostrom (2014), with permission). Revisions were based on discussions within the 'SES Club', an international network of scholars who further developed the SES framework. First-tier components are hierarchically decomposed into lower level variables. For example, second-tier variables associated with the resource system are sector, clarity of system boundaries or size. Second-tier variables associated with the governance system are government organization, nongovernment organization and network structure

among core variables. In contrast, a model constitutes a more detailed manifestation of a general theoretical explanation in terms of the functional relationships among independent and dependent variables important in a particular setting" (McGinnis and Ostrom 2014). Frameworks thus enable the comparison of theories by providing a meta-theoretical language. Overarching frameworks should identify and embrace all the elements that any theory relevant to a particular kind of phenomenon would need to include.

While I agree that it is useful to develop such frameworks and that these frameworks should be as flexible as possible to allow many different and complementary uses, I am hesitant to endorse the view that frameworks can be largely free of theory. Any framework always includes tacit assumptions. The SES-framework builds heavily upon the Institutional Analysis and Development (IAD) framework (Ostrom 2005) and is thus clearly informed by a rational choice approach. Unlike the MTF, it does not, for example, take into account the situatedness of knowledge production and human behaviour. However, the SES framework is not so internally closed as to prevent extensions or a re-interpretation of terms. Obviously such re-interpretation and extension may only jeopardize standardized comparison which has been a major argument for developing such frameworks. It would already be a significant step forward, however, if researchers conducting empirical analyses would be explicit in how they define and operationalize different terms.

In our comparison of frameworks for analysing social-ecological systems we stated: "*A framework provides a set of assumptions, concepts, values and practices that constitute the way of viewing the specific reality*" (Binder et al. 2013). This definition takes into account the fact that frameworks comprise more than a language since only the practice of a language conveys its meaning. Being armed only with a dictionary does not allow one to communicate in a new language. The comparison of frameworks also demonstrated that the whole research field would profit from a more systematic exchange of terminology including the use of the notions of framework, theory and model.

Applications of shared frameworks do not come without costs. There may be a trade-off between the number of people who use a framework and the degree of standardization a framework imposes on the user. In the application of frameworks I thus introduce a distinction between 'light' and 'deep' applications as contrasted in Table 9.2.

The term 'light' refers to a framework application which uses terms and language without predefined standards or definitions of concepts, variables, methods for data collection and practices of application and analysis. An example of the SES framework is provided by a meta-analysis on the role of leadership in the management of fisheries by Gutiérrez et al. (2011). They distinguished among Resource System, Resource Unit, Governance System, User System and Outcomes as provided by the SES framework (Fig. 9.1) and added a further category at the same level: Co-Management. For the lower-tier variables which specify all of these categories they introduced new variables that suited their research focus and allowed them to capture what was important from their conceptual point of view.

Table 9.2 Comparison between 'light' and 'deep' applications of frameworks

Criterion	Light	Deep
Effort involved for new user to apply the framework	Low	High
Use operationalized terms for overall categorization—higher tier variables	Yes	Yes
Use categories for lower-tier variables (zoom in on details)	Partly	Yes
Standardized definitions of concepts	No	Yes
Standardized definition of operational variables	No	Yes
Protocols for data collection	No	Yes
Protocols for analysis	No	Yes
Further development of the framework by individuals who apply it to their research foci	Yes	No

The use of some categories of the SES framework supported the structuring of a complex analytical task without being constrained. Such applications constitute significant progress since they at least facilitate the comparison of important insights. The paper on the SES framework is often cited.[1] However, an analysis has shown that most of the publications cited make only general reference to the framework without applying it at all for any kind of analysis.[2] The number of citations can be seen as an indication of a general interest in, and agreement on, developing such frameworks. The small number of publications that refer to applications of the SES framework indicates that significant obstacles have yet to be overcome.

A much more ambitious approach to supporting meta-analyses based on what approaches a 'deep' application of the SES framework has been undertaken by the SESMAD project (Cox 2014). The project team has initiated the development of a public database in which diverse users can enter their data. Such an undertaking is highly demanding, in particular when users are to have some freedom (e.g., specifying additional variables) when adding their own cases to the database.

A similar undertaking launched by the so-called "SES Club", an informal network of researchers from Europe and the US who embarked on further developing the SES framework, is also underway (McGinnis and Ostrom 2014). As active member of the SES Club I can confirm the enormous effort which is required among those jointly developing a framework to arrive at shared understanding. I would argue that moving towards 'deep' application is only possible when a group of researchers engages in collaborative research on a specific research theme.

An example for such a 'deep' application is the International Forestry Resources and Institutions (IFRI)[3] database which was developed based on the IAD framework. Albeit much simpler than the SES framework it took the research teams nearly two years to agree on a standardized approach. A network of research teams

[1]Ostrom (2007) had 480 and Ostrom (2009) had 733 citations according to SCOPUS, 17.01.2015.

[2]Michael Cox—personal communication and in preparation.

[3]http://www.ifriresearch.net/.

collected data for more than a decade using the same strict data collection protocols. The investment paid off in terms of unprecedented opportunities for in-depth analyses for a large number of cases (Wollenberg et al. 2007). The entire process was tightly managed and focused on a set of well-defined research questions. During the first decade the database was only accessible to the members of the network. In recent years the network has started to open up and exchange data and methods with a larger scientific community.

Experiences with the SES framework and the IFRI database largely confirm the experiences in the development and application of the MTF. In a special issue in *Environmental Science and Policy* we reported examples of various kinds of initial applications of the MTF (Pahl-Wostl and Kranz 2010). In line with a light application some teams involved in these applications used only individual concepts defined by the MTF. Kranz et al. (2010) made use of the stylized representation of policy processes in terms of stylized phases (cf. Fig. 3.7) to compare trans-boundary water management in the Orange and Mekong basins. Bisaro et al. (2010) used the representation of multi-level policy processes as sequences of connected action situations to analyse climate change adaptation and adaptive management in Lesotho. The application of standardized databases albeit for in depth analyses of individual cases in the Amudarya basin in Uzbekistan (Schlüter et al. 2010) and the Tisza basin in Hungary (Sendzimir et al. 2010), took the use of the MTF one step further. In subsequent work we extended these analyses to compare the transformations towards integrative flood management in various countries (Pahl-Wostl et al. 2013). Using standardized databases for the various data collection efforts proved to be highly useful for developing comprehensive representations of complex governance systems. However, standardization and thus deep application also increases the effort for any individuals who had not been involved in the initial development of the framework.

Nevertheless I am convinced that developing shared frameworks and applications is the only way forward in order to overcome fragmentation in research on governance systems and SES in general. Joint framework development and shared practice can also support processes of social learning across the various scientific disciplines and communicate tacit assumptions (Dewulf et al. 2007; Hovelynck et al. 2010). Social learning does not imply consensus but a basis for discussing differences constructively. Transforming fragmentation and ideological debates into pluralism and deliberation would constitute a major step forward for the research community.

9.2.2 *Standardized Databases and Analytical Protocols for the MTF*

This section summarizes efforts towards, and experiences of standardization of applications of the MTF. Figure 9.2 shows the main classes of the water system

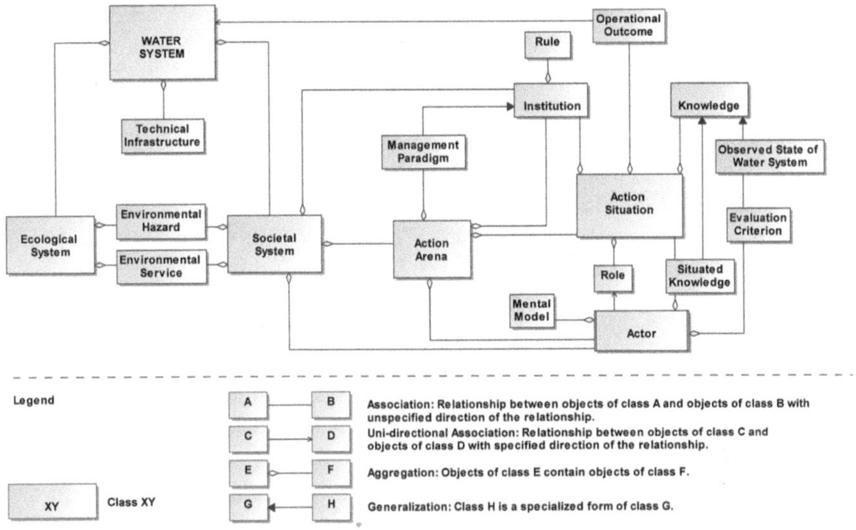

Fig. 9.2 Class hierarchy of the management and transition framework (MTF) in UML notation. The water system class comprises the ecological and societal systems and technical infrastructure. Environmental Hazard and Environmental Service constitute the interface between the ecological and societal systems. The societal system is further disaggregated into its components (Slightly modified from Fig. 1 in Pahl-Wostl et al. 2010, p. 575)

including social and ecological systems and infrastructure represented in the MTF in Unified Modelling language (UML) notation. UML is a graphical modelling language which allows the representation of object-oriented hierarchies by introducing classes and relationships between classes as explained in the captions of Fig. 9.2 and Fig. 3.5 in Chap. 3.

We decided to use UML to introduce a certain degree of stringency into the participatory development process of the MTF (Dewulf et al. 2005; Hovelynck et al. 2010). Without some formal notation it proved to be quite difficult to arrive at shared agreements on terminology. But increased formalization is always associated with some costs in terms of a reduced richness of expression and the increased effort required for the novice to become acquainted with the framework. UML imposes more formal relationships between the variables in the MTF than is, for example, the case in the SES framework. The latter has a hierarchical multi-tier structure. Relationships between variables are only indicated at the highest level (cf. Fig. 9.1). This leaves more space for interpretation by the user but also introduces ambiguities and inconsistencies that may reduce comparability of different applications of the SES by different users.

In the MTF, the UML literate immediately sees the nested structure. Others may take more time to do so. The water system is the overarching class which comprises all environmental and human components: the ecological system, technical infrastructure and the societal system. The interface between societal and ecological systems is described by environmental hazards and services. Environmental hazards embrace both hazards that human activities impose on the environment and hazards for humans that arise from the natural environment such as floods. The societal system is further decomposed. As introduced in Chap. 3 action situations (ASs) constitute core building blocks of governance processes. They provide spaces for social interactions and may be influenced and/or produce institutions, knowledge or operational outcomes. Action arenas are thematic platforms—networks of ASs constituting a societal function (Pahl-Wostl et al. 2010; Knieper et al. 2010). In addition to the hierarchical decomposition, UML also facilitates the representation of relational characteristics. The class role, for example, is defined for an actor in the social context of an action situation (cf. Chap. 4). For roles which emerge from social interactions such as leadership this is quite evident. But other roles such as the role of a scientific expert as defined in the MTF or the role of an actor refer as well to an interpretation and enactment in a specific social interaction context. Interpretation may however also be influenced by societal norms, institutions.

To translate this conceptual framework into an operational tool, relational databases and associated protocols for data collection and analysis have been developed (Pahl-Wostl et al. 2014; Knieper et al. 2010). The benefit of relational databases is manifold: first, large amounts of data can be stored and structured. Second, relational databases comprise related tables for defined datasets and thus facilitate the easy adoption of the structure provided by the MTF in general and the MTF class diagram in particular. Third, the storage of both quantitative and qualitative data is possible which allows the integration of diverse types of components and the diverse types of attributes used in the MTF. Finally, relational databases facilitate data analysis. Queries can be designed to calculate quantitative indicators of governance regimes and to provide structured input for graphical representations of management processes (Knieper et al. 2010).

In applications of the MTF the first step is to choose the classes that are relevant for addressing the questions under consideration. Let us assume that a user is interested in comparing the influence of vertical integration in dealing with water shortages during droughts in a number of case studies. Referring to Fig. 9.2, the following classes might be considered relevant: action situation, actor, role of actor, institution, knowledge and operational outcome. When the MTF is applied to an empirical case, normally several instances are identified for each class, such as specific action situations, the actors acting in these situations and the institutions influencing or resulting from these action situations (e.g., Pahl-Wostl et al. 2013). Let us again draw on the example of the influence of vertical integration in dealing with droughts. Instances of the class Action Situation might be the development of a drought policy or the implementation of a drought management plan. Moreover, classes have several attributes. For example, attributes of the class actor would be "individual" or "collective". Classes are related to each other (e.g., an actor participates in an action situation and holds a specific role there) and the stylized process representations of the MTF specify process phases to which action situations can be related. This adds up to a sizeable amount of data (classes, attributes, relations) that are required to describe a case study by means of the MTF.

Figure 9.3 shows the scheme of an encompassing relational database with classes, their attributes and relations. A comprehensive guidance document was developed which is openly available online (Pahl-Wostl et al. 2014). Nevertheless, it still proved to be helpful to provide an online introduction to new users prior to their working with the guidance documents.

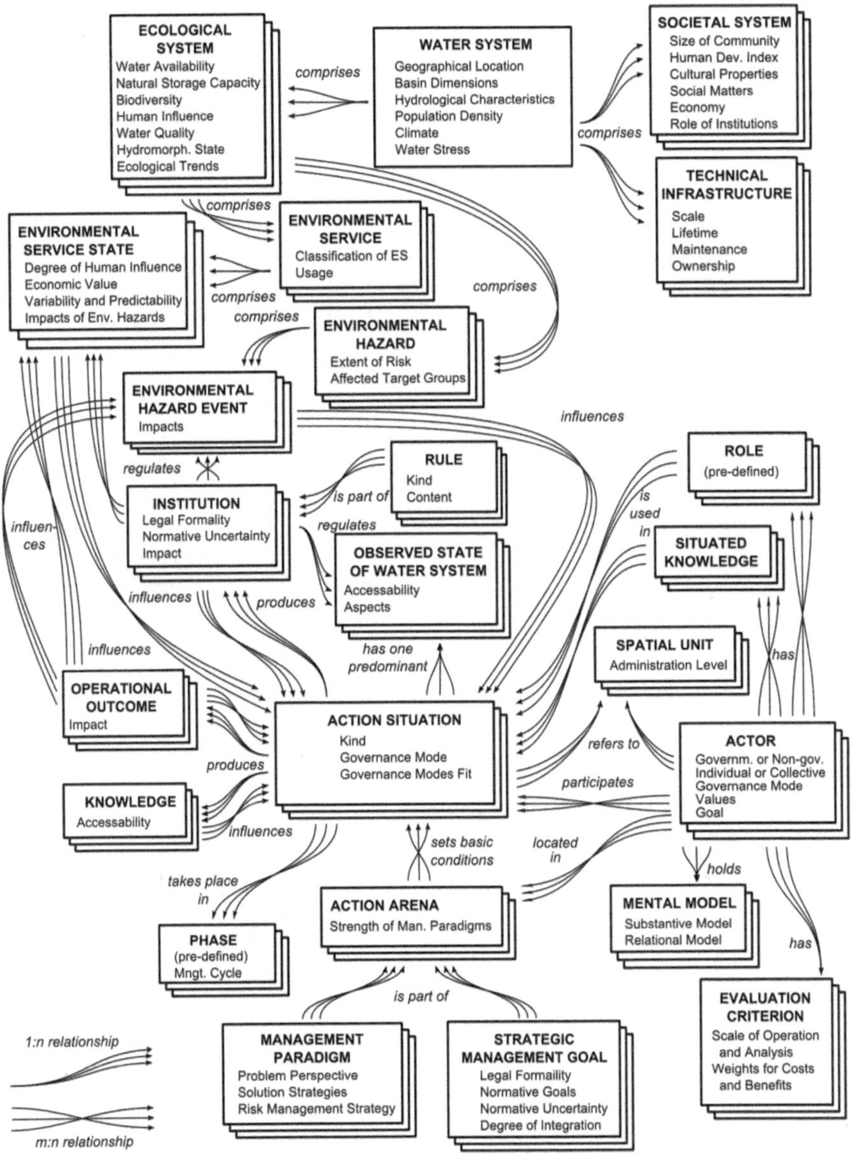

Fig. 9.3 Scheme of a relational system database which comprises all classes of the MTF (Pahl-Wostl et al. 2014)

A number of indicators and associated database queries were developed for a systematic analysis of governance systems. Table 9.3 lists for example indicators developed for vertical integration, stakeholder participation and centralization to operationalize theoretic propositions on what is relevant for these governance system properties.

Table 9.3 Examples of operationalizing governance system properties using the MTF and derived relational databases (Pahl-Wostl et al. 2014)

Governance system property	Indicator	Operationalization in MTF database
Vertical integration	Link via institutions	Number of links between two levels by institutions connecting ASs and direction of influence
	Link via knowledge	Number of links between two levels by knowledge connecting ASs and direction of influence
	Link by actors	Number of actors from different levels participating in an AS
	Actors as integrators	Single actor is active on multiple scales by participating in ASs at several levels
Stakeholder participation	Participation in rule production	For each pair of ASs (AS1, AS2) for which AS1 produces an institution which influences AS2, the number and role of actors from AS2 involved in AS1
	Participation in knowledge generation	For each pair of AS (AS1, AS2) for which AS1 produces knowledge which influences AS2, the number and role of actors from AS2 involved in AS1
	Involvement of non-gov. actors	Number of governmental and number of non-governmental Actors participating in the AS of a formal policy cycle
	Role of non-gov. actors	Number of governmental and number of non-governmental Actors with a certain role (e.g., lead) participating in an AS of policy cycle
Centralization	Dominant levels	Dominance of levels in leading AS on the same or other levels. Quantified by counting for each pair of levels (level 1, level 2) how often an AS on level 2 is led by an actor from level 1
	Dominant actor	Lead and dominant roles distributed across many or held by one or several actors
	Centralization per level	High horizontal centralisation if at one level most ASs are led by the same actor. For each level the extent of distribution of the lead role across all ASs is determined

9.2.3 *Illustration of Applications of MTF Databases*

Analyses of the application of the MTF databases to characterize governance systems have appeared in several publications (Schlüter et al. 2010; Sendzimir et al. 2010; Pahl-Wostl et al. 2013). In this section, I provide several examples to illustrate the potential of using the indicators to characterize different governance system properties introduced in Table 9.3. These examples demonstrate as well that results from database analyses can and should be complemented by qualitative in depth case study knowledge to understand and to explain the findings.

The example of flood management in the Hungarian Tisza is used to illustrate analyses of the role of stakeholder participation in the formal policy process triggered by informal learning processes. In the past few decades Hungarian flood policy underwent major reform. The formal governance of flood management has traditionally been top-down and quite technocratic. However, despite a more centralized and rigid governance structure Hungarian flood management was found to be characterised by advanced levels of learning and innovation (Huntjens et al. 2011) which contradicts expectations. This can be explained by an informal learning process carried by a shadow network that succeeded in introducing innovative ideas in the policy reform process (Sendzimir et al. 2007, 2010). Those actors involved in bottom-up-driven learning processes explored alternative flood management strategies (Sendzimir et al. 2007; Pahl-Wostl et al. 2013). A kind of shadow-network emerged and managed to influence state-led water policy processes. (cf. Box 6.2, Fig. 6.5 which provides a multi-level representation of the development of flood policy in the Hungarian Tisza). The activities of this informal actor network resulted in improved vertical integration across levels and increased participation of non-governmental actors in policy formulation. Using relational databases allowed the analysis of the participation of actors in rule production and in knowledge generation (Pahl-Wostl et al. 2013). The influence of the shadow network became evident in the change of the nature of the process of developing and implementing flood management policy. In a first phase the new policy was developed by national authorities involving only a group of technical experts. The influence of the shadow network led to the development of a revised policy which included much more innovative elements like the implementation of pilot polders. Insights gained from pilots were again taken up in policy development. Overall many more actors, in particular from a regional and implementation level, were involved than during the first phase.

Vertical integration was analysed by identifying the links between governance levels during the development of flood policy in the Hungarian Tisza via institutions and knowledge, respectively (cf. Table 9.4). Results are summarized in Table 9.4. First, the national level is the main level where institutions are established whereas knowledge production occurs mainly at regional levels. Vertical integration across institutions mainly reflects top-down governance whereas vertical integration by knowledge also includes bottom-up influence.

Another example from the Amudarya Basin in Uzbekistan illustrates how centralization of governance systems can be assessed by the indicators introduced in Table 9.3. Figure 9.4 and Table 9.5 show patterns of dominance in the lead role in ASs of two

Table 9.4 Number of links from level 1 (column) to level 2 (row) via (a) institutions and (b) knowledge identified in the development of flood policy in the Hungarian Tisza

	National	Subbasin	Regional
(a)			
National	8	4	3
Subbasin	2	3	3
Regional	–	–	1
(b)			
National	–	1	2
Subbasin	–	–	–
Regional	2	–	4

In total 15 links through institutions emanate from the national level, and seven of those influence lower levels. In total six links through knowledge emanate from the regional levels, two of those influence higher levels of governance of governance

different governance processes: one is responsible for addressing extreme events (EE) and the other to allocating water to ecosystems (WE) (Schlüter et al. 2010).

The EE process is strongly dominated by the national government which had the lead in more than 50 % of the ASs (Fig. 9.4). The majority of the leads in the WE processes, by contrast, was not with a single actor but distributed among the Cabinet of Ministers, interstate organizations, and "no lead". The latter refers to ASs that involved representatives of all five former Soviet Union Aral Sea Basin countries where one could not assign a clear leading role to any of the countries. Table 9.5 shows the most frequent lead actors in different phases of the policy process. The phases refer to the stylized phases provided by the MTF. In the EE process the national government had the lead in most phases except in the "development of measures" and "implementation" where the Cabinet of Ministers and regional authorities, respectively, had the lead in most ASs. They acted, however, mainly

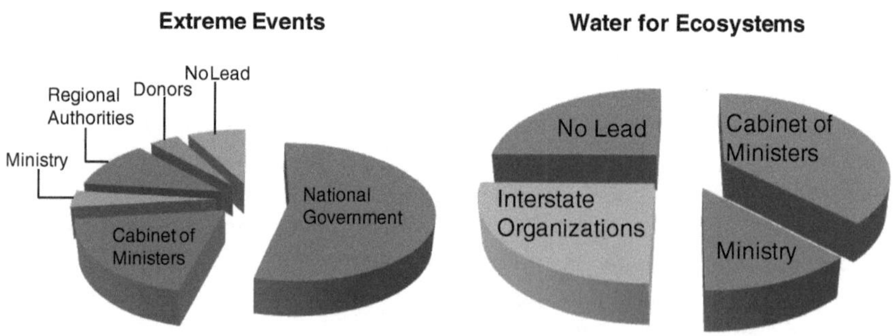

Fig. 9.4 Share of action situations in which a given actor has the lead identified for the case study in the Amudarya Basin. "No lead" in the policy process of "extreme events" refers to AS that are classified as "informal"; "no lead" in the policy process of "water for ecosystems" refers to AS that took place on an international level involving all five Central Asian heads of state (reproduced from Fig. 4 in Schlüter et al. (2010) with permission)

Table 9.5 Most frequent lead actor in each phase of the policy process for both EE and WE identified for the case study in the Amudarya Basin (reproduced from Table 1 Schlüter et al. (2010) with permission)

Phase	Lead actor EE	Lead actor WE
Strategic goal setting	National government	Cabinet of ministers
Assess current state	National government	
Policy formulation	National government	Central Asian states
Developing operational goals	National government	Central Asian states/cabinet
Developing measures	Cabinet of ministers	Cabinet of ministers
Implementation	Regional authorities	International (regional)
Monitoring	National government	

through regulations that were the enforcement of which were delegated to regional authorities, and through protocols and (inter-organizational) orders, which prescribe the exact actions and responsibilities of lower level authorities (Schlüter et al. 2010). In the WE process the representatives of the Central Asia states were dominant in policy formulation and the development of operational goals (shared with the cabinet), while strategic goal setting and the development of measures (which take place predominantly at the national and regional scales) were mainly lead by the cabinet. Implementation, however, was led by an interstate organization which was set up for project implementation financed by contributions of the member states and international donors (Schlüter et al. 2010).

Some words of caution are warranted. Using such a database and analytical protocols can produce a large number of results. Any application should be guided by specific research questions and a carefully designed protocol for how and which analyses should be used. Furthermore, any kind of analyses includes some subjective judgements. Hence applications should refrain from interpretation of small quantitative differences. For example one might find that in case A 13.7 % of actors at implementation level participated in higher level ASs producing rules that influenced their actions, versus 17.3 % of actors in case B. It would be quite meaningless to conclude from these findings that case A and B reveal major differences in the involvement of actors in rule production which should be reflected in the degree of compliance with rules. The obvious question following from such an example is: 'How can we best assess and explain a small quantitative difference?'—a question with no simple answer. The higher the level of aggregation at which an AS is defined the larger the probability that choices may be influenced by subjective assessment and the availability of data. An AS may, for example, refer to the development of a new flood policy over several years. In such a case it is more meaningful to introduce categories such as low, medium or high. The MTF supports structured representation and analyses of cases. It provides guidance but no blueprints or simplistic recipes on how to conduct the analyses. The following is a list of recommendations for what to bear in mind when applying the MTF.

1. *Define the research questions to be addressed clearly.*
2. *Select scales and resolution in space and time for the analysis.* The overall time frame should be determined by boundaries that are appropriate for addressing the research question. If one wants to compare, for example, the development of flood policies in different countries the analyses may comprise two decades in one country and three decades in another. The resolution in time should be determined by the types of events and outcomes one intends to include in the analyses.
3. *Select classes and attributes required for the analysis.* Only classes and attributes of classes relevant for addressing the research question should be included in the analyses.
4. *Develop a data collection protocol.*
5. *Determine action situations (ASs) with inputs and outputs.* This is a critical step since it is not a trivial task to decide what should be aggregated within an AS. In process representations it may be useful to focus on important outcomes—institutions, knowledge, operational outcomes—and identify the social interaction context which led to their production. If several largely independent social interaction contexts (e.g., a bottom-up learning process at the regional level and a formal policy process at national level) contribute to an outcome (e.g., set of measures) these should be represented by different ASs.
6. *Identify actors, their attributes and their role in various action situations.*

9.2.4 Methodological Diversity

Different analyses can complement each other. Even when I promote the MTF in general, I acknowledge it might not always be the method of choice. Collecting primary data is labour-intensive. The MTF does not constitute an exception in this respect. It may be useful to combine exploratory analyses comprising a large number of cases that look at a large number of variables with in-depth studies of selected cases that focus on a reduced number of variables only. Findings from exploratory analyses might guide more in-depth studies. In referring to a large number of cases I mean dozens rather than hundreds. Still, if one wants to collect data from 20 to 30 cases capturing the full complexity of the water governance system and the context in which it is embedded one easily ends up with dozens of indicators for each individual case. Here meta-analyses and expert scores may be the method of choice and often the only feasible way to go.

Table 9.6 illustrates this approach by showing an excerpt of a questionnaire developed for the Twin2Go project.[4] Twin2Go synthesized insights from individual case studies conducted in the context of twinning projects, i.e. projects with case studies in Europe as well as in Latin America, Africa, and Central, South and

[4]Twin2Go (Coordinating twinning partnerships towards more adaptive governance in river basins) was a project funded under the 6th EU Framework Program. More information and access to the data from the different river basins can be found at: www.twin2go.uos.de.

Table 9.6 Excerpt of guidance for questionnaire developed under the umbrella of the Twin2Go project for standardized data collection on properties of water governance systems (Pahl-Wostl and Lebel 2011a, pp. 15–16)

No.	Indicator	Definition	Hypothesis/statement on relationship	Scoring scheme	How to assign (i.e. on which basis are scores allocated)	Comment on data source
34.	Vertical coordination (governmental)	Formal provisions to support coordination in water sector among governmental organisations across administrative levels	Clear and distinctive allocation of tasks and functions avoiding overlaps and coordination increases performance	– A	(A) Cooperation and clear allocation of tasks	Expert judgement
				– B	(B) Clear allocation of tasks, and coordination	
				– C	(C) Task overlap, but coordination	
				– D	(D) Clear allocation of tasks, but no coordination	
				– E	(E) No coordination, much overlap	
36.	Role of local governments	Involvement of local governments in the creation of water-related institutions at higher levels, if the institutions affect the local level	Involving of local government at higher levels improves the performance of water governance	– A	(A) Local government are involved in the creation of institutions at higher levels and participate in decision-making, if they will be affected by these institutions	Expert judgement
				– B	(B) Local government are consulted in the creation of institutions at higher levels, if they will be affected by these institutions	
				– C	(C) The role of local government is basically restricted to the implementation of institutions from higher levels	

Southeast Asia. The main goal was to analyse factors that determine adaptive capacity of water governance systems. Considering the need to work with a highly heterogeneous knowledge base (results from previous research projects in river basins conducted largely independently, with secondary datasets and expert judgement) a methodological approach was developed to bring the knowledge from the different case studies into a consistent and comparable format. To achieve data standardization, Twin2Go developed a questionnaire with 81 indicators and applied it during a series of case study review workshops involving practitioners and researchers from the twinning projects in the river basins as well as additional case study experts. Table 9.6 lists two indicators for vertical integration. By using such guidance documents we tried to make the scoring by the participating experts as comparable as possible across the large number of cases in different cultural and political contexts. Nevertheless, expert judgment is not the same as primary data analysis. This would imply analyzing a range of policy processes based on interviews of participants in those processes, carrying out document analyses or undertaking direct observation. The resulting data would allow the development of more accurate indicators and more sophisticated comparisons. Increased accuracy comes at the expense of additional effort. It would have been entirely unfeasible to collect primary data for 29 cases on dozens of indicators! Despite the pragmatic approach chosen, Twin2Go provided highly interesting insights (Pahl-Wostl et al. 2012; Pahl-Wostl and Knieper 2014; Lebel et al. 2013). One can only guess at what kind of outstanding analyses would have been possible had all projects chosen a standardized approach for data collection from the outset—another argument for improving data standardization and developing more shared data sets.

9.3 Conclusions

The universes of what is desirable and what is realistic may not always coincide. I have made a strong plea for the need for more standardized databases and analytical frameworks as a foundation for systematic comparative analyses and cumulative knowledge production. It is unrealistic to assume that a wider scientific community in the social sciences would adopt and use a highly standardized framework for analysis and data collection protocols. This might even be undesirable and lead to orthodoxy if it suppresses pluralism. However, it would be desirable if various communities of practice that have developed and use different standardized approaches exchange and engage in comparative exercises using a shared terminology. Such practice would pave the way for moving from the prevailing fragmentation towards constructive pluralism. I have used a wide array of methods in the numerous empirical analyses I have led over the past two decades. The reasons have been in part pragmatic. However, in part they were also influenced by the context in which the research was conducted. Chapter 10 provides a summary of a decade of empirical research and how it has contributed to improving our understanding of water governance and management.

References

Binder, C. R., Hinkel, J., Bots, P. W. G., & Pahl-Wostl, C. (2013). Comparison of frameworks for analyzing social-ecological systems. *Ecology and Society, 18*(4), 26. doi:10.5751/ES-05551-180426

Bisaro, A., Hinkel, J., & Kranz, N. (2010). Multilevel water, biodiversity and climate adaptation governance: Evaluating adaptive management in Lesotho. *Environmental Science and Policy, 13*(7), 637–647. doi:10.1016/j.envsci.2010.08.004

Blomquist, W., Dinar, A., & Kemper, K. (2005). Comparison of institutional arrangements for river basin management in eight basins. World Bank Policy Research Working Paper 3636. Washington D.C.: The World Bank.

Borgatti, S. P., Everett, M. G., & Johnson, J. C. (2013). *Analyzing social networks*. London: Sage.

Borowski, I., Le Bourhis, J., Pahl-Wostl, C., & Barraqué, B. (2008). Spatial misfit in participatory river basin management: Effects on social learning. A comparative analysis of German and French case studies. *Ecology and Society, 13*(1), 7. URL:http://www.ecologyandsociety.org/vol13/iss11/art17/

Chenoweth, J. L., Ewing, S., & Bird, J. (2002). Procedures for ensuring community involvement in multijurisdictional river basins: A comparison of the Murray-Darling and Mekong River basins. *Environmental Management, 29*(4), 497–509.

Cox, M. (2014). Understanding large social-ecological systems: Introducing the SESMAD project. *International Journal of the Commons, 8*(2), 265–276.

Dewulf, A., Craps, M., Bouwen, R., Taillieu, T., & Pahl-Wostl, C. (2005). Integrated management of natural resources: Dealing with ambiguous issues, multiple actors and diverging frames. *Water Science and Technology, 52*(6), 115–124.

Dewulf, A., François, G., Pahl-Wostl, C., & Taillieu, T. (2007). A framing approach to cross-disciplinary research collaboration: Experiences from a large-scale research project on adaptive water management. *Ecology and Society, 12*(2), 14.

Foran, T., & Manorom, K. (2007). Pak Mun Dam: perpetually contested? In F. Molle, T. Foran, & M. Käkönen (Eds.), *Contested waterscapes in the Mekong Region: Hydropower, livelihoods and governance* (pp. 55–80). London, UK: Earthscan.

Gutiérrez, N., Hilborn, R., & Defeo, O. (2011). Leadership, social capital and incentives promote successful fisheries. *Nature, 470*, 386–389.

Hovelynck, J., Dewulf, A., François, G., & Taillieu, T. (2010). Interdisciplinary knowledge integration through group model building: Recognizing dualities and triadizing the conversation. *Environmental Science and Policy, 13*(7), 582–591. doi:10.1016/j.envsci.2010.04.002

Krugman, P. (2009). How Did Economists Get It So Wrong? (2009, October, 02). *The New York Times*.

Huitema, D., & Meijerink, S. (2010). Realizing water transitions: The role of policy entrepreneurs in water policy change. *Ecology and Society, 15*(2), 26.

Huntjens, P., Pahl-Wostl, C., Rihoux, B., Schlüter, M., Flachner, Z., Neto, S., et al. (2011). Adaptive water management and policy learning in a changing climate: A formal comparative analysis of eight water management regimes in Europe, Africa and Asia. *Environmental Policy and Governance, 21*(3), 145–163. doi:10.1002/eet.571

Knieper, C., Holtz, G., Kastens, B., & Pahl-Wostl, C. (2010). Analysing water governance in heterogeneous case studies—experiences with a database approach. *Environmental Science and Policy, 13*(7), 592–603. doi:10.1016/j.envsci.2010.09.002

Kranz, N., Menniken, T., & Hinkel, J. (2010). Climate change adaptation strategies in the Mekong and Orange-Senqu Basins: What determines the state-of-play? *Environmental Science and Policy, 13*(7), 648–659. doi:10.1016/j.envsci.2010.09.003

Lebel, L., Dore, J., Daniel, R., & Koma, Y. S. (Eds.). (2007). *Democratizing water governance in the Mekong Region*. Chiang Mai, Thailand: Mekong Press.

Lebel, L., Nikitina, E., Pahl-Wostl, C., & Knieper, C. (2013). Institutional fit and river basin governance: A new approach using multiple composite measures. *Ecology and Society, 18*(1), 1. doi:10.5751/ES-05097-180101

Mayntz, R. (2005). Forschungsmethoden und Erkenntnispotential Natur- und Sozialwissenschaften im Vergleich. *MPIfG Discussion Paper*. Cologne: Max Planck Institute for the Study of Societies.

McGinnis, M. D., & Ostrom, E. (2014). Social-ecological system framework: Initial changes and continuing challenges. *Ecology and Society, 19*(2), 30. doi:10.5751/ES-06387-190230

Molle, F., Foran, T., & Kakonen, M. (Eds.). (2009). *Contested waterscapes in the Mekong Region: Hydropower, livelihoods and governance*. London, UK: Earthscan.

Mostert, E., Pahl-Wostl, C., Rees, Y., Searle, B., Tàbara, D., & Tippett, J. (2007). Social learning in European River-Basin management: Barriers and fostering mechanisms from 10 river basins. *Ecology and Society, 12*(1).

Ostrom, E. (2005). *Understanding institutional diversity*. Princeton: Princeton University Press.

Ostrom, E. (2007). A diagnostic approach for going beyond Panaceas. *Proceedings of the National Academy of Sciences of the United States of America, 104*, 15181–15187.

Ostrom, E. (2009). A general framework for analyzing sustainability of social-ecological systems. *Science, 325*(5939), 419–422. doi:10.1126/science.1172133

Ostrom, E. (2010). Beyond Markets and States: Polycentric governance of complex economic systems. *American Economic Review, 100*, 641–672. doi:10.1257/aer.100.3.641

Pahl-Wostl, C., Becker, G., Sendzimir, J., & Knieper, C. (2013). How multilevel societal learning processes facilitate transformative change: A comparative case study analysis on flood management. *Ecology and Society, 18*(4), 58. doi:10.5751/ES-05779-180458

Pahl-Wostl, C., Holtz, G., Kastens, B., & Knieper, C. (2010). Analyzing complex water governance regimes: The Management and Transition Framework. *Environmental Science and Policy, 13*(7), 571–581. doi:10.1016/j.envsci.2010.08.006

Pahl-Wostl, C., & Knieper, C. (2014). The capacity of water governance to deal with the climate change adaptation challenge: Using fuzzy set qualitative comparative analysis to distinguish between polycentric, fragmented and centralized regimes. *Global Environmental Change, 29*, 139–154.

Pahl-Wostl, C., Knieper, C., & Holtz, G. (2014). Introduction to the MTF total system database (version 11). University of Osnabrueck. http://www.newater.uni-osnabrueck.de

Pahl-Wostl, C., & Kranz, N. (2010). Water governance in times of change. *Environmental Science and Policy, 13*(7), 567–570. doi:10.1016/j.envsci.2010.09.004

Pahl-Wostl, C., & Lebel, L. (2011a). Guidance on Twin2Go Questionnaire V3: Appendix to report on methods for comparative analysis. *Twin2Go Deliverables* Twin2Go project.

Pahl-Wostl, C., & Lebel, L. (2011b). Methods for comparative analysis. *Twin2Go Deliverables* Twin2Go project.

Pahl-Wostl, C., Lebel, L., Knieper, C., & Nikitina, E. (2012). From applying panaceas to mastering complexity: Toward adaptive water governance in river basins. *Environmental Science and Policy, 23*, 24–34.

Scheierling, S. M., Loomis, J. B., & Young, R. A. (2006). Irrigation water demand: A meta-analysis of price elasticities. *Water Resources Research, 42*(1), W01411. doi:10.1029/2005WR004009

Schlüter, M., Hirsch, D., & Pahl-Wostl, C. (2010). Coping with change: Responses of the Uzbek water management regime to socio-economic transition and global change. *Environmental Science and Policy, 13*(7), 620–636. doi:10.1016/j.envsci.2010.09.001

Scott, J. (2000). *Social network analysis: A handbook* (2nd ed.). London, UK: Sage.

Sendzimir, J., Flachner, Z., Pahl-Wostl, C., & Knieper, C. (2010). Stalled regime transition in the upper Tisza River Basin: The dynamics of linked action situations. *Environmental Science and Policy, 13*(7), 604–619. doi:10.1016/j.envsci.2010.09.005

Sendzimir, J., Magnuszewski, P., Flachner, Z., Balogh, P., Molnar, G., Sarvari, A., et al. (2007). Assessing the resilience of a river management regime: Informal learning in a shadow network in the Tisza River Basin. *Ecology and Society, 13*(1), 11.

van der Brugge, R., Rotmans, J., & Loorbach, D. (2005). The transition in Dutch water management. *Regional Environmental Change, 5*, 164–176.

Wollenberg, E., Merino, L., Agrawal, A., & Ostrom, E. (2007). Fourteen years of monitoring community-managed forests: Learning from IFRI's experience. *International Forestry Review, 9*(2), 670–684. doi:10.1505/ifor.9.2.670

Chapter 10
Empirical Analyses—From Single Case Studies to Comparative Analyses

Theoretical propositions need testing against empirical evidence. Theoretical breakthroughs have always been informed by an in-depth understanding of the "real" world. This chapter summarizes and synthesizes major results from a number of empirical studies that were conducted over the past decade under the umbrella of the research programme on water governance and management I developed. The studies range from single case studies to comparative analyses of nearly 30 cases. The chapter introduces the overall logic behind the classification of the various analyses under three thematic clusters: characterization of dynamic state, analysis of social learning processes and analysis of transformative change. Then the results and conclusions on major insights with respect to the guiding theme of the book, the transformation of water governance towards sustainability, are discussed for each thematic stream.

10.1 Overview on Different Thematic Clusters

Based on the theoretical framework represented in Fig. 8.1, three different streams of empirical analyses are distinguished as schematically represented in Fig. 10.1. Figure 10.1 links a graphical representation of the major conceptual frameworks that were used to each stream.

(A) Characterization of dynamic state

The first research stream, characterization of the dynamic state, implements a diagnostic approach by analysing the relationships among the three characteristics of a water governance system, its functional performance and the influence of context on these relationships. The insights gained allow the identification of desirable characteristics of future governance systems.

(B) Analysis of social learning processes

Research on social learning focuses on governance processes and social interactions in multi-party settings. It draws on and tests the validity of a relational concept for social learning (Pahl-Wostl et al. 2007a). Insights gained from such analyses are of

C. Pahl-Wostl, *Water Governance in the Face of Global Change*,
Water Governance—Concepts, Methods, and Practice,
DOI 10.1007/978-3-319-21855-7_10

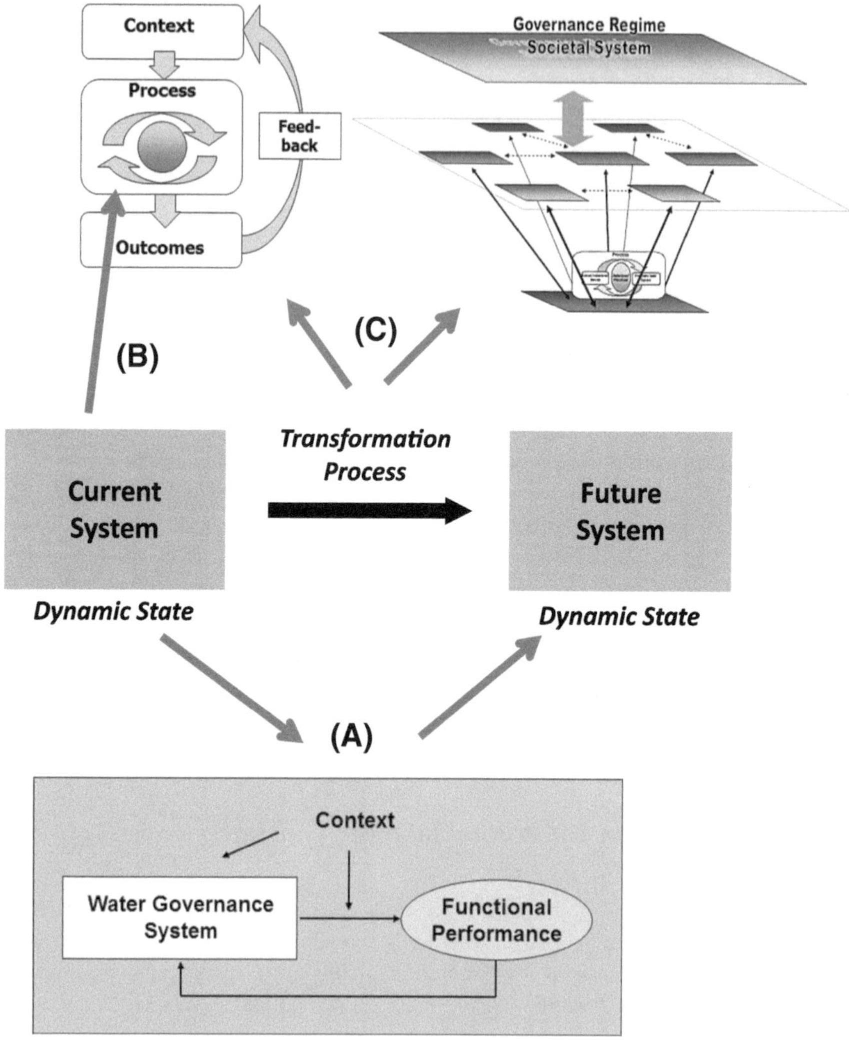

Fig. 10.1 Schematic representations of streams of empirical analyses (*green arrows*) and underlying conceptual frameworks based on the theoretical framework represented in Fig. 8.1. The three streams refer to (**A**) characterization of dynamic state, (**B**) analysis of social learning, (**C**) analysis of transformative change

importance for understanding the requirements for both adaptive and transformative capacity.

(C) Analysis of transformative change

This research stream analyses characteristics of governance systems and of societal learning processes that support transformative change. It adopts a multi-level perspective to assess the relative importance of bottom-up and top-down pathways of influence in sustainability transformations.

Table 10.1 provides an overview of publications on comparative analyses conducted in recent years by myself and collaborators. Most of these analyses were realised under the umbrella of large coordinated projects funded by the European Commission. Each empirical study is assigned to one of the three different thematic streams. The analyses differ with respect to the number of cases, geographical regions covered, spatial and temporal scales of analysis, methods used and issues addressed. In the following section, the analyses conducted under the umbrella of each of the three thematic streams are introduced and discussed in more depth with respect to overall research objectives, approaches selected and major insights gained.

10.2 Characterization of Dynamic State

The thematic stream on the characterization of the dynamic state of water governance systems addresses the overarching research questions: How are characteristics of governance systems related to their performance? What is the influence of context on these relationships? The conceptual framework underpinning such analyses is based on a diagnostic approach which makes a distinction between characteristics of the governance system, context and performance (c.f. Sects. 3.1.2 and 8.2). Table 10.2 provides an overview of empirical analyses which contribute to this approach.

10.2.1 Moving from Panaceas to a Diagnostic Approach—the Twin2Go Project

The most comprehensive of these analyses (Pahl-Wostl et al. 2012; Pahl-Wostl and Knieper 2014; Lebel et al. 2013) were based on the Twin2Go data set[1] that comprised data from governance systems of 29 national river basins. The EU-funded project Twin2Go began with the observation that numerous recommendations often relying on simplistic 'standard' panaceas were put forward for water governance reform without testing their appropriateness in diverse contexts. We conducted the first comprehensive comparative analysis of complex water governance and

[1]The Twin2Go '*Coordinating Twinning partnerships towards more adaptive Governance in river basins*' project (www.twin2go.uos.de), funded under the 6th framework program of the European Commission, ran from 2009 to 2011. The dataset includes 29 case studies—domestic river basins, parts of transboundary basins—in Europe, Latin America, Africa, as well as Central, South and Southeast Asia.

Table 10.1 Overview of case study analyses presented in this chapter

Reference—original title of publication	No of cases[a]	Regions covered	Spatial scale	Methods used
Characterization of dynamic state				
The capacity of water governance to deal with the climate change adaptation challenge: Using fuzzy set qualitative comparative analysis to distinguish between polycentric, fragmented and centralized regimes (Pahl-Wostl and Knieper 2014)	27 TG	Europe, Africa, Asia, Latin America	River basin—national basins in the case of transboundary rivers	Fuzzy-set-qualitative comparative analysis (QCA) Expert scoring
From applying panaceas to mastering complexity: toward adaptive water governance in river basins (Pahl-Wostl et al. 2012)	29 TG	Europe, Africa, Asia, Latin America	River basin—national basins in the case of transboundary rivers	Linear regression Qualitative Expert scoring
Institutional fit and river basin governance: a new approach using multiple composite measures (Lebel et al. 2013)	28 TG	Europe, Africa, Asia, Latin America	River basin—national basins in the case of transboundary rivers	Semi-quantitative
Climate change adaptation in European river basins (Huntjens et al. 2010)	4 NW	Hungary, Ukraine, Portugal, Netherlands	Sub-basins within national basins of transboundary rivers	Semi-quantitative Expert scoring
Institutional design propositions for the governance of adaptation to climate change in the water sector (Huntjens et al. 2012)	3 NW	Netherlands, Australia, South Africa	National basins of transboundary rivers	Qualitative Expert scoring
Cross-Comparison of Climate Change Adaptation Strategies Across Large River Basins in Europe, Africa and Asia (Krysanova et al. 2010)	6 NW	Europe, Africa, Central Asia	Transboundary river basins	Qualitative Expert scoring
Requirements for adaptive governance of groundwater ecosystem services: insights from Sandveld (South Africa), Upper Guadiana (Spain) and Spree (Germany) (Knüppe and Pahl-Wostl 2013)	3	Spain (Guadiana), Germany (Elbe), South Africa	Sub-basins within national basins of transboundary rivers	MTF Primary data

(continued)

Table 10.1 (continued)

Reference—original title of publication	No of cases[a]	Regions covered	Spatial scale	Methods used
Analyses of social learning processes				
Social learning in European river-basin management: barriers and fostering mechanisms from 10 river basins (Mostert et al. 2007)	10 HC	Europe	Sub-basins—regions within countries	Semi-quantitative Primary data
Spatial misfit in participatory river basin management: effects on social learning, a comparative analysis of German and French case studies (Borowski et al. 2008)	2 HC	Germany (Elbe), France (Dordogne)	Sub-basins within countries	Qualitative Primary data
Making framing of uncertainty in water management practice explicit by using a participant-structured approach (Isendahl et al. 2009)	5 NW	Germany (Rhine, Elbe), Netherlands (Rhine), Czech (Elbe), Spain (Guadiana)	Sub-basins within countries	Qualitative Primary data
Using framing parameters to improve handling of uncertainties in water management practice (Isendahl et al. 2010a)	2 NW	Germany (Rhine), Netherlands (Rhine)	Sub-basins within countries	Qualitative Primary data
Analyses of transformative change				
How multilevel societal learning processes facilitate transformative change: a comparative case study analysis on flood management (Pahl-Wostl et al. 2013a)	3 NW	Germany, Netherlands, Tisza	Sub-basins within national basins of transboundary rivers	MTF Primary data
Continuity and change in social-ecological systems: the role of institutional resilience (Herrfahrdt-Pähle and Pahl-Wostl 2012)	2 NW	South Africa Uzbekistan	National	Narrative Primary data
Stalled regime transition in the upper Tisza river basin: the dynamics of linked action situations (Sendzimir et al. 2010)	1 NW	Hungary	National basin of transboundary river	MTF Primary data
Coping with change: responses of the Uzbek water management regime to socio-economic transition and global change (Schlüter et al. 2010)	1 NW	Uzbekistan	National basin of transboundary river	MTF Primary data

(continued)

Table 10.1 (continued)

Reference—original title of publication	No of cases[a]	Regions covered	Spatial scale	Methods used
Understanding the development of flood management in the middle Yangtze River (Xia and Pahl-Wostl 2012c)	1 NW	China (Yangtse)	Sub-basins of national river	MTF Primary data
The process of innovation during transition to a water saving society in China (Xia and Pahl-Wostl 2012b)	1 NW	China	Province	Qualitative Primary data
The development of water allocation management in the Yellow River basin (Xia and Pahl-Wostl 2012a)	1 NW	China (Yellow River)	Sub-basins of national river	Qualitative Primary data
Adaptive water management and policy learning in a changing climate: formal comparative analysis of eight water management regimes in Europe, Africa and Asia (Huntjens et al. 2011)	8 NW	Europe, Africa, Central Asia	Sub-basins within national basins of transboundary rivers	Multi-value-QCA Expert scoring

[a]The acronyms under the numbers refer to large European projects under the umbrella of which the respective study was conducted. *TG* Twin2Go, *NW* NeWater, *HC* HarmoniCOP. The projects are explained in more details where results are presented in this chapter

Table 10.2 Overview of empirical analyses contributing to thematic stream (A) characterization of dynamic state (cf. Fig. 10.1)

Reference	Major research questions	System characteristics	Functional performance
Pahl-Wostl et al. (2012)	How are characteristics of governance systems related to their performance? What is the influence of context on these relationships?	Legal frameworks Basin principle Polycentricity Vertical and horizontal coordination Knowledge and information management	Climate change adaptation Good governance Achievement of sustainability targets Water management practice
Pahl-Wostl and Knieper (2014)	What is the role of polycentricity in increasing the ability of governance systems to deal with emerging challenges?	Coordination (horizontal and vertical) Degree of decentralization	Climate change adaptation
Lebel et al. (2013)	Can multi-dimensional measures of institutional fit support a diagnostic approach?	Allocation, integration, conservation, basinization, participation, adaptation	Ratio of institutional capacity to level of challenges arising from biophysical and socio-economic conditions
Huntjens et al. (2010)	What are the requirements for adaptive and integrated water management (AIWM)? Does AIWM facilitate climate change adaptation?	Agency, awareness raising and education, type of governance and cooperation structures, information management and—exchange, policy development and—implementation, risk management, and finances and cost recovery	Climate change adaptation
Huntjens et al. (2012)	Can institutional design propositions (principles) be derived for the governance of climate change adaptation in the water sector?	Ostrom design principles Policy learning Robustness of process	Climate change adaptation
Krysanova et al. (2010)	What facilitates and what are barriers to climate change adaptation?	Horizontal cooperation, Vertical cooperation, Financial and human resources	Climate change adaptation
Knüppe and Pahl-Wostl (2013)	Which factors influence trade-offs between human and environmental water needs? What are	Vertical (hierarchical) and horizontal (cross-sectoral) integration	State of groundwater ecosystem services trade-offs

(continued)

Table 10.2 (continued)

Reference	Major research questions	System characteristics	Functional performance
	requirements for a sustainable governance of groundwater ecosystem services		

management systems in national river basins, compiling insights from basins in developed and developing/emerging countries (Pahl-Wostl et al. 2012).

Compiling insights was not a trivial task, however. Twin2Go was developed primarily on insights from a wide range of individual case studies conducted in the context of twinning projects, i.e. projects with case studies in Europe as well as developing and transition countries. The projects differed in their research emphasis, in scale and methods used.

To support a diagnostic approach we developed an analytical framework that makes a distinction among water governance regime, regime performance and environmental and socio-economic context (Pahl-Wostl et al. 2012). The operationalization of the framework required nearly 100 indicators which were used to derive aggregated measures. The establishment of a comprehensive data base had to build on highly heterogeneous knowledge (results from previous research projects in river basins conducted independently, secondary datasets and expert judgement). A methodological approach was developed to bring the knowledge from the different case studies and heterogeneous sources into a consistent and comparable format (Pahl-Wostl and Lebel 2011). Data standardization was achieved by developing a questionnaire for a set of indicators, and applying it during a series of Case Study Review Workshops involving practitioners and researchers from the twinning projects in the river basins, as well as additional case study experts. Additional data were collected in the various basins on what stakeholders considered to be best practices, and if and how they could be transferred to other locations.

Combining Statistical Analyses and Qualitative Assessments

Comprehensive comparative analyses were guided by a set of hypotheses relating regime characteristics to performance. These hypotheses reflected major assumptions discussed in research on effective and adaptive water and resource governance (Pahl-Wostl et al. 2011). Table 10.3 lists hypotheses which provided robust and meaningful results and were discussed in Pahl-Wostl et al. (2012). The first column lists hypotheses. The second column lists the regime measures to which the hypotheses on the left refer. An X in a cell indicates an expected positive relationship between a regime characteristic (row) and a performance measure (column). For example, regime characteristic R1, presence of comprehensive legal frameworks, is expected to have a positive influence on performance measures P1, implementation of MDG targets as measure of social sustainability goals, on P2, adherence to good governance principles, on P4, environmental conditions, and on P5, provisions for environmental management. The hypotheses examined in

Table 10.3 Hypotheses on the relationship between regime characteristics R_i and different performance measures P_i

	Performance => Regime ⇓	P1 MDG targets	P2 good governance	P3 CC adaptation	P4 environ. conditions	P5 environ. management
Legal frameworks regulating water management improve performance with regard to good governance principles and the achievement of sustainability goals	R1 Legal frame-works	**X**	**X**		**X**	**X**
Legal frameworks prescribing the basin principle improve performance with regard to the achievement of sustainability goals and increase adaptive capacity	R2 Basin principles	**X**		**X**	**X**	**X**
Polycentricity and multi-level arrangements combining decentralization and coordination increase effectiveness and adaptive capacity of governance regimes	R3 Poly-centricity	**X**	**X**	**X**	**X**	**X**
Lack of vertical integration leads to disconnection of governance levels and gap between policy process and operational implementation. Vertical integration and cooperation increase adaptive capacity and performance	R4 Vertical integration	**X**		**X**	**X**	**X**
Lack of horizontal integration leads to disconnection of sectors or regions and impedes integration in policies and implementation. High vertical integration and effective cooperation increase adaptive capacity and performance	R5 Horizontal integration			**X**	**X**	**X**
Open access to information and integration of different kinds of knowledge support higher levels of learning and increase adaptive capacity	R6 Knowledge			**X**		**X**
Adaptive capacity increases if various types of uncertainty are taken into account and addressed in an appropriate way	R7 Handling Uncertainty			**X**		

Table 10.3 did not yet specify the influence of context. Not all regime measures are independent. As defined, polycentricity embraces elements of both vertical and horizontal integration.

Two complementary approaches were applied to the analysis of case study data. (1) Statistical investigation: Linear regression analyses were applied to the sets of composite performance, regime and context measures to assess the significance and strengths of associations before and after adjusting for context. (2) Qualitative analysis: Cases were clustered in groups that either supported or contradicted the assumptions made in the hypotheses, or in groups that did not allow a conclusion to be drawn either way. In the event that contradictions in the assumptions appeared, context factors were examined as potential explanations. Table 10.4 represents results from statistical analyses with and without taking context into account.

Results support some of the general patterns stated in the hypotheses (cf. Table 10.3) without supporting simplistic prescriptions. In most relationships between regime characteristics and performance, considerable variation could be detected. In general, taking context into account helped to explain additional variation, but overall, adjusting for context did not invalidate patterns of associations of regime characteristics with performance. Only for performance measure P1, achievement of water related MDGs, was the socio-economic context (C3) found to have an overriding influence (Pahl-Wostl et al. 2012). What is striking though is the absence of correlations between governance characteristics and the state of the environment (P4). There is a weak correlation between the degree of watershed modification (C3) and environmental conditions. These findings suggest the need for more differentiated analyses that classify basins according to the degree and kind of human pressure on the environment.

The analyses clearly confirmed the hypotheses on the importance of polycentricity which is already a more complex regime measure embracing horizontal and vertical integration. Analyses taking into account context suggest that the importance of polycentricity is not simply a spurious association that can be explained by the state of economic and institutional development (c.f. Table 10.4b). Even though polycentric regimes were predominantly found in European countries, developing countries and transition economies worldwide may also display such characteristics.

The presence of legal frameworks as well as the adoption of the basin principle (including the development of basin management plans)—hallmarks of water governance reform—are weakly associated with the overall performance of water governance (Table 10.4a). Both are associated with the adoption of good governance principles. The presence of legal frameworks is not a sufficient condition for overall high performance. The qualitative case assessments revealed that effective legal frameworks require the capacity to implement and to follow good governance principles, i.e. respect for the rule of law.

No strong correlation could be detected between the regime measure R6 Knowledge (open access to information, consideration of expert and local/traditional knowledge), and the performance measure P3 Climate Change Adaptation (Table 10.4). In contrast, there is a strong and context-independent

Table 10.4 Results from statistical analyses

Performance regime	P1 MDG goals	P2 good governance	P3 adaptation policies	P4 Environ. conditions	P5 environ. manage.	P All aggregation over all
(**a**) *Schematic representation of associations between performance and regime measures*						
R1 legal frameworks	~	+++	++	~	~	++
R2 basin principles	~	+++	~	~	~	+
R3 poly-centricity	~	+++	+++	~	~	+++
R4 vertical integration	~	+++	++	~	~	++
R5 horizontal integration	~	++	+++	~	~	+
R6 knowledge	~	+++	~	~	+	+
R7 handling Uncertainty	~	+++	+++	~	~	+++

Performance regime	P1 MDG goals	P2 good governance	P3 adaptation policies	P4 Environ. conditions	P5 environ. manage.	P All aggregated
(**b**) *Schematic representation of associations between performance and regime measures adjusted for context*						
R1 legal frameworks	~ **C1 +++**	+++ C1 + C4 +	++ ~	~ C3 +	~ **C1 +++**	+ **C1 +++**
R2 basin principles	~ **C1 +++**	+++ C1 +	~ C1 +	~ C3 ++	~ **C1 +++**	++ **C1 +++**
R3 poly-centricity	~ **C1 +++**	+++ ~	+++ ~	~ C3 +	~ **C1 +++**	++ **C1 +++**
R4 vertical integration	~ **C1 +++**	+++ C4 ++	++ ~	~ C3 +	~ **C1 +++**	+ **C1 +++**
R5 horizontal integration	~ **C1 +++**	++ C1 ++ C4 +	+++ ~	~ C3 +	~ **C1 +++**	+ **C1 +++**
R6 knowledge	~ **C1 +++**	+++ C1 +	~ C1 +	~ C3 +	~ **C1 +++**	++ **C1 +++**
R7 handling uncertainty	~ **C1 +++**	+++ ~	+++ ~	~ C3 ++ C1 +	~ **C1 +++**	++ **C1 +++**

+ symbols refer to positive correlations with different degrees of significance: +++ -> $P < 0.001$, ++ -> $P < 0.01$, + -> $P < 0.05$; P refers to regression coefficient. ~ implies that no significant correlation could be identified. C1–C4 Prefixes indicate regression coefficients for context variables that were included in the regression: C1: State of economic and institutional development, C2: Water availability, C3: Degree of watershed modification, C4: Basin size

correlation between the regime measure R7, Handling Uncertainties, and P3. Practices for handling uncertainties can be seen as a manifestation of the paradigm shaping water management practice. These findings suggest that open access to information and taking into account different kinds of knowledge do not lead to

Table 10.5 Characteristics of different governance regime types (Pahl-Wostl et al. 2012)

	Polycentric	Fragmented	Centralized
Distribution of formal power	High	High	Low
Multi-level distribution of functions and resources	High	High	Low
Coordination vertical	High	Low	Low
Coordination horizontal	High	Low	Low
Typical countries—cases	Netherlands	India	Uzbekistan

increased capacity to respond to the emerging challenge, viz. climate change, unless the reigning paradigm and modes of coordination provide an enabling environment.

Results suggest that combinations of, rather than individual regime characteristics are decisive in accounting for differences in performance. Multivariate statistical analyses, however, could not detect any meaningful combinations of regime measures. Multi-factorial causation seems to be more complex than what can be captured by linear combinations.

Set Theoretic Approaches—Fuzzy Set QCA

To capture the importance of different regime architectures we developed a classification which makes a distinction among three ideal-typical configurations: polycentric, centralized, and fragmented regimes (Table 10.5). A polycentric regime is characterized by a high distribution of formal power, functions and resources combined with effective vertical and horizontal coordination. A regime with a high distribution of power, functions and resources but lacking coordination is referred to as a fragmented regime. A centralized regime is characterized by a centralization of power and resources, and lack of effective coordination.

This classification was applied in the qualitative comparative analyses of the case studies in the Twin2Go data set (Pahl-Wostl et al. 2012). The cases with highest performance averaged across all performance measures could all be classified as polycentric regimes. With regard to the low performing group, both centralized (e.g., Uzbekistan) and fragmented (e.g., India) regimes were identified. The qualitative assessment provided interesting insights into the importance of regime architectures which could not be captured by statistical analyses. Hence, we decided to go one step further and use a set-theoretic approach, i.e. fuzzy set-QCA (Qualitative Comparative Analysis), to analyse which regime configurations lead to high or low performance with a focus on the importance of polycentricity (Pahl-Wostl and Knieper 2014). Regarding performance, we focused in this study on the ability of water governance systems to deal with the climate change adaptation challenge. Polycentricity has been claimed to be an essential characteristic of adaptive governance systems (cf. Sect. 6.2.2).

Polycentricity has been and is nowadays often identified with decentralization only, ignoring thereby, an essential criterion characterizing polycentric governance systems—the coordination including a shared set of rules (c.f. Sect. 6.2.2 in Chap. 6).

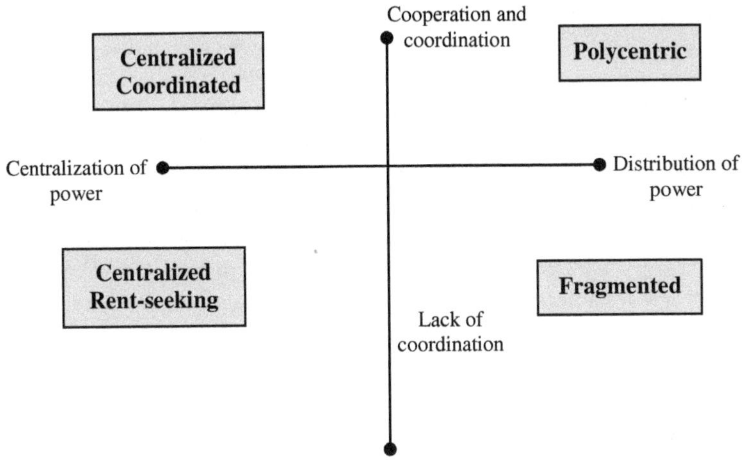

Fig. 10.2 Categorization of governance regimes in a two-dimensional grid of the distribution of power and degree of coordination/cooperation. The shaded boxes in the four corners denote the ideal-typical configurations (reproduced from Fig. 1 in Pahl-Wostl and Knieper 2014, p. 141, with permission)

In order to clarify terminology and support more systematic analyses, we proposed a categorization of governance regimes that are based on their degrees of coordination and centralization as shown in Fig. 10.2. This two-dimensional classification constitutes a more comprehensive derivation of the types introduced in Table 10.4. Based on this two-dimensional classification, four kinds of configurations which can be understood as Weberian ideal types are distinguished: Polycentric, Fragmented, Centralized Coordinated, and Centralized Rent-seeking regimes.

Polycentric regimes combine distribution of power and authority with effective coordination among various centres and across spatial levels. The modular structure characterizing polycentric systems increases resilience and the capacity for dealing with shocks and disturbances. Coordinated centres for decision making that have a certain degree of autonomy can support experimentation and learning. Therefore, polycentric regimes are assumed to have high performance, in particular, with respect to adaptive capacity and to dealing with emerging challenges such as climate change (Ostrom 2001, 2010; Folke et al. 2005; Pahl-Wostl 2009; Pahl-Wostl et al. 2012). Fragmented regimes lack coordination. Without coordination, the distribution of power and authority and overlapping responsibilities of the various decision making centres, may lead to uncoordinated and contradicting actions with loss of effectiveness and efficiency, as highlighted by the study by Lieberman (2011) of the South African health system. In a comparison of climate change adaptation strategies in large river basins in Europe, Africa, and Asia, Krysanova et al. (2010) identified a lack of horizontal (i.e. across sectoral and administrative boundaries) cooperation as an important barrier to climate change adaptation.

Centralized regimes operate under a hierarchical governance mode. A dominant governmental actor at the national level holds all power and authority. Centralized regimes lack response capacity and flexibility, which is associated with a more modular and decentralized system configuration even when top-down coordination is in place. The argument of increased flexibility has been a core argument in support of the decentralization of government functions (Hooghe and Marks 2003). Regarding coordination, a distinction is now made between centralized rent-seeking and centralized coordinated regimes. Centralized regimes without coordination are assumed to be characterized by rent-seeking. A prevalence of rent-seeking behaviour impedes effective coordination, and the lack of coordination encourages rent-seeking behaviour. Rent-seeking implies that governmental representatives and bureaucrats abuse their power and role in the hierarchy to increase their own benefits rather than caring for the provision of public goods (Tullock 2008). In centralized rent-seeking regimes, the reigning elite has little incentive to deal with emerging problems, and adaptive capacity is assumed to be low. In centralized coordinated regimes, actors at lower levels may be consulted during the decision-making process. But sub-ordinate centres have little autonomy and mainly implement decisions made at the top level. This reduces the capacity to deal with complex, potentially conflictual governance problems, and to take into account regional context. This leads to the conclusion that centralized regimes in general have less adaptive capacity, and their ability to deal with emerging challenges is lower than in polycentric regimes.

Set-theoretic methods such as Qualitative Comparative Analysis (QCA) are appropriate tools to analyse the empirical relevance of, and test hypotheses related to, different regime configurations. QCA is a powerful method to analyse causal relationships between a set of conditions and an outcome (Ragin 1987). It is based on set-theoretic logic formalism where conditions are either TRUE or FALSE. In contrast to statistical regression methods, QCA is particularly powerful at analysing multiple causation (equifinality) for the phenomenon i.e., that more than one path (or set of conditions) may lead to a particular outcome.

The different governance regime ideal types, and their assumed performance properties, can now be expressed in a formal logical notation as represented in Table 10.6. The hypotheses that link governance regime types to performance are shown in the last row of Table 10.6.

Anyone interested in further details of the analyses using fsQCA should consult (Pahl-Wostl and Knieper 2014). In the following remainder of this section I summarize the significant insights.

The analyses confirmed the hypothesis that polycentric governance regimes, defined as the combination of decentralized power with effective coordination, are characterized by high adaptive capacity. In this study, adaptive capacity referred to the ability of a water governance system to respond to challenges arising from, and triggered by, (expected) climate change. Polycentric regimes were found predominantly, but not exclusively, in European countries with high levels of institutional development and high material standard of living. The presence of integrative regulatory frameworks in the water sector proved to be a necessary but not

Table 10.6 Characteristics of different regime ideal types

Regime characteristics	Polycentric	Fragmented	Centralized rent-seeking	Centralized coordinated
Distribution of power = DIS	High/DIS	High/DIS	Low/dis	Low/dis
Coordination = COR	High/COR	Low/cor	Low/cor	High/COR
Adaptive Capacity = ADAP	High/ADAP	Low/adap	Low/adap	Low/adap
Hypotheses in logical terms[a]	DIS * COR → ADAP	DIS * cor → adap	dis * cor → adap	dis * COR → adap

Characteristics are described in both qualitative terms (High and Low) and as conditions in Boolean Logic. Uppercase letters represent a Boolean TRUE value for a binary variable (e.g. COR). Lowercase letters represent a FALSE value for a binary variable (e.g. cor) (Pahl-Wostl and Knieper 2014)

[a]The multiplication symbol [*] represents a logical "AND", whereas the addition [+] symbol represents a logical "OR". The arrow [→] represents the assumed association between conditions and outcome. The expression "X * Y → Z" can be read as: the presence of condition X in combination with the presence of condition Y is sufficient for the presence of outcome Z

sufficient condition for high performance. Even when the state of economic and institutional development was not identified as necessary conditions, it proved helpful to explain the performance characteristics of countries in a transition phase, in particular, regarding the effectiveness of policy implementation.

The state of institutional development measured in these analyses by the effectiveness of formal institutions (Corruption Perception Index) was identified as a central condition for explaining poor performance. It proved to be more important than economic development. These findings support claims that water governance reform with the aim of decentralization does not address systemic governance problems (Brown and Cloke 2004, 2005; Soliman and Cable 2011). This became apparent in the cases in Latin America. Most of these countries have been characterized by centralized governance. The water governance systems in Latin American countries in the study belonged to the centralized rent-seeking regimes (Ecuador, Chile), the fragmented (Colombia, Brazil) or the centralized coordinated (Nicaragua) regime types, and all had low adaptive capacities. Efforts towards decentralization does not seem to have increased effectiveness or supported the development of more polycentric regimes.

It is evident that resource governance needs to be entwined with capacity development, and the implementation of ambitious policies needs to proceed in a realistic and stepwise process. The structure of polycentric governance offers this potential, but it is no guarantee of success. Steps towards improved resource governance must take into account context-dependent constraints and opportunities.

Assessing Institutional Fit

Lebel et al. (2013) developed and applied an approach to take context-specific constraints into account in the context of the Twin2Go project. The point of departure for this work was that effective environmental governance depends in part on achieving a reasonable fit between institutional arrangements on the one hand, and ecosystem and social processes on the other. An approach for measuring the fit was developed that could inform diagnostic analysis. The dimensions of allocation, integration, conservation, 'basinization', participation, and adaptation were identified for this purpose based on views of experts on best practices and their transfer. Quantitative indicators of fit were defined as the difference between measures of institutional capacity of a water governance regime and the degree to which social-ecological conditions or context were challenging to that aspect of governance. A good fit corresponded thus to high institutional capacity relative to challenging conditions, and vice versa.

Figure 10.3 displays the 'fit profiles' for selected basins where the overall water governance system can be categorized as polycentric, fragmented or centralized (rent-seeking). A larger shaded polygon indicates an overall better fit to challenging conditions than a smaller polygon. Polycentric systems display a better overall fit for many dimensions. The centralized system in the Amudarya basin is characterized by damn low fit in participation and conservation, but still gives attention to basinization. Similar shapes were also found for other basins of this type

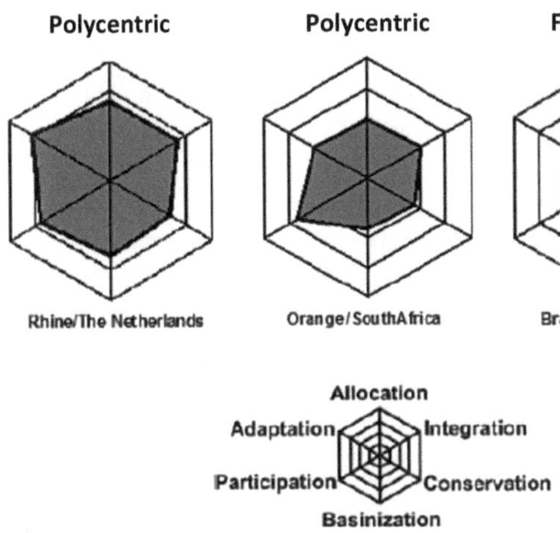

Fig. 10.3 Profiles of multidimensional fit for selected national basins based on Fig. 2 of Lebel et al. (2013). The overall water governance system can be categorized as polycentric, fragmented or centralized (rent-seeking)

(Lebel et al. 2013). Comparing multidimensional profiles give a sense of how well water governance regimes are equipped to deal with a range of conditions associated with natural resource and use-related. The analyses demonstrated that it is seemingly difficult, but not impossible, to simultaneously achieve a high level of institutional fit in the face of many challenging conditions. Fit profiles can be used to identify and prioritize interventions. A poor institutional fit in many dimensions indicates systemic problems, and may most likely require systemic transformation transcending the water governance system.

10.2.2 Requirements for Adaptive Governance and Management

The EU funded project NeWater (New approaches to adaptive water management under uncertainty) was founded upon the normative claim that sustainable water management requires a transition to integrated and adaptive water governance and management systems (Pahl-Wostl 2007). Although no blue-print was proposed for the design of such systems, NeWater proposed a set of system characteristics that were deemed to be required for realising integrated and adaptive governance and management (cf. Table 8.1). Several studies were conducted to test these propositions, to analyse requirements and develop new methods for adaptive and integrated water management that take into account the complexity of the river basins to be managed, and the difficulty in forecasting the future development of factors influencing them. NeWater had case studies in Europe (Rhine, Elbe, Guadiana,

Tisza), Africa (Orange, Nile) and Central Asia (Amudarya). At a later stage, case studies in China (Yellow River, Yangtze) and Australia (Murray Darling) were added to the project. Correspondingly a rich knowledge base could be established and several comparative analyses were conducted. The MTF (Management and Transition Framework—cf. Chaps. 3 and 9) was developed during the NeWater project in a concerted effort involving many project partners. As ought be expected, finding common conceptual ground among diverse groups of scientists and turning this into an operational framework required a considerable amount of effort and time, i.e. several years. This was counter to the constraints imposed by the funding body that expected that framework development, application and analyses (in participatory settings) can be accomplished within four years. Hence the MTF did not guide data collection from the outset. However, we made a virtue out of necessity and developed complementary approaches for standardized data collection for the purpose of conducting comparative analyses.

The first exploratory analysis of Newater focused on the role of adaptive and integrated water management (AIWM) in coping with the impacts of climate change on floods and droughts. A comparative assessment of the current water management regimes was conducted in four case studies in three European river basins: the Hungarian part of the Upper Tisza, the Ukrainian part of the Upper Tisza, the Alentejo region in the Lower Guadiana in Portugal, and Rivierenland in the Upper Rhine in the Netherlands (Huntjens et al. 2010). The analysis tested the assumption that regimes with a higher level of AIWM consider and implement a more advanced and a more diverse set of structural and non-structural measures to deal with climate change. AIWM was characterized by several regime elements: agency, awareness raising and education, type of governance and cooperation structures, information management and exchange, policy development and implementation, risk management, and finances and cost recovery (c.f. Table 8.1). A calibrated approach using standardized questionnaires, expert judgement and reinterpretation of outcomes with support of the relevant literature, supported the development of a quantitative scoring scheme as a basis for comparing the water management regimes in the selected case studies.

The results provided evidence for a strong interdependence of the elements within water governance and management systems, and as such, this interdependence is a stabilizing factor. For example, results showed that a lack of joint knowledge is an important obstacle to cooperation, and vice versa. Moreover, insights from the analyses suggested that bottom-up governance is not a straightforward answer to water management problems in large-scale, complex, multiple-use systems such as river basins. Instead, all the regimes being analysed were in a process of balancing between bottom-up and top–down governance, and thus moving towards polycentric governance regimes. One reason for the key role of joint knowledge production for cooperation may be that the four case studies analysed were at sub-basin level with a focus on policy implementation and the operational level of governance. The findings suggest a further interpretation of results on the role of knowledge obtained under the umbrella of the Twin2Go project (Table 10.4). It is not only the different kinds of knowledge taken into account that matter, but also the modes of knowledge production and use in collaborative processes of social learning.

In light of such insights, the question arose: could the importance of learning processes be captured by institutional design principles? Huntjens et al. (2012) analysed the applicability of the institutional design principles developed by Ostrom for local common pool resources systems (Ostrom 2005) in order to develop institutional design propositions for climate change adaptation in water governance systems. As Ostrom developed her design principles for local common pool resource systems, one cannot assume that these principles can also be applied to climate change adaptation in multi-level governance settings. Furthermore climate change adaptation has a stronger emphasis on learning. We used the term proposition instead of principle to emphasize the fact that propositions (as much as principles) should not be seen as blueprints for institutional design. Based on analyses of climate change adaptation policies in the Netherlands, Australia and South Africa, the conclusion was that for dealing with complexities and uncertainties related to climate change impacts (e.g., increased frequency and intensity of floods or droughts), additional or adjusted institutional design propositions were necessary to facilitate learning processes. This seemed to be the case especially for dealing with complex, cross-boundary and large-scale resource systems such as river basins and delta areas in the Netherlands and South Africa, or groundwater systems in Western Australia.

Table 10.7 shows the set of eight refined and extended institutional design propositions for the governance of adaptation to climate change in the water sector. Together they capture structural, agency and learning dimensions of the adaptation challenge, and they provide a strong initial framework for exploring key institutional issues in the governance of adaptation to climate change. These institutional design propositions support a 'management as learning' approach to dealing with complexity and uncertainty. This implies that they do not specify blueprints for institutional settings, but incorporate processes that encourage adaptation tuned to the specific features of local geography, ecology, economies and cultures.

Furthermore, Huntjens et al. (2010) demonstrated that in a basin where one type of extreme is dominant—such as droughts in the Alentejo (Portugal) and floods in Rivierenland (Netherlands)—the potential impacts of other extremes are somehow ignored or not perceived with the degree of urgency they might deserve. A difference was also identified regarding the responsiveness to drought and low-flow problems versus flood problems. The responsiveness to challenges posed by droughts seemed to be considerably lower than to challenges posed by floods. These findings underpin the need for a more systematic classification of hazards from the perspective of governance and management (c.f. Sect. 7.3). In comparison to floods, droughts and their impacts develop over longer periods of time. Unless immediately catastrophic, they tend to convey less of a sense of urgency to policy makers.

Responsiveness to droughts and other emerging problems seems to be particularly low if groundwater resources are affected. Groundwater is largely invisible, boundaries are not well defined and monitoring is difficult. In a comparative study,

Table 10.7 Institutional design propositions for climate change adaptation in complex water governance systems (derived from Table 2 in Huntjens et al. (2012))

Design principle	Institutional proposition
Clearly defined boundaries	Clearly defined water use rights in the event of droughts. Clear allocation of responsibilities and resources in the event of floods
Equal and fair (re-)distribution of risks, benefits and costs	Engagement with, and strong representation of, groups likely to be highly affected or especially vulnerable
Collective choice arrangements	To enhance the participation of those involved in making key decisions about the system, in particular on how to adapt
Monitoring and evaluation of the process	Providing a basis for reflexive social learning and supporting accountability
Conflict prevention and resolution mechanisms	Including timing and careful sequencing, transparency, trust-building, and sharing of (or clarifying) responsibilities
Nested enterprises/polycentric governance	(In a multi-level context), modular governance systems balancing top-down and bottom-up processes
Robust and flexible process	Institutions and policy processes that are resilient when confronted with social and physical challenges and disturbance, but which at the same time are capable of changing
Policy learning	Settings that foster policy and institutional adjustments based on a commitment to addressing uncertainties, deliberating on alternatives, and reframing problems and solutions

we analysed requirements for adaptive governance with a focus on groundwater ecosystem services (Knüppe and Pahl-Wostl 2013). The ecosystem services concept has up to now received little attention with respect to groundwater governance and management. However, for groundwater the ESS concept may be particularly useful in increasing awareness of complex interdependencies. It could be instrumental for overcoming a lack of responsiveness to emerging groundwater problems if the governance system provides enabling and supportive conditions such as stakeholder participation and horizontal and vertical coordination. We used the MTF to develop and calculate operational indicators for vertical and horizontal coordination. Case studies in Germany, Spain and South Africa were located at sub-basin level and were all characterized by severe groundwater problems.

Stakeholder participation in groundwater management was identified as relatively low and unpopular compared, for example, with surface water management. One reason is that the link between users and the resource is often not apparent, and because many benefits are public goods, and therefore the economic value of groundwater and its services goes unrecognized (especially for its regulating, supporting and cultural services). At all levels, groundwater governance was found to be challenged by geographic and political boundaries. Therefore, it proved to be essential that the knowledge and experience of actors from different levels be taken into account when shaping management of groundwater ecosystem services. A higher degree of sectoral integration is difficult to achieve in regions that are dominated by a single sector such as agriculture. Although agriculture is a key sector for sustaining livelihoods, it is also crucial that governance and management activities take into account all sectors, including different goals of groundwater ecosystem services, which do not have an explicit market (e.g., baseflow, aesthetic beauty).

Results indicated that higher degrees of integration during management activities were not directly linked to an improvement in groundwater ecosystem services. However, evidence highlighted that integration (1) opens up the political arena for environmental perspectives, (2) increases the quality of groundwater and conservation plans, (3) accelerates the implementation of policies, (4) mitigates conflicts between different groundwater users and (5) increases the awareness of various ecosystem services. One simple explanation for the lack of a visible effect of integration on groundwater ecosystem services could be that more time was required for the effective implementation of promising plans. However, despite this, integration may also be hampered by a lack of financial and human resources and a lack of political will.

10.2.3 Insights on Dynamic States

In conclusion, one can state that empirical analyses provided clear evidence for the importance of the following characteristics of water governance systems for increasing adaptive capacity and the ability of a water governance system to deal with emerging challenges:

- Polycentric regime architectures combining distribution of authority with effective horizontal and vertical coordination and balancing top-down and bottom-up processes.
- Integrative and flexible regulatory frameworks combined with institutional capacity to implement them.
- Respect for good governance principles which is strongly influenced by the state of institutional development at the level of society as a whole.
- Opportunities for informal learning and joint knowledge production taking into account uncertainties and different perspectives of actors involved in the process.

These characteristics are not simple recipes but rather they are guiding principles that need to be tailored to environmental, socio-economic, cultural and political contexts. Water governance systems and requirements for water governance reform cannot be seen as detached from the state of institutional and economic development in a country. Institutional capacity seems to be an essential requirement—even more than economic development—for implementing sustainable water governance and management.

10.3 Analysis of Social Learning Processes

The research stream addressed in this section focuses on processes of social learning which are assumed to be essential for increasing both adaptive and transformative capacity of water governance and management systems (cf. Chap. 4, Sect. 4.2.3). The research questions that were addressed in empirical studies are: Which factors facilitate or hinder social learning in multi-party interactions? What is the role of social learning processes in (a transformation to) adaptive and integrated water governance and management? Table 10.8 provides an overview of the more specific issues addressed in the various publications contributing to this thematic stream.

10.3.1 HarmoniCOP—Social Learning in Multi-party Interactions

The EU funded project HarmoniCOP (Harmonizing Collaborative Planning) investigated the role of social learning in water management with a focus on the implementation of the European Water Framework Directive (WFD). The WFD prescribed the participation of stakeholders in the development of river basins management plans. However, the responsible authorities were not trained in the design and implementation of stakeholder participation, nor did they see its real benefit. HarmoniCOP forwarded the hypotheses that participatory processes could and should promote social learning to build capacity among stakeholder groups in order to address the complex challenges inherent in integrated water management. HarmoniCOP developed a relations concept of social learning (Pahl-Wostl et al. 2007a). Social learning in the context of addressing resource governance and management problems can be defined as a process of multiparty interactions where actors engage in relational practices to assess and generate knowledge about a problem domain. Social learning manifests itself at the level of participating individuals and the group as a whole through change in and development of shared practices, change in individual mental models, individual and collective perceptions or cultural cognitive institutions, in the development of trust and the capacity for collective action.

A comparative analysis of the case studies focused on the influence of characteristics of processes and of context factors on social learning in participatory river-basin management (Mostert et al. 2007). Most cases were linked to participatory processes conducted in the context of the implementation of the WFD. The national teams chose different case study approaches, comprising literature reviews of a completed process, interviews with stakeholders involved in completed process, observation of an ongoing process, participation in an ongoing process, and design of and participation in an ongoing process. Case studies and comparative analyses were guided by the framework developed in the context of the

Table 10.8 Overview of analyses contributing to thematic stream (B) processes of social learning (cf. Figs. 10.1 and 8.3)

Reference	Major research questions	Process characteristics	Outcomes	Context factors
Mostert et al. (2007)	Which factors (process, context) hinder or foster social learning in river basin management?	The role of stakeholder involvement Opportunities for interaction Motivation and skills of leaders and facilitators Openness and transparency Representativeness Framing and reframing Resources	Increased understanding of key issues Reframing, building trust, and improving relationships Development of new organizations Substantive outcomes	Politics and institutions
Borowski et al. (2008)	How does spatial misfit in participatory river basin management affect social learning?	Process boundary Diversity of exchange Information flow	Nature and implementation of outcome	Spatial scale characteristics of institutions and actors Availability of multi-party interaction
Isendahl et al. (2009)	How are uncertainties framed in water manage-ment practice?	Framing of uncertainties	Structured approach for management of uncertainties and risk	Institutional context
Isendahl et al. (2010a)	How can improved understanding of framing be used to improve the management of uncertainties in water management practice?	Reframing of uncertainties	Improved structuring and handling of uncertainties and risk	

HarmoniCOP project, which makes a distinction between social learning processes in multi-party interactions, substantive and relational outcomes and institutional and environmental context (c.f. Fig. 8.3) (Pahl-Wostl et al. 2007a).

The case studies showed that social learning in river-basin management is not an unrealistic ideal (Mostert et al. 2007). Resistance to social learning was encountered, but many instances of social learning were nonetheless found. Table 10.9 shows the ten most significant factors that were identified to foster or hinder social learning. Factors identified could be grouped into eight categories: the role of

Table 10.9 The ten highest scoring factors fostering or hindering social learning (based on Tables 2 and 3 in Mostert et al. (2007))

Factors fostering social learning
Continued, high motivation and engagement with high technical competence: personal qualities that establish and maintain the legitimacy of the organizer
Independent technical mediator or facilitator
High level of commitment of the leaders
Establish and maintain legitimacy and openness of project, continuous feedback, dissemination of minutes, questionnaires, comprehensive language, presentations and background documents
Flexibility from both sides to do common work and willingness to move from original position
Crisis moments or issues of high concern, e.g., flooding
Good exchange of information
Limited number of participants to enable in-depth discussions
Sufficient time and resources
Joint planning of approach
Factors hindering social learning
Lack of clarity about role of stakeholder involvement (form, timing and aims)
Stakeholders' lack of resources
Lack of adequate time and resources for the process
Lack of stakeholder belief that their inputs will make a difference
Lack of clarity of status and aims of initiative
Failure to include all stakeholders
Difficulties in moving to a multiparty approach because of a reluctance to change governance structure
Differences in scale of the project and scale of interest of the stakeholder
Omission of important aspects, e.g., costs
Overly technical language

stakeholder involvement; opportunities for interaction; motivation and skills of leaders and facilitators; openness and transparency; representativeness; framing and reframing; adequate resources; and politics and institutions. Seven categories refer mainly to process characteristics, one category, politics and institutions, mainly to context. The emphasis on process factors does not imply that context does not matter. In particular, the role of power, the impact of formal procedures on collaboration, possibilities to link local collaboration processes with institutional change, and factors explaining political support for collaboration, were identified as important topics which would be treated with more in depth analyses. It became evident that the understanding of the role of social learning processes in the context of transformative change, required analysing such processes embedded in a multi-level governance system, in order to take into account the influence of context on social learning processes and the feedback from the outcomes of social learning processes to context conditions (c.f. Fig. 8.3).

The importance of institutional context was also revealed in analyses on the spatial misfits and their influence on social learning (Borowski et al. 2008). The introduction of the basin principle implies that water resources should be governed at the spatial scale of a river basin to improve the fit between institutional and biophysical scales. However, improving the fit between environmental and water governance scales has often led to spatial misfits between different institutions (Moss 2003). For example, river basin management operates then at different spatial scales than regional planning or agricultural policies.

As a consequence of the introduction of river basin management, participatory structures in water management have frequently been introduced at the hydrological scale (e.g., sub-basin) without fully adapting them to the established decision-making structure. This may result in parallel structures and spatial misfits within the institutional settings of river basin governance systems. Actors and institutions regulating other sectors (e.g., agriculture, regional planning) may operate at provincial scales. A comparative analysis of French and German case studies revealed how social learning is impeded by such misfits (Borowski et al. 2008). It could also be demonstrated that river basin-scale institutions or actors that link parallel structures were essential for promoting river basins as management entities, and for encouraging social learning among actors at the river basin scale. In the multi-scale, multi-level settings of river basin governance, it seems to be difficult or even impossible to fully exclude spatial misfits. Therefore, boundary organizations become much more important for fostering effective coordination and avoiding fragmentation.

10.3.2 Framing of Uncertainties

Risk governance and strategies for dealing with uncertainties are important elements of water governance and management systems. They are strongly influenced by the prevailing water management paradigm and the mental models actors hold. Adaptive water governance and management is characterised by acknowledgement of different perspectives and world-views, conscious decision-taking under (irreducible) uncertainties and innovative approaches to managing uncertainty and risk (cf. Table 8.1 and Pahl-Wostl et al. 2007b; Pahl-Wostl 2009). Empirical analyses have provided evidence that such advanced practices for handling uncertainties are strongly associated with higher performance in responding to the emerging challenge of climate change (Table 10.4). Social learning in the context of handling uncertainties can thus be assumed to be an essential requirement for change in the direction of more integrated and adaptive water governance systems.

Under the umbrella of the Newater project, several analyses were undertaken to discover more about the handling of uncertainty in water management practice (Isendahl et al. 2009, 2010a). These analyses aimed at identifying the important parameters for the framing of uncertainties (Isendahl et al. 2009). Knowing such parameters provides an important foundation for defining requirements for social learning with the goal of developing and adopting more advanced approaches for addressing uncertainties. The analyses built on a series of "Uncertainty Dialogues" carried out with water managers in the Rhine, Elbe and Guadiana basins. During these dialogues, decision-makers in water management described a diverse range of uncertainty situations encountered in water management practice. A participant-structured approach was used to make the framing of uncertainties explicit (Isendahl et al. 2010a). Several important parameters could be identified on how uncertainties are framed: positioning, urgency, responsibility and trustworthiness.

Positioning towards uncertainty refers to the evaluative quality people attach to the uncertainties, in other words, whether they frame the uncertainty as something positive or negative (Levin et al. 1998). A slight predominance of negative framing and only a few instances of positive framing was observed. Positioning seems to be linked as well to the other framing parameters.

Urgency is related to the point of time at which a decision is taken on the uncertainty situation, or to the time frame within which a decision is supposed to have an effect. Concerns about situations with high levels of urgency and exerting pressure over the short-term prevailed. They were often outweighing and pushing to the background long-term goals associated with high uncertainties due to the difficulty of making predictions for the distant future.

Responsibility issues refer to various aspects of an uncertain situation. They may relate to the question of who is perceived as responsible for solving an uncertain situation, and they also provide an indication of the perceived range of options upon which decisions in a situation marked by uncertainty can be taken. As water management is, in general, a governmental responsibility, authorities must mainly enforce rules and regulations. It is not expected that authorities have responsibility for addressing uncertain situations and the authority to make autonomous decisions, and this is, often not wanted by governmental employees themselves. Abiding by the rules is usually perceived as the safest option with respect to formal or legal accountability, even when it may prevent the identification of a solution in response to an unusual or new problem characterised by uncertainty.

Trustworthiness refers to how much trust actors have in components of an uncertain situation. Components of an uncertain situation, such as actors or data, are framed as trustworthy (or not). Trust or lack thereof, in the relationships between actors b, as well as in the reliability of data, can decrease or enhance uncertainty. In some situations, the perception of untrustworthiness can even be a source of uncertainty.

The results of the analyses revealed that the handling of uncertainties in water management practice was quite pragmatic, and could be described as 'muddling through' rather than as a reflective and systemic approach. An analysis of the handling of uncertainties in the implementation of the EU Water Framework Directive came to similar conclusions (Sigel et al. 2010). However, an explicit awareness for the importance of and willingness to adopt a systematic approach for dealing with different kinds of uncertainties is an important condition for moving towards more adaptive and integrated water management approaches.

Table 10.10 Strategies developed in stakeholder workshops to improve the handling of uncertainties (Isendahl et al. 2010b)

Framing parameter	Strategy to improve handling of uncertainties
Positioning	Revalue uncertainty—move away from a negative attitude towards uncertainties and accept the fact that uncertainties may be irreducible and may even offer opportunities
Trustworthiness	Enhance trust—increase trust among actors involved in dealing with an uncertain problem situation. Ensuring open and transparent communication, involving all relevant actors and making expectations clear can enhance trust
Type of uncertainty	Reconsider the nature of uncertainties—increase awareness of the fact that uncertainties differ in types and may, according to type, require different strategies for dealing with them
Unpredictability	Deal with unpredictability—be open to the development of strategies for dealing with unpredictability
Incomplete knowledge	Improve incomplete knowledge
Multiple knowledge frames	Take into account multiple knowledge frames—acknowledge that people make sense of issues in different ways and appreciate the beneficial effect of different perspectives on the problem solving capacity of a collective of actors
Urgency	Reconsider timelines and setting priorities—keep deadlines and make decisions despite uncertainties if needed, but develop adaptive strategy to revisit and improve decisions
Responsibility	Reconsider responsibilities—clarify and assign responsibilities for dealing with uncertainties. Clarify the role of participants in a stakeholder process

To develop and assess strategies to improve the management of uncertainties, the parameters identified for the framing of uncertainty were applied in two sub-basins of the Rhine, in the Dutch Kromme Rijn and in the German Wupper, through a series of stakeholder interactions (Isendahl et al. 2010b). As an additional framing parameter deemed to be important, a distinction was made among different types of uncertainty: unpredictability, incomplete knowledge and multiple knowledge frames.

Table 10.10 summarizes strategies that were identified in stakeholder meetings to improve the handling of uncertainty with respect to the different framing parameters.

The strategies identified in Table 10.10 suggest that the first step necessary for improving the management of uncertainty is making the established framing of uncertainty and limitations of current practices for addressing it explicit. The analyses also revealed the limited capacity of the basin authorities in question to deal with uncertainties (Isendahl et al. 2010a, b). Barriers lie in the influence of cultural cognitive institutions, prevailing paradigms, but also in the formal constraints imposed by regulatory frameworks. Accountability is an essential point that requires careful consideration in institutional design for adaptive governance. Water management is a sector where being innovative and taking risks has been penalized rather than rewarded. This is due to the fact that water management has been largely dominated by a hierarchical governance mode, and an expert culture with a focus on safety and secure provision of water-related services. Being risk-averse is of course not a negative trait when dealing with vital services for the benefit of society. However, aversion to risk taking is an impediment in situations where uncertainties can neither be controlled nor avoided. The strategies identified can be valuable guides for supporting a reframing in processes of social learning. Reframing of uncertainties is an essential contribution to double loop learning, and supports the shift to adaptive and integrated water governance and management (Pahl-Wostl 2009).

10.3.3 Insights into Social Learning Processes

Empirical analyses confirm the importance of adopting a relational perspective of social learning. The cognitive processing of (factual) knowledge cannot be separated from social relational issues such as framing or roles of actors in a social context. It provides as well an improved process based understanding of why effective cooperation and knowledge production are identified as interdependent factors in comparative analyses of governance systems (cf. Sect. 10.2).

The characteristics of the social process itself strongly influence the effectiveness of social learning. Analyses have shown that transparent and open communication, inclusiveness, clear roles and rules, and sufficient resources are important requirements for successful social learning processes. Results from the HarmoniCOP analyses also confirmed that the governance context has an important influence on social learning. Problem framing imposed by the governance context, asymmetric power constellations, and rigid bureaucratic rules are major obstacles to learning processes.

The important role of social learning was also demonstrated in analyses of strategies developed by water managers to deal with uncertainties. A negative attitude towards uncertainties seems to prevail. Understandably, uncertainties are seen as an impediment to the management of water resources that needs to be overcome or at least reduced. This attitude has been reinforced by water governance systems that are built on the basis of the command and control paradigm.

A change in attitude towards uncertainties is essential for adopting the more integrated and adaptive management approaches that are now also promoted in regulatory frameworks such as the European Framework Directive. Perceiving uncertainty as an opportunity rather than a threat is a prerequisite for promoting innovative approaches in general. Developing trust in a collective of actors was identified as one key element for developing and embracing new approaches for dealing with uncertainties.

Social learning processes thus provide an interpretative space for challenging established norms and engaging in double-loop learning. They build social capital which is essential for moving from double to triple-loop learning, and ultimately for transformative change.

10.4 Analysis of Transformative Change

The overarching research questions addressed in empirical analyses of transformative change are: What supports and what hinders transformative change? What is the explanatory power of the triple-loop learning concept with respect to understanding transformative change? Table 10.11 provides an overview of the more specific issues addressed in the various publications contributing to this thematic

Table 10.11 Overview of analyses contributing to (C) transformative change thematic stream (cf. Fig. 10.1)

Reference	Major research questions	Concept
Huntjens et al. (2011)	Do adaptive and integrated water governance and management systems support higher levels of policy learning in coping with floods and droughts?	Triple-loop learning
Herrfahrdt-Pähle and Pahl-Wostl (2012)	How do institutional change and institutional continuity interact and build or reduce institutional resilience?	Panarchy
Schlüter et al. (2010)	Which characteristics of the current water governance and management system in Uzbekistan prevent or support transformative change leading to integrated and adaptive water resources management?	Social learning
Sendzimir et al. (2010)	What factors explain the resistance of regimes to becoming adaptive in the face of uncertainty?	Triple-loop learning
Pahl-Wostl et al. (2013a)	How do multi-level societal learning processes facilitate transformative change?	Triple-loop learning
Xia and Pahl-Wostl (2012a)	What it the role of "windows of opportunity for transition" in triggering a transition in the water allocation management in the Yellow River basin? How did informal learning processes contribute to the regime change?	Kindgon's multiple stream model
Xia and Pahl-Wostl (2012b)	What is the role of policy experiments in supporting innovation processes during the transition to a Water Saving Society in China? What can the role of policy experiments be in supporting transformative change in general?	Transition management
Xia and Pahl-Wostl (2012c)	To which extent is a transition towards integrated flood management in the middle Yangtze River basin taking place? What it the role of informal learning processes in supporting such a transition?	Triple-loop learning

stream. The analyses range from single case studies to comparative analyses of up to eight basins. Comparative analyses of transformative change pose considerable challenges due to the need for analysis over longer time scales and the limited comparability of transformation processes. Different approaches have been chosen in the various studies examined to deal with these challenges.

10.4.1 Factors That Facilitate Transformative Change

Prospects of climate change have been a trigger for a paradigm shift leading to more adaptive and integrated water governance and management (AIWGM), due to the increased awareness of uncertainties and of interdependencies and feedbacks in the complex social-ecological systems to be managed. At the same time, AIWGM is expected to support transformative change and higher levels of policy learning.

In another study under the umbrella of the NeWater project, Huntjens et al. (2011) investigated whether higher levels of AIWGM revealed enhanced capacity for coping with floods and for policy learning when addressing the climate change adaptation challenge. Policy learning was defined as a deliberate attempt to adjust policy goals or techniques of policy development and implementation in light of the consequences of past policies and new information, so as to better attain the ultimate objectives of governance. The concept of triple-loop learning was used to operationalize the different levels of policy learning that were being reflected and/or consolidated in the adaptation strategies for either floods and/or droughts. Formal comparative analysis (multi-value QCA) was applied to identify configurations of characteristics of water governance and management systems that led to higher levels of policy learning. The comparative analysis was based on eight case studies in Europe, Africa and Central Asia at the sub-basin level, but taking the embedding in a multi-level governance system into account.

Results confirmed insights from first exploratory analyses into factors influencing climate change adaptation in four European sub-basins (Huntjens et al. 2010). Key factors influencing adaptation proved to be also central to supporting transformative change. Better integrated cooperation structures in combination with advanced information management proved to be highly important for contributing to higher levels of policy learning. Furthermore, the analyses highlighted interdependencies between different system elements and the necessity to fine-tune centralized control with bottom-up approaches. Table 10.12 shows results for three selected case studies. Rivierenland in the Netherlands revealed the highest level of policy learning of all the cases, and clear signs of transformative change in dealing with the climate change adaptation challenge. The Upper Vaal, a sub-basin of the Orange in South Africa, revealed that an initial step had been taken towards higher levels of policy learning. The Amudarya in Uzbekistan was characterized by ad hoc problem solving revealing no signs of policy learning at all. In the analyses conducted under the umbrella of the Twin2Go project, the Netherlands was one of the countries approaching the ideal type of polycentric water governance system (c.f.

Table 10.12 Scoring of dimensions of AIGWM in case studies (based on Huntjens et al. (2011))

Case	Scoring of dimensions of AIGWM					Policy learning
	Cooperation structures	Policy development and implemen.	Information management	Finances and cost recovery	Risk management	Dominant type of learning
Rivierenland (Netherlands)	High	Med	High	Med	Med	Double-/triple-loop learning
Upper Vaal (South Africa)	Med	Med	High	Med	Med	Single-/double-loop learning
Amudarya (Uzbekistan)	Low	Low	Low	Low	Low	Single-loop learning (ad-hoc problem solving)

Table 10.5), and revealed a good fit profile with respect to institutional capacity (Rhine/Netherlands Fig. 10.3). South Africa was also classified as polycentric in its basin governance, and the fit profile showed moderate capacity along most dimensions (Orange/South Africa Fig. 10.3). Uzbekistan in contrast, was the country that came closest to the ideal type of centralized water governance system (c.f. Table 10.5), and showed a poor fit profile with respect to institutional capacity (Amudarya/Uzbekistan Fig. 10.3). Some countries seem to be particularly resistant to change. In this regard, it may be interesting to compare South Africa and Uzbekistan, two countries that have experienced serious societal crises and political changes within the past few decades, but wherein exhibit quite different capacity for policy learning.

Herrfahrdt-Pähle and Pahl-Wostl (2012) posed the question: When do major societal crises lead to long-term transformative change? Political and economic frameworks seem to be characterised by deeply-rooted resistance to fundamental change. Confronted with a major crisis, paradigms, mechanisms, and structures that led into the crisis may perpetuate in an effort to repair the system as fast as possible. However, instead of preserving conventional patterns and focusing on continuity, crises could be used as an opportunity for learning, adapting, and embarking on more sustainable pathways. An example of these contrasting responses is given by South Africa and Uzbekistan, which were both locked into persistent regimes over many decades. Faced with the challenge of transformation, Uzbekistan followed the path of institutional continuity, while South Africa opted for comprehensive reforms and a high level of change.

The Uzbek case mirrors what may be called pathological resource management. Soviet water managers were successful in achieving the narrowly-defined goal of providing enough water for cotton monoculture in the Central Asian republics, which, in turn encouraged the rapid enlargement of irrigated agriculture in the region. As a result, Uzbekistan is highly dependent on agricultural production today, especially cotton, and the economically, socially and environmentally unsustainable use of its water resources. The Republic of Uzbekistan which was established after the breakdown of the Soviet Union, did not use the crisis to adapt or transform its institutional system. The authoritarian political system prevailed, and unsustainable water use continued.

In the South-African case, the crisis and the end of the apartheid system have been used to introduce paramount change in the social, economic, and political domains. In particular, South African water governance has undergone fundamental change. The National Water Act includes many innovative elements and was developed with broad stakeholder consultation. South Africa is struggling, however, with the implementation of its ambitious new water legislation. The South African system seems to be experiencing extensive change, and shows clear signs of triple loop learning at the level of constitutional rules, while collective choice and operational rules lag behind.

These cases illustrate that successful transformation seems to be critically dependent on finding the right balance between continuity and change (Herrfahrdt-Pähle and Pahl-Wostl 2012). Elements of institutional continuity during times of transformation include preserving key institutions, which define how the rules are made; maintaining social memory; providing transparency of reform processes and allowing them time to take effect. Elements of institutional change required during phases of consolidation include flexible legislation; regular reviews; and adaptation of legislation during and after implementation.

10.4.2 Informal Institutions—Curse or Blessing?

The Uzbek case illustrates in particular the problematic role of persistent informal institutions preventing institutional change. The Uzbek governance system is a prototype of what I termed a 'rent-seeking centrally coordinated system' (Pahl-Wostl and Knieper 2014). The Uzbek government initiated reforms in the agricultural and water sectors to steer the socio-economic transition, and to address the threats of increasing water scarcity and decreasing agricultural productivity. However, despite the urgency of the problems and massive international assistance, changes to the water management regime have only been minimal so far. Schlüter et al. (2010) identified structural barriers to change by analysing in more depth, two policy processes. The MTF was used for a multi-level representation of policy development (cf. Sect. 3.3) in the domains 'coping with extreme events (EE)' and 'providing water for ecosystems'. MTF analyses were complemented with group model building exercises with stakeholders on the national, regional and local levels. Figure 10.4 is a simplified graph of the ASs and their interrelationships in the EE processes (ibid Fig. 5). The figure includes only links between ASs based on institutions and operational outcomes. Only the names of most important institutions are included. Two ASs are linked when one AS produces an institution or an operational outcome that influences another AS. There are two parallel strands: the general water sector reform process on the right and the actual process of coping with extreme events on the left. The figure demonstrates that both processes operate to a large extent separately from each other. Institutional reform is highly fragmented being based on individual decrees. The newly established water saving council operated largely in isolation of other elements of reform and the actual water allocation process. The analyses also identified a high degree of centralization (cf. Fig. 9.3 and Table 9.6) and top-down control. Figure 10.4 also highlights the overriding influence of informal institutions.

The main types of informal institutions identified were clientelism, patronage and corruption, as well as codes of conduct that regulate interaction within and among clans and regional elites (Schlüter et al. 2010). The impact of those informal institutions on the policy process takes place in all policy phases; from policy design to implementation. The pervasiveness of informal institutions throughout the formal policy process highlights their important role in determining policy outcomes. The interference of informal institutions and networks with the formal policy process, is, one factor explaining the large discrepancy between the formal rules and reality. Informal interactions provide links across administrative levels

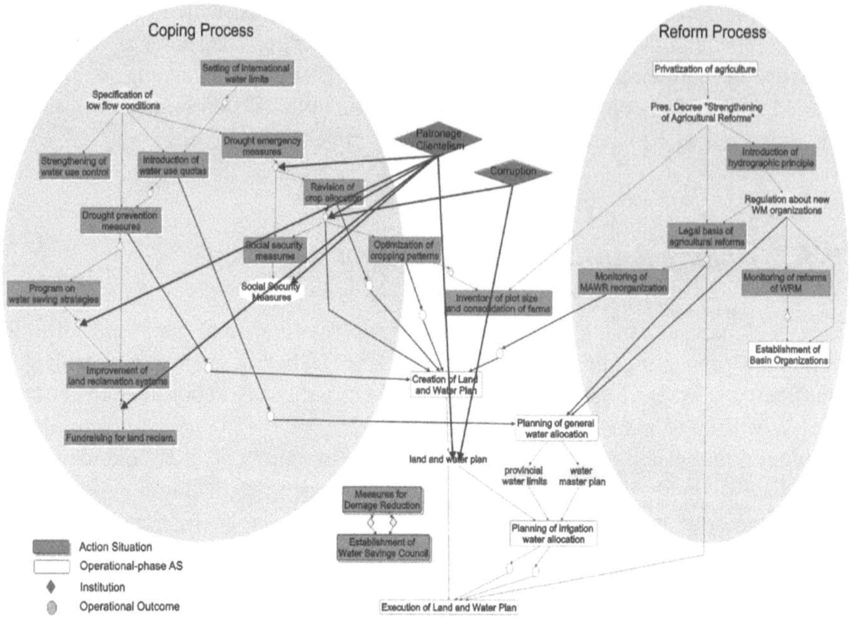

Fig. 10.4 Action situations of the "Coping with extreme events" policy process and their linkages through institutions and operational outcomes including the role of informal institutions (patronage and corruption). (reproduced from Fig. 5 in Schlüter et al. (2010) with permission)

largely missing in the formal structure. Water allocation decisions are negotiated in informal networks rather than through a process of following regulations and official management plans. The detailed analyses confirm the findings of the comparative analyses based on fsQCA. Comparative analyses based in fsQCA identified the lack of institutional capacity as a central condition for paths towards low performance in water governance systems of the centralized rent-seeking ideal-type, where Uzbekistan was the most typical representative (cf. Sect. 10.2.1 and Pahl-Wostl and Knieper (2014)).

Overall one can conclude that the analyses by Schlüter et al. (2010) revealed a lack of vertical integration across administrative levels of the formal system, and the prevalence of strongly centralized water management systems. Informal institutions compete with ineffective formal institutions and shape the outcomes of policy processes. Informal networks and the social capital embedded prevent changes that would be urgently needed to improve economic, social and environmental sustainability of water management. The combination of top down institutional change initiated by socio-economic transition and bottom-up consolidation of the existing status quo, via informal processes and networks, inhibits social learning.

Be that as it may, informal networks can also be drivers of change. In an exploratory analysis of the relationship between the structural characteristics of water governance systems and different levels of learning, the Hungarian Tisza basin showed signs of double-, and even triple-loop learning and innovation, despite being under a more rigid and centralized national regime (Huntjens et al. 2011). Most likely such learning can be attributed to the influence of an informal shadow network. This was shown by Sendzimir et al. (2010) who analysed in depth the development of flood management in the Hungarian Tisza. The paper reviewed hypotheses on the factors that act as bridges or barriers to transformation, and then used the MTF to examine how the interactions linking action situations, operational outcomes, knowledge and institutions, influenced the river management policy debate in Hungary from 1997 to 2009. The analysis focused on how factors characteristic of conventional control versus progressive adaptive management regimes influenced policy interactions in ways that contributed to or hindered transformation.

An informal shadow network had important influence in triggering innovation in the formal policy process. Its development was triggered by joint knowledge production and experimentation with new ideas in flood management. Its participation in the national policy dialogue effectively established a bridge between local and national actors that integrated, even if temporarily, these levels, and thereby filled a gap in the formal policy structure. This was partly accomplished by allowing public access to different sources of information that could be integrated into the dialogue, and, eventually, policy. However, a more detailed analysis in Sendzimir et al. (2010) showed that double loop learning (reframing), did not translate to effective triple loop learning (transformation).

The EU played an important role by providing financial support for the educational and research activities of the shadow network. The EU had long recognized the value of informal, participatory policy-science processes in revealing novel and innovative ideas, and it enshrined that principle in the Water Framework Directive. However, the quality of innovative knowledge produced by the shadow network in the Hungarian Tisza basin had to be recognized by some party and integrated into the formal policy process. The critical role of leadership in this respect was demonstrated first by its presence, and then by its absence. A parliamentary committee chair was instrumental in securing the input of the shadow network in national policy debate. But his departure created an opening for conservative groups to block any innovative approaches. The established informal networks dating back to communist rule seemed to compete for dominance with a novel, emerging informal network making use of opportunities offered by international exchange and which mobilized support at the local and regional levels. This case demonstrates that transformative change must be endorsed by many parties, and firmly tied in with established institutional structures. Links to formal policy that depend on individuals are fragile.

Structural transformations moving from command and control towards integrated and adaptive flood management are long-term processes that take place over decades. Change can be non-linear and fast at times, followed by periods of almost stasis. Pahl-Wostl et al. (2013a) analysed and compared transformative change in flood management in three case studies in Europe with a long history of severe flooding: the Hungarian Tisza and the German and Dutch Rhine. The triple-loop learning concept was applied to identify drivers for and barriers to change. And as described in Chap. 6 for the Hungarian Tisza (Box 6.2), the MTF was used for a multi-level representation of flood policy and informal learning processes and their linkages.

In all the three cases, severe floods provided windows of opportunity, since public awareness and political pressure increased. Such periods offered opportunities to promote alternative strategies developed in years preceding a disaster, often triggered by earlier extreme events. This happened in all case studies after the severe floods in the 1990s, which supported the inclusion of integrated approaches and ecological knowledge in policy development and implementation. Informal settings proved to be important for generating new knowledge and innovative policy approaches. The effectiveness of innovation diffusion depends on the links between formal and informal spaces. The three countries show substantial differences in the overall governance structure in this respect. Table 10.13 summarizes major results of the comparative analyses of the three cases.

The shadow network in Hungary was effective in integrating different kinds of knowledge, and bridging the different levels from local to regional to national. It had a strong influence on the policy process during phases of policy development and implementation. However, the role of actors from the shadow network remained informal, and their influence on the policy process depended on the political climate and contingent factors (catastrophes, influential individuals), rather than developing more formal and mature contractual relationships.

A similar, largely autonomous bottom-up process could not be identified in either the Dutch or German Rhine basins. The strength of the informal shadow network in the Hungarian case seemed to have resulted from the weakness of the government and its absence in bridging levels, e.g., by engaging stakeholders from lower levels in policy development. On the other hand, adopting innovative approaches in long-term strategic thinking and supporting their implementation has been pronounced in the Netherlands. One reason may be that the country is also the most exposed to flood risk. But the Netherlands has also had a substantial reliance on a technical control paradigm and a highly sophisticated technical infrastructure aimed at controlling floods. With respect to identified instances of informal learning, a clear influence has been exerted on setting strategic goals and during implementation. However, it seems that further informal expert networks are more closely embedded in formal policy processes (Nooteboom 2006), which was not fully captured by Pahl-Wostl et al. (2013a). Knowledge integration and links to formal policy seemed to be quite effective. This might reflect the influence of an expert network that operates across multiple sectors, which is also indicated by the large number of collaborative actors that could be identified in the policy processes. These collaborative actors referred to commissions with representatives from policy, science, and business established by government to revisit existing policies.

Germany and its federal state of Baden-Wuerttemberg, which was explored in more depth, proved to be less advanced in moving towards an integrated, long-term flood management paradigm than the Netherlands. More advanced approaches within the traditional (conventional engineering) paradigm were pursued, in

Table 10.13 Characteristics of multi-level societal learning processes in flood management: comparison of cases in Hungary, The Netherlands and Germany (based on Table 3 Pahl-Wostl et al. (2013a))

	Tisza HU	Rhine NL	Rhine D
Informal learning process	Driven by informal bottom-up process, shadow network led by NGOs developing around shared mission and new management paradigm Influence on formal policy process in strategic and operational goal setting and implementation phases	Expert communities with actors from government, NGOs, science, and business develop alternative approaches Ad hoc Advocacy Coalitions oppose implementation projects and trigger policy change. Influence on formal policy in phases of strategic goal setting/policy formulation and implementation	Expert communities with actors from science and government develop alternative approaches Ad hoc Advocacy Coalitions oppose implementation projects. Influence on formal policy process in operational goal setting and implementation phases
Knowledge integration in actor networks	Effective integration of expert and traditional, local ecological knowledge in shadow network	Knowledge integration in the expert community—ecological expert knowledge	Knowledge integration in the expert community—ecological expert knowledge
Multi-level structure and vertical coordination	National dominance Shadow network effective in bridging levels—national, regional, local	National dominance Key governmental organization (RWS) links levels	Federal system with autonomy at state level. National level comparatively weak
Learning process outcome—change in flood management paradigm	Discourse advanced and coordinated by shadow network Yet weak implementation in formal policy process and management practice	Discourse advanced, long-term strategic planning Increasing implementation in formal policy and management practice	Discourse emerging but barely coordinated across levels or actor groups. Initial steps towards long-term strategic planning Part implementation in policy and weak coordination in management practice

particular during policy implementation (i.e. combining polder construction with ecological considerations and management systems). Learning from local experiences had an influence on policy processes during the setting of operational goals and implementation. Germany is a federal system with considerable autonomy at the state level. Such a system has the potential advantage that various federal states can test different approaches and hence promote and collect experience on various innovations in parallel. However, limited knowledge exchange, and missing coordination across federal states, seemed to counteract the potential benefits of such parallel innovation processes.

In both Rhine case studies, the ecological issues in flood protection were considered, and integrated, without the trigger of a wider public and stakeholder participation process. One reason may be that the integration of new kinds of ecological knowledge is still dominated by an expert-centred approach to planning. Despite the Dutch consensus culture in water policy, this did not imply wide stakeholder involvement in policy formulation and the design of operational measures. In both the Netherlands and Germany, opposition from groups that had been consulted only at a late stage of the planning process triggered governmental efforts in promoting wider public participation.

It is evident that informal institutions and agency in informal actor networks can have quite different effects on resources management. They may be drivers of societal learning and knowledge production. They may also be reasons for institutional inertia and block transformative change. Another quite unique development can be observed in China as described in the next section.

10.4.3 Policy Experiments in China in Support of Transformative Change

With regard to water management, China has a somewhat ambivalent reputation, namely the implementation of infrastructure projects of unprecedented dimensions. The Three Gorges Dam is a gigantic reservoir serving mainly as a means of flood control and hydropower production. The world's biggest water diversion scheme, the recently opened South-North transfer from the Yangtze to the Yellow River, serves to combat water scarcity in the North Chinese plains. Northern China is scarce in water resources but is home to two thirds of China's agricultural land. The capital of Beijing suffers from a chronic water shortage. Technical megaprojects such as the Three Gorges Dam or the South-North Diversion are quite controversial due to social (e.g., thousands of people are relocated) and environmental (e.g., disruption of flow regimes) impacts. Furthermore, shipping water around the country does not solve the water scarcity problem in the long term. To the contrary, it may provide incentives to not improve the efficiency and effectiveness of water use (The Economist 2014). Despite severe water shortages, the efficiency of water use in China's industrial and agricultural production remains low.

In parallel to such efforts to solve problems with a command and control paradigm, one also observes the first signs of a paradigm shift towards more integrated and adaptive approaches. In the 10th National Five-Year Plan (2000–2005), the central Chinese government promoted the strategy for the 'Construction of a Water-Saving Society'. This strategy was underlined by a paradigm shift from an engineering-centred approach to water supply management towards an integrated approach aimed at achieving the effective allocation and efficient use of water (Xia and Pahl-Wostl 2012b). It is guided by the principle that socio-economic development should take into account the limited availability of water resources. The central element of 'constructing a water-saving society' is institutional innovation which builds on the water use rights concept. Implementing such a strategy would constitute a clear shift from supply to demand management. The fact that a project such as the South-North Diversion is realized at the same time as the proclamation of this new strategy may raise some doubts about the sincerity of and political support for a new policy. In 2012, Li Keqiang, Vice-premier at that time, even stated that the South-North-Diversion represented a 'key to a water-saving society' (Moore 2014). For centuries, China's rulers, from emperors to Communist leaders, have followed a command and control paradigm in ruling the country and in managing its water supply. Eight of the nine members of the previous Politburo's standing committee were engineers, and former president, Hu Jintao, was also a water engineer (The Economist 2013). Hence it would be quite unrealistic to expect change towards a new water management paradigm to be fast, and to expect that it would truly be implemented from the top. Nevertheless, one can also observe other indications of change over the past decade with respect to a move to more integrated approaches in flood management at different levels, even when an explicit guiding strategy at the national level is as yet missing (Xia and Pahl-Wostl 2012c). To find out the extent to which a paradigm shift in policy implementation can be observed, we conducted case studies dealing with water scarcity (Xia and Pahl-Wostl 2012a, b) and with flood management (Xia and Pahl-Wostl 2012c). Attention was devoted to how transformative change was achieved and what a paradigm shift to more integrated and adaptive approaches means in the Chinese context.

In China, transformative change is not a bottom-up process but is largely controlled by the central government. Economic reform has relied solidly on experiment-based policymaking. What may be called 'experimentation under hierarchy' has been a potent approach in bringing about transformative change in a rigid authoritarian, and bureaucratic environment, regardless of strong political opposition (Heilmann 2008). One could also argue that such experiments constitute 'controlled' innovation where the government tries to never lose control over the transformation process. Policy experiments have also been a major instrument in assessing the institutional innovations in water governance. Policy experiments in China are usually initiated by the government. They are implemented at the operational level in regions to test the feasibility of new policies and policy instruments. Regions that are chosen by the central government as pilots have little opportunity to refrain from participating. However, such policy experiments do not aim at legally binding outcomes and are not formal policy implementation processes. They provide the freedom to explore innovative policy implementation. Despite central political control, regions have some degree of autonomy due to financial decentralization, which implies that regions dispose of their own financial resources.

Experimental policies have been prominent in the implementation of a water saving society, with the Government encouraging experimentation at different levels. One policy was the water use rights exchange. The experiment analysed was carried out in the Yellow River basin in Mongolia, and explored the possibility of transferring water use rights between different sectors to increase the efficiency of water use. The industrial sector was supposed to provide funding for water-saving infrastructure for irrigation, and would receive the quantity of water saved that was originally assigned to the agricultural sector (Xia and Pahl-Wostl 2012c). Other experiments encouraged cities to analyse the potential of institutional innovations (water rights, water pricing, and water user associations) in supporting the move towards a water-saving society. The analyses focused on experiences in two major cities, Tianjin in the Haihe and Zhangye in the Heihe basin (Xia and Pahl-Wostl 2012a, b).

Overall, one can observe a kind of co-evolution of governance activities between national and local, and between strategic and operational levels. However, more systemic assessments of experiences from pilot experiments did not take place. Although aggregated lessons from local experiments were fed into the national strategy for the construction of a water saving society, regular feedback and exchange among pilots and between regional and national level were lacking (Xia and Pahl-Wostl 2012a, b). The ability of experiments to support learning has been distorted by the failure to define clear indicators, and to regularly incorporate the results of monitoring into ongoing experiments. Furthermore, uncertainties were not explicitly identified at any stage of the experiments. Such findings suggest that initiating experiments from the top does not provide strong incentives for the regional level to learn from experience, and to monitor successes and failures. The incentives are stronger to report what the government wants (or expects) to hear, and to report successes only.

Policy experiments were also conducted in the implementation of new flood policies. These supported the creation of flood retention and storage areas, and the introduction of flood insurance schemes. Experimental pilots were analysed in the Dongting Lake area in the Yangtze River basin (Xia and Pahl-Wostl 2012c). A pilot on wetland restoration was implemented in collaboration with a research project supported by international organizations such as the WWF. Despite progress being made, one could still note major incompatibilities between what was on paper in terms of innovative plans, and real implementation.

Top-down driven adaptive management can sometimes be highly effective, for example, the rapid implementation of floodplain restoration directly after the 1998 flood in China. The government has the power to relocate people if required for polder construction. This is rather different to what happened in, for example, the Netherlands, where the ecological approach has become mainstream through a series of learning processes initiated by government agencies and researchers (Pahl-Wostl et al. 2013a). However, top-down driven learning can also easily block the transition towards more integrated flood management approaches that emphasize non-structural measures. The top level's prevailing preference for investing in structural measures could further reinforce the mind-set of local water managers in favour of the traditional flood control regime. Local policy-makers may prefer structural solutions over non-structural measures, because the former are more visible to the central government. Water managers comply with expectations rather than endorsing a new paradigm.

It is evident that the Chinese government has its own interpretation of transformative change and innovation in water management. Moore (2014) argues that the authoritarian Chinese government is capable of employing a range of strategies, persuasive as well as coercive, to pursue its goals. Overall, China's experience with policy experimentation is intriguing, and numerous policy experiments have been conducted to date. Unfortunately, more systematic analyses of lessons learned have not been conducted from either the perspective of policy or research. This casts some doubt on the interest of the central government in using this instrument for a systematic improvement of policy, according to transparent criteria. Nevertheless, the Chinese approach may be appropriate for the specific circumstances of the country. Given the specific political, cultural, historical and socio-economic context of China, the transferring of experiences from China to other countries or vice versa is not straightforward. Still, I would like to argue that insights gained up to now suggest that even in the Chinese context, enduring transformative change requires a balance between top-down and bottom-up processes. Adopting a more experimental attitude towards public policy rather than portraying a chosen policy as the ultimate solution to a problem might be beneficial for dealing with complex societal problems in many countries.

10.4.4 Insights on Transformative Change

The validity of the triple-loop learning concept in capturing the essential features of transformative change can be largely confirmed. Transformative change can be depicted as an evolutionary search process where actors experiment with innovation until they encounter new constraints and boundaries. Making a distinction between reframing at the level of discourse and transformation implying structural change appears to be useful. However, transformative change is not necessarily a stepwise process; that is, moving from one learning stage to the next. It can be more appropriately described as an iterative process, which may also be differentially advanced in different domains of the water governance and management systems.

Overall, one can conclude from empirical evidence that enduring transformative change depends on:

- A balance between top-down and bottom-up processes and a flexible combination of governance modes: It seems that neither hierarchical top-down nor bottom-up network governance can provide the balance between change and stability, between diverging and converging developments required for enduring transformative change. A flexible combination of governance modes seems to be required to achieve such a balance.

- Joint knowledge production in social learning contexts: Such knowledge production serves not only to produce innovative knowledge, but it also builds social capital and the capacity for collective action in a group of actors.
- Nature and effectiveness of links between informal settings and formal policy processes: Informal spaces and diverse actor networks are important for supporting the integration of knowledge and experimentation with innovative approaches. However, the connections between learning and policy processes (e.g., hinging on individual actors) are fragile if innovative approaches are not codified in formal institutions and widely-shared practices.
- Crises and disasters as windows of opportunity for policy change: The immediate response to a crisis requires rapid action and does not allow for much reflection on the most appropriate means that should be employed. However, crises may open windows of opportunity to reflect on the need for transformative change and to develop innovative practices which would then be available when the next crisis occurs.

10.5 Overall Conclusions and the Way Forward

Empirical analyses have largely confirmed the hypotheses on the requirements for adaptive and integrated water governance and management. Similar characteristics are shared by both the adaptive as well as the transformative capacity of water governance and management systems. Polycentricity, the combination of governance modes, and the integration of informal learning processes into formal policy settings contribute to the increased flexibility of governance systems and their capacity to respond to emerging challenges. Such broad guiding principles can and must be tailored to the context of individual countries. Emphasis needs to be placed on understanding the processes of change and policy implementation, to social and societal learning rather than creating blue-prints for system architectures that often end up as simplistic panaceas for governance reform.

The relationship between formal and informal institutions deserves special attention. Helmke and Levitsky (2004) introduced a typology to describe this relationship and its effect on the governance in a country. On the one hand, they made a distinction between countries with effective and ineffective formal institutions. On the other, they distinguished between formal and informal institutions that have either compatible or conflicting goals. Hence one can derive four possible relationships as depicted in Fig. 10.5.

Complementary relationships are desirable for transformative change and, in general, from the perspective of adaptive governance and management. Informal settings support innovation and learning, and formal regulations provide a stabilizing environment in which actors can develop long-term expectations. Examples were encountered in the Dutch and, to some extent, the German cases. One often finds a competing relationship between informal and formal institutions in countries

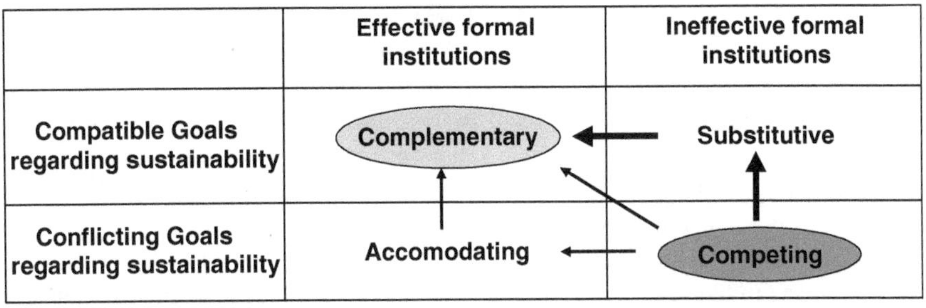

Fig. 10.5 Depending on the effectiveness of formal institutions and compatibility of goals between formal and informal institutions, four relationships between formal and informal institutions can be derived (Helmke and Levitsky 2004). The arrows denote potential pathways from competing to complementary relationships. The former are the least, the latter are the most desirable kind of relationship regarding the achievement of sustainability goals

with high corruption levels and thus ineffective formal institutions, which leads to a decline in economic, social and environmental sustainability and water security. A pertinent example is the Uzbek case. However, in such countries one may also find a substitutive relationship with effective patterns of local self-organisation (Ostrom 2005). The Hungarian case can be classified as a combination of substitutive and complementary relationships. Even though corruption in Hungary is moderate compared to Uzbekistan, formal institutions in flood management were not effective. The network substituted formal institutions at the local level and supported policy reform at the national level in a complementary mode. China's is a combination of accommodation and competition. The central government can enforce environmental regulation. However, there is considerable conflict between maximizing short-term economic profit and aiming at long-term environmental and social sustainability. The central government itself had for a long time mainly promoted the economic goal. Corruption is a problem and the role of informal networks is substantial. During policy experiments in the environmental realm, informal institutions may accommodate rather than comply with and support formal policy goals, and thus undermine the effectiveness of the overall process.

The (dynamics of the) relationships between formal and informal institutions clearly requires more analysis. It is a characteristic of governance systems as a whole, and decisive for understanding the (lack of) effectiveness and efficiency of water governance and water governance reform.

Analysing multi-level transformative change is a major challenge that has not yet fully been met by the empirical analyses we have conducted. The challenge arises not only from the tremendous complexity of transformation processes but also from the boundary conditions for such research. The funding periods for projects typically cover three to four years. This is a particularly short period in the time scale of societal transformations and does not allow the analysis of cross-level interactions, the role of innovation and learning platforms and the influence of societal discourse on transformative change. It is scarcely possible to reconstruct evidence for such processes from historical data. Such adverse conditions can only partly be compensated for by using systematic and integrative frameworks that facilitate the comparisons and reuse of data across case studies. What is required are networks of long-term (at least decadal) studies that analyse and build the capacity of regions to sustainably govern and manage their water resources in the face of global environmental change and its unexpected developments (Pahl-Wostl et al. 2013b, c). Chapter 11 elaborates in more depth the importance of real world experimentation. But realizing such trans-disciplinary research programmes would require transformative change in the scientific community and in the funding structure of science as well.

References

Borowski, I., Le Bourhis, J., Pahl-Wostl, C., & Barraqué, B. (2008). Spatial misfit in participatory river basin management: Effects on social learning. A comparative analysis of German and French case studies. *Ecology and Society, 13*(1), 7. (online) URL: http://www.ecologyandsociety.org/vol13/iss11/art17/

Brown, E., & Cloke, J. (2004). Neoliberal reform, governance and corruption in the south: Assessing the international anti-corruption crusade. *Antipode, 36*(2), 272–294. doi:10.1111/j.1467-8330.2004.00406.x

Brown, E., & Cloke, J. (2005). Neoliberal reform, governance and corruption in Central America: Exploring the Nicaraguan case. *Political Geography, 24*(5), 601–630. doi:10.1016/j.polgeo.2005.01.008

Folke, C., Hahn, T., Olsson, P., & Norberg, J. (2005). Adaptive governance of social-ecological systems. *Annual Review of Environment and Resources, 30*, 441–473.

Heilmann, S. (2008). Policy experimentation in China's economic rise. *Studies in Comparative International Development, 43*(1), 1–26. doi:10.1007/s12116-007-9014-4

Helmke, G., & Levitsky, S. (2004). Informal institutions and comparative politics. *Perspectives on Politics, 2*, 725–740.

Herrfahrdt-Pähle, E., & Pahl-Wostl, C. (2012). Continuity and change in social-ecological systems: The role of institutional resilience. *Ecology and Society, 17*(2), doi:10.5751/ES-04565-170208

Hooghe, L., & Marks, G. (2003). Unraveling the Central State but how? Types of multi-level governance. *American Political Science Review, 97*, 233–243.

Huntjens, P., Pahl-Wostl, C., & Grin, J. (2010). Climate change adaptation in European river basins. *Regional Environmental Change, 10*(4), 263–284. doi:10.1007/s10113-009-0108-6

Huntjens, P., Pahl-Wostl, C., Rihoux, B., Schlüter, M., Flachner, Z., Neto, S., et al. (2011). Adaptive water management and policy learning in a changing climate: A formal comparative analysis of eight water management regimes in Europe, Africa and Asia. *Environmental Policy and Governance, 21*(3), 145–163. doi:10.1002/eet.571

Huntjens, P., Lebel, L., Pahl-Wostl, C., Camkin, J., Schulze, R., & Kranz, N. (2012). Institutional design propositions for the governance of adaptation to climate change in the water sector. *Global Environmental Change, 22*(1), 67–81. doi:10.1016/j.gloenvcha.2011.09.015

Isendahl, N., Dewulf, A., Brugnach, M., François, G., Möllenkamp, S., & Pahl-Wostl, C. (2009). Assessing framing of uncertainties in water management practice. *Water Resources Management, 23*(15), 3191–3205. doi:10.1007/s11269-009-9429-y

Isendahl, N., Dewulf, A., & Pahl-Wostl, C. (2010a). Making framing of uncertainty in water management practice explicit by using a participant-structured approach. *Journal of Environmental Management, 91*(4), 844–851. doi:10.1016/j.jenvman.2009.10.016

Isendahl, N., Pahl-Wostl, C., & Dewulf, A. (2010b). Using framing parameters to improve handling of uncertainties in water management practice. *Environmental Policy and Governance, 20*, 107–122.

Knüppe, K., & Pahl-Wostl, C. (2013). Requirements for adaptive governance of groundwater ecosystem services: insights from Sandveld (South Africa), Upper Guadiana (Spain) and Spree (Germany). *Regional Environmental Change, 13*(1), 53–66. doi:10.1007/s10113-012-0312-7

Krysanova, V., Dickens, C., Timmerman, J., Varela-Ortega, C., Schlüter, M., Roest, K., et al. (2010). Cross-comparison of climate change adaptation strategies across large river basins in Europe, Africa and Asia. *Water Resources Management, 24*(14), 4121–4160. doi:10.1007/s11269-010-9650-8

Lebel, L., Nikitina, E., Pahl-Wostl, C., & Knieper, C. (2013). Institutional fit and river basin governance: A new approach using multiple composite measures. *Ecology and Society, 18*(1), doi:10.5751/ES-05097-180101

Lieberman, E. S. (2011). The perils of polycentric governance of infectious disease in South Africa. *Social Science & Medicine, 73*, 676–684.

Levin, I. P., Schneider, S. L., & Gaeth, G. J. (1998). All frames are not created equal. *Organizational Behavior Human Decision Processes, 76*(2), 149–188. doi:10.1006/obhd.1998.2804

Moore, S. M. (2014). Modernisation, authoritarianism, and the environment: the politics of China's South-North water transfer project. *Environmental Politics, 23*(6), 947–964. doi:10.1080/09644016.2014.943544

Moss, T. (2003). Solving problems of 'Fit' at the expense of problems of 'interplay'? The spatial reorganisation of water management following the EU water framework directive. In H. Breit, A. Engels, T. Moss, & M. Troja (Eds.), *How institutions change. Perspectives on social learning in global and local environmental contexts* (pp. 85–121). Opladen: Leske + Budrich.

Mostert, E., Pahl-Wostl, C., Rees, Y., Searle, B., Tàbara, D., & Tippett, J. (2007). Social learning in European river-basin management: Barriers and fostering mechanisms from 10 river basins. *Ecology and Society, 12*(1).

Nooteboom, S. (2006). Adaptive Networks – The Governance for Sustainable Development. Delft, The Netherlands: Eburon Academic Publishers.

Ostrom, E. (2001). Vulnerability and Polycentric Governance Systems. IHDP (International Human Dimensions Programme on Global Environmental Change) Newsletter UPDATE, *3*(1), 3–4.

Ostrom, E. (2005). *Understanding institutional diversity*. Princeton: Princeton University Press.

Ostrom, E. (2010). Polycentric systems for coping with collective action and global environmental change. *Global Environmental Change, 20*, 550–557.

Pahl-Wostl, C. (2007). Transition towards adaptive management of water facing climate and global change. *Water Resources Management, 21*(1), 49–62.

Pahl-Wostl, C. (2009). A conceptual framework for analysing adaptive capacity and multi-level learning processes in resource governance regimes. *Global Environmental Change, 19*, 354–365.

Pahl-Wostl, C., & Knieper, C. (2014). The capacity of water governance to deal with the climate change adaptation challenge: Using fuzzy set qualitative comparative analysis to distinguish between polycentric, fragmented and centralized regimes. *Global Environmental Change, 29*, 139–154.

Pahl-Wostl, C., & Lebel, L. (2011). Methods for comparative analysis. *Twin2Go Deliverables* Twin2Go project.

Pahl-Wostl, C., Craps, M., Dewulf, A., Mostert, E., Tabara, D., & Taillieu, T. (2007a). Social learning and water resources management. *Ecology and Society, 12*(2), 5. (online) URL:http://www.ecologyandsociety.org/vol12/iss12/art15/

Pahl-Wostl, C., Sendzimir, J., Jeffrey, P., Aerts, J., Berkamp, G., & Cross, K. (2007b). Managing change toward adaptive water management through social learning. *Ecology and Society, 12* (2), 30. (online) URL:http://www.ecologyandsociety.org/vol12/iss32/art30/

Pahl-Wostl, C., Lebel, L., Knieper, C., & D'Haeyer, T. (2011). Synthesis report. Context-sensitive comparative analysis of associations between water governance properties and performance in water management. *Twin2Go Deliverable No. 2.3.*

Pahl-Wostl, C., Lebel, L., Knieper, C., & Nikitina, E. (2012). From applying panaceas to mastering complexity: Toward adaptive water governance in river basins. *Environmental Science & Policy, 23*, 24–34.

Pahl-Wostl, C., Becker, G., Sendzimir, J., & Knieper, C. (2013a). How multilevel societal learning processes facilitate transformative change: A comparative case study analysis on flood management. *Ecology and Society, 18*(4), 58. doi:10.5751/ES-05779-180458

Pahl-Wostl, C., Giupponi, C., Richards, K., Binder, C., de Sherbinin, A., Sprinz, D., et al. (2013b). Transition towards a new global change science: Requirements for methodologies, methods, data and knowledge. *Environmental Science & Policy, 28*, 36–47.

Pahl-Wostl, C., Vörösmarty, C., Bhaduri, A., Bogardi, J., Rockström, J., & Alcamo, J. (2013c). Towards a sustainable water future: Shaping the next decade of global water research. *Current Opinion in Environmental Sustainability, 5*(6), 708–714. doi:10.1016/j.cosust.2013.10.012

Ragin, C. (1987). *The comparative method. Moving beyond qualitative and quantitative strategies*. Berkeley, Los Angeles and London: University of California Press.

Schlüter, M., Hirsch, D., & Pahl-Wostl, C. (2010). Coping with change: responses of the Uzbek water management regime to socio-economic transition and global change. *Environmental Science & Policy, 13*(7), 620–636. doi:10.1016/j.envsci.2010.09.001

Sendzimir, J., Flachner, Z., Pahl-Wostl, C., & Knieper, C. (2010). Stalled regime transition in the upper Tisza River Basin: The dynamics of linked action situations. *Environmental Science & Policy, 13*(7), 604–619. doi:10.1016/j.envsci.2010.09.005

Sigel, K., Klauser, B., & Pahl-Wostl, C. (2010). Conceptualising uncertainty in environmental decision-making: The example of the EU water framework directive. *Ecological Economics, 69*, 502–510.

Soliman, H., & Cable, S. (2011). Sinking under the weight of corruption: Neoliberal reform, political accountability and justice. *Current Sociology, 59*(6), 735–753. doi:10.1177/0011392111419748

The Economist. (2013). Desperate measures: Rivers are disappearing in China. Building canals is not the solution. *The Economist.*

The Economist. (2014). Water consumption: A canal too far. *The Economist.*

Tullock, G. (2008). *Public goods*. Redistribution and Rent Seeking: Edward Elgar Publishing.

Xia, C., & Pahl-Wostl, C. (2012a). The development of water allocation management in the Yellow River basin. *Water Resources Management, 26*(12), 3395–3414. doi:10.1007/s11269-012-0078-1

Xia, C., & Pahl-Wostl, C. (2012b). The process of innovation during transition to a water saving society in China. *Water Policy, 4*(3), 447–469.

Xia, C., & Pahl-Wostl, C. (2012c). Understanding the development of flood management in the middle Yangtze River. *Environmental Innovation and Societal Transitions, 5*, 60–75. doi:10.1016/j.eist.2012.10.001

Chapter 11
Virtual and Real World Experimentation

Understanding human behaviour and complex societal dynamics is essential for understanding and supporting transformative change. Such change may often require investigations beyond the realm of observed patterns of behaviour. This chapter elaborates on the potential of virtual and real world experimentation to broaden the scope of analyses, in order to foster creativity and innovation and to explore new terrains that are beyond current experience. Simulation models are a tool whose potential has only recently started to be exploited in the social sciences. The chapter discusses the role of models for exploratory analyses in this field, but also for supporting communication and social learning that contribute to or stimulate transformative change. It elaborates on the role of virtual and real world laboratories to build knowledge and capacity for transformative change.

11.1 Simulation Models as Virtual Laboratories

As abstract, simplified representations of real-world phenomena models serve an important purpose in scientific analysis. I would venture to say that this applies to all scientific disciplines[1] even when perceptions of what a model is and how it should be used may differ. One specific approach to using models is to develop formalized representations that can be implemented and simulated as computer models. Those models offer the possibility for systematic manipulation in order to test for example the plausibility of assumptions or to explore a model's behaviour in situations that have not yet been encountered in a real-word context. Until now the potential of such simulation models has not been fully exploited in the social sciences. The reluctance to make more extensive use of simulation models as a research method may derive from the perception that simulation models are tools for quantitative prediction only and that simulation modelling is a natural science or engineering approach. Indeed, the using of simulation models for prediction has a strong and highly visible tradition in natural sciences and engineering. However,

[1]All scientific disciplines, as referred to in this book, embrace natural and social sciences (including economics), engineering and humanities.

C. Pahl-Wostl, *Water Governance in the Face of Global Change*,
Water Governance—Concepts, Methods, and Practice,
DOI 10.1007/978-3-319-21855-7_11

such a tradition cannot and must not be the standard for how simulation models can and should be used in other areas of scientific inquiry. With the advent of an improved understanding of complex systems and the increasing interest in complex adaptive systems, a much richer understanding of the role of models has developed (Epstein 2008; Brugnach et al. 2008). New techniques like agent-based modelling also opened up a new frontier for the application of simulations models in the social sciences. Furthermore, these models may be used for diverse purposes. Brugnach et al. (2011) distinguished between four different model purposes: prediction, exploratory analysis, communication and learning.

Using models for prediction is intended to forecast system dynamics. Such forecasts may refer to individual variables such as changes in the population of fish species in a specific marine ecosystem resulting from different harvesting strategies. However, in complex, adaptive systems prediction of trajectories of individual variables in a specific system may not be meaningful due to inherent and irreducible uncertainties. Instead models may be applied to produce general insights about regularities in system behaviour. Prediction refers then to the ability to foresee properties and relationships at the level of overall system behaviour. This could, for example, be an analysis of the influence of network structure on a system's adaptive capacity.

In exploratory analysis models are used for detecting different types of system behaviour or potential development trajectories. This could, for example, imply an investigation into the possibility of undesirable regime shifts based on the current understanding of the social-ecological system. In contrast to prediction, exploratory analysis does not mimick reality and predict the most likely future states of a system. Rather, the goal is to explore the potential patterns of a system's dynamics and the implications of making certain assumptions about model components and their interdependencies. The distinction between using models for exploration and for prediction is not always clear cut. Exploratory analysis is particularly useful if empirical knowledge about a certain phenomenon or system is limited.

Models may also serve to illustrate and communicate knowledge about complex systems to decision makers, stakeholder groups and/or the general public. Such models can be seen as communication and educational tools. In addition, they also offer ways to challenge and make transparent assumptions that ignore or even contradict substantiated scientific understanding such as ignoring complex feedbacks.

Models can also be used as tools of intervention and to support social learning processes. In such cases the modelling process is as important or even more important than the final product, the model, in order to achieve the purpose of learning. Participatory model building processes can make different perspectives of a complex problem situation and tacit assumptions about it explicit.

The following sections mainly address the modelling purposes of prediction and exploratory analyses. The focus is on agent based modelling which is the most interesting approach for representing social systems in simulation models (Gilbert and Troitzsch 1999; Balke and Gilbert 2014). Section 11.2 then elaborates in more detail on models used for learning and for communication.

11.1.1 Supporting Theory Integration

A systems analyst who, for the first time, intends to develop an agent-based model of a certain phenomenon such as consumer behaviour is initially puzzled by the bewildering diversity of theories on human behaviour. Beginning with the task of representing a phenomenon and being armed with a methodology without embedded assumptions about the theoretical foundations underpinning that phenomenon are excellent prerequisites for theory integration and development.

An interesting example of theory integration is provided by the "Consumat Model" developed by Jager (2000). The model integrates theories from cognitive and social psychology and economics to develop a comprehensive model of consumer behaviour. Individual theories address and explain only parts of the factors and processes which determine how consumers behave. Developing an integrative conceptual framework can be highly valuable for identifying interfaces between different theories and gaps in theorizing. A representation of a consumer in a computer model must be sufficiently complete, i.e. include specifications of all relevant processes, in order to simulate behavioural patterns typically encountered in the real world. Consumat agents can choose between different heuristics (see next section for a more in depth explanation of heuristics) to decide what their actions will be and make choices in their role as consumers including: repetition, imitation, improving, social comparison, satisficing or deliberation (Jager and Janssen 2003). Heuristics differ in cognitive effort where repetition requires the least and deliberation the highest effort. The model contains numerous feedbacks. The choice of decision strategies is determined by an agent's level of "needs satisfaction" and by uncertainty, and furthermore by an agent's perception of behavioural control based on abilities and opportunities. The social and physical environment influences the perception of opportunities available to the agent. Agents can learn and gather experience which is changing their perceptions.

The Consumat model is much more appropriate for simulating realistic, context and path-dependent behaviour of real consumers (as opposed to agents) than the simplistic assumption of utility maximization of the representative consumer in neo-classical economics. However, moving closer to meaningful behaviour also comes at a cost. The Consumat model is quite complex and needs many parameters and assumptions to run. Drawing generalizable conclusions from the results of model simulations becomes more demanding.

11.1.2 An Evidence-Based Simulation Model Using Heuristics

It is much easier to identify the limitations of the behavioural model of a utility-maximizing rational actor in the context of neo-classical economics than to develop convincing alternatives. One approach that may explain behaviour in many real world situations is based on heuristics (Gigerenzer and Selten 2001). Heuristics are simple behavioural patterns that are triggered by certain characteristics in the decision context or environment that are also called cues. Heuristics require little information or even exploit the informational structure of the environment. Using simple heuristics instead of maximisation procedures is plausible within the realm of psychology. These heuristics work with the cognitive, emotional, social, and behavioural repertoire that humans actually have. Being simple, these heuristics do not work for any and all decisions. Instead, they are domain specific. Some work in some decision environments and others in other environments. The match between heuristics and environmental structures is precisely what makes them work. Having a range of such simple heuristics at their disposal is an asset rather than a weakness of real human actors. It provides humans with an 'adaptive toolbox' that endows them with the capability of navigating in complex and changing environments (Gigerenzer and Selten 2001).

Ebenhöh (2006) used the adaptive toolbox approach to develop a modular and flexible representation of agents that allows the reproduction of a wide range of behavioural patterns. We were particularly interested in analysing conditions for cooperation in common pool resource contexts (Pahl-Wostl and Ebenhöh 2004). Understanding cooperation is essential for understanding collective decision making and network governance which builds on cooperation among actors and the interaction of governance modes.

Instead of using theory as a point of departure, the modelling framework was developed based on data from experimental economics. Experimental approaches have become an increasingly popular approach in behavioural economics. At the time the model was developed the set of available data was still more limited but rich enough to ground the model in experimental evidence. We used data from three types of games: the Dictator, the Ultimatum and a common pool resource game. The main features of these games are summarized in Box 11.1. Results from experiments using such games have shown that human beings are more cooperative than would be expected based on economic theory (Chaudhuri 2011; Fehr and Gächter 2000, 2002; Oosterbeek et al. 2004; Engel 2011). Social control promotes cooperation whereas anonymity reduces cooperation. Compliance with a social norm of cooperation seems to be path and context dependent.

The three games represent different decision situations. The Dictator game involves only one asymmetric interaction. The Ultimatum game encourages strategic considerations by the first player to move based on expectations about the behaviour of the other player. The common pool resource game mimics the typical decision making context that encourages free-riding behaviour. It involves repeated albeit only anonymous interactions. When they have the opportunity, players try to enforce a norm of "fairness" by punishing other players for their not being "fair" where "fairness" is based on subjective judgement (Fehr and Gächter 2002).

Box 11.1 Economics Games addressing social preferences and cooperative behaviour

Dictator Game (Engel 2011):

- The first of two players divides a sum of money between the two players.
- The other player can only accept the decision and sum of money given to her.

A considerable portion of first players give some of the money to the other player. According to the neo-classical paradigm of a rational utility-maximizing actor the first player should give nothing to the second player.

Ultimatum Game (Oosterbeek et al. 2004):

- The first player's task is as in the Dictator game.
- Payment of the allocated money is conditional on the second player's acceptance of the first player's decision.
- If the second player accepts both receive a sum of money according to the first player's decision. If the offer is rejected, both get nothing.

The average sum of money given by the first mover to the second player is about 50 % of the overall sum received. A considerable number of second players decline the offer, and the lower the offer, the greater the likelihood of this decision. A rational actor would give a small sum in the role of first player and would not decline any offer in the role of second player.

Allocation—Common Pool Resource Game (Fehr and Gächter 2002):

- Four players in a game receive 20 money units (MU) of assets and could contribute between 0 and 20 MU to a common project.
- The common investment is increased by the experimenter by 60 % and divided evenly among the four.
- Punishment for investment decisions: Players can invest in punishment of players whose investment is considerably below the average investment. For each MU invested in punishment the punished player had to pay 3 MUs.

Three experimental settings:

- Common pool resource game without punishment.
- Start with six games with the possibility of punishment and conclude with six games without punishment.
- Start without the possibility of punishment and conclude with punishment.

In games without punishment the presence of a single free rider leads to a breakdown of cooperative behaviour by other players and the investment in the common pool declines to zero. Punishment leads to an increase in common investment and stabilizes cooperative behaviour in the team of players.

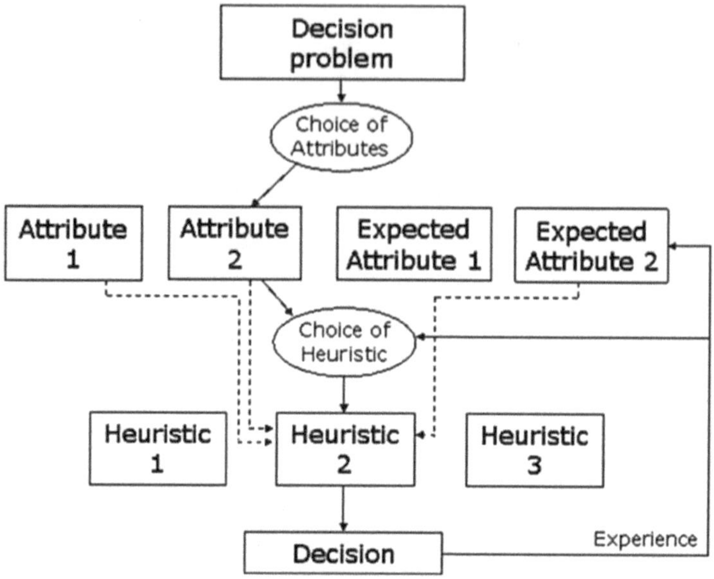

Fig. 11.1 Major elements of a decision making process as represented in the agent-based modelling framework (based on Fig. 1.1 in Ebenhöh 2006)

The agents in our adaptive toolbox approach are characterised by a set of attributes. Attributes such as cooperativeness are not perceived as immutable personal traits but can change over time. Box 11.2 provides more details on the whole set of attributes chosen. Agents have expectations about their environment and attributes of the other agents. They are characterized by bounded rationality using heuristics and exhibiting satisficing behaviour.[2] A behavioural pattern is judged to be satisfactory if it meets an individual aspiration level of the agent; if not, the behaviour is changed with simple search rules. The search stops, when the aspiration level is met. If a search continues for too long, aspiration levels may be adapted. Using such a process, bounded rationality bases decision making upon simple, psychologically plausible heuristics.

[2]The concept of bounded rationality was originally introduced by Simon (1982) as an alternative and more realistic model for human decision making in comparison with the rational actor model of neo-classical economics. Bounded rationality takes into account the fact that actors only have access to a limited amount of information and only have a limited amount of time available to evaluate alternative options. Bounded rational actors are satisficers who search for satisfactory rather than optimal solutions.

Figure 11.1 is a representation of the major elements of a decision making process in the agent based modelling framework. The choice of attributes is influenced by the framing and the nature of the decision problem. Attributes influence the choice of heuristics and may be used within heuristics. Box 11.2 provides examples of heuristics employed by the first player in the Ultimatum game.

Box 11.2 Characteristics of the Modelling Framework (Ebenhöh 2006)

The model captures individual characteristics of agents by introducing attributes which refer to dispositions to behave one way or another. These dispositions can be altered by experiences. During typical simulation runs, however, the dispositions modelled as attributes were not changed, because of the short time horizon of experimental games. Expectations about other agents on the other hand change quickly. Such expectations can refer to an anonymous other or to a specific agent depending on the experience collected by the agent. The attribute set that is chosen comprises those seven attributes that were required to reproduce the results from the three experimental game settings:

Cooperativeness: Cooperativeness expresses the disposition of an agent towards group rationality, to invest individual resources and engage in collective action to increase group resources and achieve joint outcomes.

Fairness concerning me and fairness concerning others: Fairness refers to judgements made in relationships between one agent and a single other agent. Fairness refers to a disposition to equalize outcomes. There may be quite a difference if agents judge fairness for themselves with respect to others compared to judging fairness for others as compared to themselves.

Positive and negative reciprocity: Reciprocity defines how much an agent's behaviour depends on the previous behaviour of other agents. Positive reciprocity refers to friendly reactions to the act of another agent who is perceived as nice. Negative reciprocity refers to retaliation for acts perceived as hostile.

Conformity: Conformity refers to the disposition of an agent to comply with social norms and strategies pursued by the majority of agents.

Risk aversion: Risk aversion expresses the reluctance of agents with respect to risky actions.

Examples of heuristics for the first player's decision in the Ultimatum game:
gift = 10 MU
gift = expected trustworthiness * 10 MU
if (expected trustworthiness < low limit) gift = 0 MU else if (expected trustworthiness > high limit) gift = 10 MU else gift = expected trustworthiness * 10 MU

The agent based modelling framework developed by Ebenhöh (2006) allows for comparison among different individual and collective decision situations. This is a prerequisite for analysing social embeddedness and to contextualize the influence of institutions on cooperative behaviour in a group. Cooperativeness is not an immutable personal trait. Nevertheless one finds often typical categories of players in experiments. These can be characterized as maximizers, as cooperators, and "waverers" with quickly changing behaviour due to changing circumstances and recent experiences (Ebenhöh and Pahl-Wostl 2008). The latter group can be drawn into mutual cooperation or mutual defection. It is this kind of meta-stable behaviour resulting from social embeddedness that causes most social systems to be inherently unpredictable. Furthermore, cooperative behaviour can break down in a group due to the presence of free-riding behaviour and the influence of norms of fairness even when all agents have an initial disposition towards high levels of cooperativeness. The impact of institutions on stabilisation of cooperation may differ depending on the mutual esteem among group members or other contextual variables. Context and path dependence such as this has consequences for the management of natural resources that builds upon cooperation or trust processes in groups of stakeholders. The pluralistic and flexible agent based modelling framework constitutes a virtual laboratory that allows an exploration of the influence of path dependencies associated with the introduction of institutions and interrelationships among institutional settings. In using such a pluralistic approach there will always remain ambiguities in the interpretation of situations and actions. What seems to be a cooperative situation to one actor may appear competitive to another. Such a modelling framework hence does not provide unique and best solutions for the institutional design of water governance systems. However, it would be inappropriate to provide simple recipes for complex decision situations by ruling out such ambiguities. Furthermore, variability and ambiguities in interpretation are also prerequisites for broadening the interpretation space of existing institutions and thus for institutional change and higher levels of learning. Higher levels of social learning and transformative change require changes in dominant cultural cognitive institutions that determine mental models and the framing of problem situations and potential solutions (cf. Chap. 4, Sect. 4.2.2).

11.1.3 Exploratory Analysis of Social Learning Processes

Despite an increased focus on social learning in participatory resource governance and management there is still a lack of overall agreement on definitions and systematic analyses of conditions for social learning (cf. Chap. 10, Sect. 10.3). In such a situation agent based modelling can support integration, synthesis and the focusing of empirical research questions. Making a significant contribution to this has been the goal of the doctoral dissertation of Scholz (2014) who developed a conceptual and methodological framework for analysing moderated social learning processes in natural resource management. One focus of the research was the analysis of conditions for convergent rather than divergent learning as a basis for developing a shared and broadened understanding of a problem situation among a group of actors. Shared understanding refers here to a degree of similarity between the mental models of the participants involved in a participatory exercise (process).

Knowledge about the social dynamics in processes of social learning is rather limited. Apart from the lack of a common framework which impedes the cumulative development of a larger knowledge base, empirical studies of social learning are notoriously difficult to conduct. Developing observable measures for individual and social learning poses considerable challenges to the analyst. The agent-based model developed by Scholz (2014) had thus two major objectives: a synthesis of knowledge from different strands of research and exploratory analysis of the social dynamics during social learning processes. Two important features of the social dynamics analysed were the tendency of individuals to conform to what they perceive as the views of the majority, and the influences of power relations. The conceptual framework that was developed to represent such features of social dynamics was already introduced in Chap. 4 (cf. Sect. 4.3.2). Box 11.3 summarizes the most important elements and assumptions of the CollAct simulation model. Actors participate with a certain role in a collective practice (Scholz et al. 2013). Each actor holds a mental model with relational and substantive components. This mental model shapes the perception of the actor with respect to the social interaction context (relational part) and with respect to the problem situation under consideration (substantive part). One can talk of convergent learning if actors agree upon a shared representation of the problem situation under consideration and also integrate new elements in their individual mental models (Scholz et al. 2015a). Such shared representations are developed from individual mental models during collective practices, e.g., participatory modelling. The model thus captures not only social contagion but also individual learning. It allows a distinction to be made between mere consensus on one hand and real shared understanding on the other.

Box 11.3 An Agent-based Model of Social Learning (Scholz et al. 2014, 2015b)

The model, CollAct (simulating collaborative activities) was built to explore how people gain a shared understanding and reach consensus in an interactive

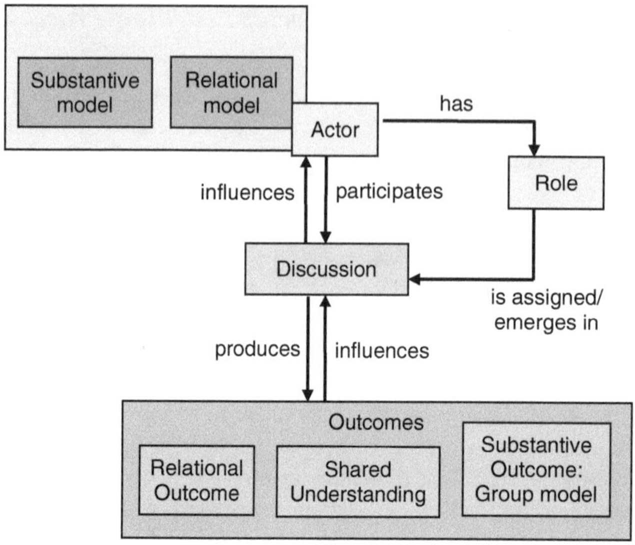

Fig. 11.2 Conceptual framework underlying CollAct (reproduced from Fig. 1, Scholz et al. 2015b)

group setting. Agents are modelled in a cognitive way, including substantive and relational knowledge of mental models, which may change through learning. Figure 11.2 and Table 11.1 summarize the most significant elements and concepts of the model.

The agents in CollAct discuss with each other and produce a group model which is an indicator for consensus. They develop a shared understanding if mental models of individual agents converge. Furthermore, two psychological biases are implemented in CollAct: the Asch and the halo effect. The Asch effect describes how people agree on a judgement that turns out to be incorrect as a result of perceived group pressure. This effect is known as conformity. The halo effect describes how a positive judgment of one characteristic of a person creates a positive bias in the judgment of other characteristics of this person.

Factors identified to have an important influence on the results of a group discussion include group size, the level of controversy within the discussion, cognitive diversity, social behaviour in form of cognitive biases (Asch and halo effect), and, depending on group size, the existence of a leading role at the outset of the discussion.

In the first version of CollAct elements included in mental models are not interdependent. Analysing the influence of reigning paradigms requires the inclusion of a more complex internal hierarchical structure in the network of beliefs that actors hold.

Table 11.1 Description of main elements of the CollAct Model (reproduced from Table 1, Scholz et al. 2015b)

Element	Description
Discussion	The discussion provides the social interaction context, leading to specific outcomes. Because we use group model building as inspiration, the substantive outcome is called the group model. The resulting group model may be seen as reflecting a consensus. Outcomes produce feedback to the discussion in which actors interact
Participant	Participants are stakeholders interacting in a discussion. Every participant has a role and a mental model
Mental model	The mental model comprises a substantive model (e.g., perceived state of the environmental system and relevant causalities) and a relational model (representation of other actors, including representations of personal characteristics, skills, preferences, and knowledge)
Role	Participants can take on roles based on their function (e.g., convener). Other roles (e.g., leader) can emerge during a process. Roles may thus exist prior to the implementation of a participatory method or emerge during the discussion process. Aspects of roles can change through engagement or through attribution by other participants. The emergence of or shift in roles at the group level is based on changes in the relational models of actors
Relational outcome	Relational outcome refers to the outcome associated with actors' relationships. An example is the creation of trust. Relational outcomes can support or hamper the further development of the process
Shared under-standing	Shared understanding refers to the convergence in mental models of participants. Interactions may possibly lead to shared concepts and a shared understanding of the topic under consideration, representing convergent learning. Convergence in learning means that participants integrate concepts gleaned from one another and/or develop new, shared concepts
Substantive outcome	The group model produced in CollAct is a substantive outcome. Further substantive outcomes of action situations include actions, rules, and knowledge. Of these, knowledge and rules can in turn impact on the original action situation

The elements are in part derived from MTF classes—see also Sect. 4.3

Results from model simulations suggest that substantive (cognitive) learning is an essential requirement for building a shared understanding that goes beyond consensus (Scholz 2014; Scholz et al. 2015b). Conformity and relational influences may lead to overall consensus on a group result that is not accompanied by the development of a shared understanding and thus individual learning by all participants. This does not imply, however, that relational learning has no influence on developing a shared understanding. High levels of mutual esteem and the building of a shared understanding were found to reinforce each other. The findings support results from empirical studies by Sol et al. (2013) who found that trust, mutual learning and reframing are emergent and interdependent processes of social learning. According to Sol et al. (2013) social learning can thus be depicted as a dynamic process in which trust, mutual learning and reframing are produced and reproduced.

CollAct constitutes a unique modelling tool for exploring the interdependence of variables affecting social learning and the influence of the application of participatory methods. It illustrates the strength of virtual laboratories and experimental computer simulations. Insights gained from model simulations can inspire and further guide further empirical research and have important implications for the facilitation of participatory social learning processes.

CollAct can and has been extended to embed group learning processes in wider social networks (cf. Chap. 8, Fig. 8.4). Actors in multi-party settings often act as representatives of their constituencies and learning is influenced by and may extend to social interactions in a wider social network. Future simulations will explore conditions for radical innovations which might require instances of divergent and convergent learning.

11.1.4 Potential of Model Simulations as Virtual Laboratories

The challenge anyone developing a model faces is to choose an appropriate level of complexity for the representation of the system under consideration. Complexity in this case, refers to the closeness of the chosen representation to the real system. Figure 11.3 illustrates a hypothesized relationship between complexity arising from closeness to reality and the potential to enhance the understanding of system behaviour through an agent-based model. Highly abstract, extremely simplified models can be useful for demonstrating some interesting phenomena. An example is provided by the model of Schelling (1971) on racial segregation. He showed that a cellular automaton model with a simple set of rules expressing preferences for one's neighbours to be of the same colour can already produce spatial segregation

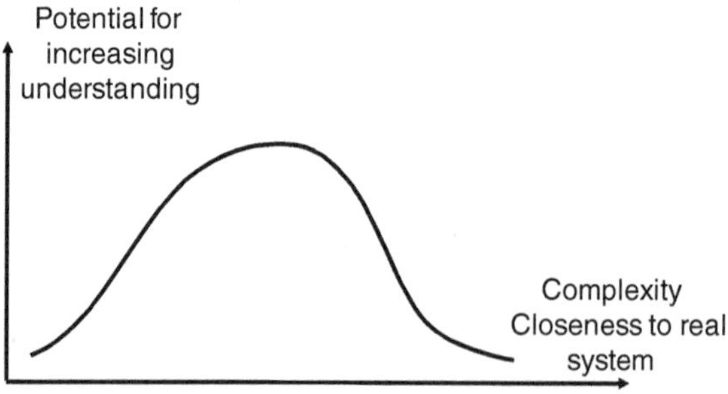

Fig. 11.3 Hypothesized relationship between complexity and closeness to reality and potential to increase understanding of system behaviour of an agent based model

of different types of agents. However, the model is limited in terms of an improved understanding of behavioural aspects that underlie the formation of such patterns. The other extreme would be a spatially-explicit model of a region where each household is represented and details like income, number of people, and education level, etc. are taken into account. Given the increase in computing power and available data the use of modelling approaches such as these have become feasible for anyone with sufficient knowledge. However, it would be rather difficult to derive from the results of such a model more general conclusions that go beyond the specific setting of the individual case in question. Hence the art of designing an appropriate model is to include sufficient details to make it meaningful without including the level of detail that makes the model context-specific and reduces the potential for systematic exploration of model behaviour. I would argue that the agent-based modelling frameworks presented in the previous sections represent such a modelling approach with an intermediate level of complexity.

11.2 Participatory Modelling to Support Social Learning

Models and in particular the model building process can also serve as a means to stimulate social learning processes. In such cases, the modelling purpose is not exploratory analysis but learning. In participatory modelling stakeholders are involved in the design of a model of a system that they are a part of. In such a situation it is inevitable that the process of developing a representation of the system also becomes a process of reflection and learning. Actors may reconsider their own role in the system, their beliefs about system elements and interdependencies, and the rules under which the system operates. By using facilitation techniques such learning processes become the explicit goal of the whole modelling exercise.

The origins of participatory modelling lie in attempts to stimulate double-loop learning to overcome deeply-held beliefs about the functioning of a system. Consultants working in management science pioneered this approach after they realized that their expert models and the unexpected and worrying results from the exploratory analyses resulting from these were largely ignored by decision makers and dismissed as implausible (Vennix 1996, 1999). There are different reasons why human beings tend to ignore and dismiss what does not agree with their experiences and assumptions about how the world around them functions. Human beings tend to ignore complexity and assume simple linear relationships where complex feedbacks loops are more appropriate. An illustrative example is provided by Sterman (2000, Sect. 5.6) with respect to strategies for dealing with traffic congestion problems. The common belief that building more roads will reduce congestion has been proven wrong. More roads increase the attractiveness of driving with a car, encourage people to move to the (then) more popular outskirts of a city, reduce the relative attractiveness of public transport and lead, after a time lag, to once-again congested roads. In the end regions are caught in lock-in situations with congested roads, unfavourable settlement patterns, degraded public transport and major barriers to transforming the overall transportation system. Engaging stakeholders in developing a conceptual model with complex feedback loops can render the consideration of complex systems less daunting, stimulate higher levels of learning and encourage people to reflect about the beliefs underlying their framing of a problem situation and how to deal with it.

Transformative change towards more sustainable water resource governance and management in general requires overcoming complex lock-in situations stabilized by a reigning paradigm. Collective action may be impeded by contradicting and even conflictual perspectives. In such a situation, participatory modelling can help to make those transparent. However, if actors hold different mental models which prevent collective decision making or even communication in more extreme conflict situations, can one decide on the basis of scientific reasoning which mental model is 'correct' and which is 'false'? For a long time scientific assessment grounded in expert knowledge was guided by the assumption that such decisions would be possible based on factual evidence and scientific knowledge only and that these ingredients would be a requirement for problem solving. This is an inappropriate assumption when dealing with wicked problems. As illustrated in Fig. 11.4 wicked problems are characterized by low consensus on the validity of both factual knowledge and values and goals related to a problem situation (Rittel and Webber 1973). Wicked problems can only be tackled by engaging actors with a different

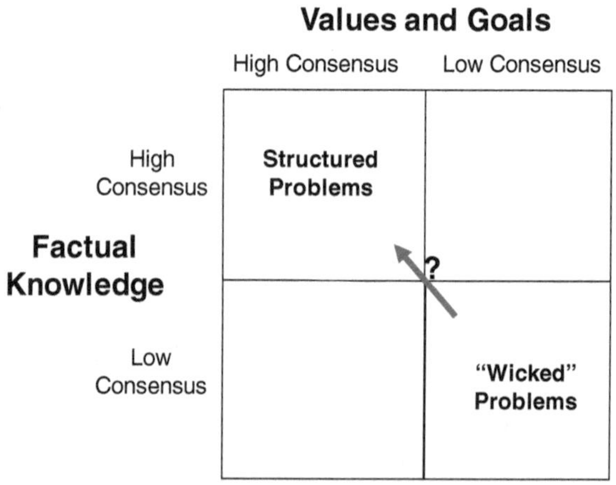

Fig. 11.4 Classification of problems in a two-dimensional space of consensus on factual knowledge and consensus on values and goals. Problems characterized by low levels of consensus on both factual knowledge and values and goals are called wicked problems. Attempts to convert wicked problems into structured problems are doomed to failure and/or constitute the inappropriate simplification of a complex problem situation since wickedness is inherent in the nature of the problem rather than being caused by insufficient knowledge and/or lack of communication

framing of the problem situation and potential solutions in constructive dialogue. Any attempt to move wicked problems to the "structured problems" (upper left in Fig. 11.4) by accumulating more knowledge is futile. Climate change is a case in point (Hulme 2009). Even when factual knowledge about anthropogenic influences on climate change may be less contested than it was a decade ago there is still little consensus on the severity of the problem and even less so on what should be done and who should act. Wicked problems are too complex and multifaceted to ever achieve this consensus. Appropriate means need to be developed to address the intricate nature of wicked problems.

Regarding evidence-based knowledge there must first be agreement that evidence can be collected empirically and second, the different parties have to accept the method selected for establishing the empirical knowledge base. The goal of the deliberations should be that evidence is taken into account and that arguments that are factually wrong are corrected. Regarding value-laden perceptions there may be more than one legitimate interpretation of a given body of empirical knowledge, resulting in ambiguity. In such a situation, accumulating more knowledge will not resolve ambiguity (Dewulf et al. 2005). The two dimensions of Fig. 11.4 are not independent. If actors associate the acknowledgment of an environmental problem and proposed or expected solutions with undesirable consequences for themselves and/or if those contradict their beliefs of what is right or wrong they might be inclined to dispute factual evidence. Relational and substantive aspects of addressing a problem situation influence and may reinforce each other.

Participatory modelling can now be seen as a relational practice which combines task-oriented actions with the relational qualities of reciprocity and reflexivity (Pahl-Wostl et al. 2007). The concept of relational practices takes the fact that substantive and relational aspects of a problem solving process cannot be separated explicitly into account. The influence of relational aspects on substantive knowledge can be intentional and conscious such as a strategic use of uncertainties to undermine the validity of factual knowledge which seems to threaten one's own interests. However, the influence of world views and deeply hold beliefs on substantive knowledge may often happen unconsciously. Participatory modelling is an important tool for helping to make implicit assumptions explicit. It builds on the concept of experiential social learning by promoting learning through engaging in shared practices (Pahl-Wostl 2002a, b).

Participatory modelling can, but does not necessarily need to be combined with the development of computer-based simulation models (Pahl-Wostl and Hare 2004; Pahl-Wostl 2002a, b). The first phase of a participatory modelling exercise normally embraces the development of conceptual models with graphical representations often combined with narratives. This phase is instrumental in initiating double-loop learning and for developing the capacity in a collective of actors for constructive dialogue. Computer-based simulation models support experimentation in a combined real and virtual laboratory setting. The setting is, on the one hand, real since the actors involved will experiment with options that affect their own future and that involve them as subjects in participatory process. On the other hand, the setting is virtual since experiments and exploratory scenarios are simulated in a digital environment. Outcomes depend on the assumptions that are used to construct the simulation model. Assumptions can be altered to test their plausibility, to identify knowledge gaps and uncertainties and to discuss the desirability of possible future developments.

The purposes of learning in participatory modelling exercises that support transformative change thus encompass a variety of aspects:

- Improving the understanding of complexity and feedbacks in social-ecological systems;
- Identifying different kinds of uncertainties and developing appropriate strategies for dealing with them;
- Promoting double-loop learning by making the influence of world views explicit and encourage reflection on them;
- Encouraging reflection on one's own role in a social setting and on perceptions of other actors' roles and expected behavioural patterns;
- Identifying options and requirements for collective action.

Various methods exist that focus more on one or the other aspect of learning. Group model building has a stronger emphasis on substantive aspects of mental models whereas agent-based approaches have a stronger focus on relational aspects. Box 11.4 summarizes in more detail the different aspects of these participatory modelling approaches.

Box 11.4 Participatory Modelling Approaches

Participatory modelling approaches engage stakeholders in model building processes with the explicit goal of supporting social learning (Pahl-Wostl 2002a, b; Pahl-Wostl and Hare 2004; van den Belt 2004; Etienne 2013). Furthermore, the knowledge held by stakeholders may sometimes be the only source of knowledge that is needed to develop a model of the system under consideration. Participatory modelling combines substantive and relational aspects of problem-solving processes. Methods differ with respect to assessing the substantive or relational parts of the mental models of actors as summarized below for two complementary approaches. Various knowledge elicitation methods may be used to elicit representations of the mental models of actors either in individual interviews or group processes. Alignment and non-alignment among mental models of the different actors are identified and may be discussed in a group setting. Scientific experts may be part of such a process by presenting the current state of scientific knowledge and debating the assumptions behind the mental models that are highly inconsistent with what is considered clearly established scientific evidence. One goal of the overall process is for the actors to agree on a group model which may consist of one or several representations of the system under consideration. The goal must not be to achieve consensus on all aspects of a group model. Different representations may also reflect different problem framing and scenarios of approaches to dealing with a problem. But it is a clear goal of such a process to establish a constructive discussion on these differences and strive for mutual respect for different perspectives.

System Dynamics Approach with a focus on substantive mental model

Representation of knowledge about a system: Knowledge is represented in terms of causal loop diagrams. The approach is derived from system dynamics.

Type of learning: People update their knowledge and assumptions about causal relationships in a specific system. This does not imply that they are capable of making appropriate judgements about future developments. People develop a general understanding of certain types of dynamics and thus extend the overall repertoire of what they consider as possible developments.

Knowledge elicited: Knowledge about important variables in the system and their causal relationships.

Computer based simulation model: System dynamics models based on differential equations.

Agent Base Approach with a focus on relational mental model

Representation of knowledge about a system: Knowledge is represented in terms of roles in the social network, and expectations about the behaviour of other actors in the network. This implicitly involves a judgement about other actors' goals and motives, and about important characteristics such as trustworthiness, willingness to cooperate or reliability.

Type of learning: People may change their expectations about other actors' behaviour. They may change the perception of their role in the social network. They may develop an understanding for the role of other actors (e.g., role playing games).

Knowledge elicited: Social network, goals and strategies of other actors.

Computer based simulation model: Agent based models using rule-based approaches often combined with role playing games.

Participatory modelling also has implications for the role of science and, the role of the system analyst within the whole model-building process. When simulation models are used for prediction or exploratory analysis they mainly serve as tools for representation and analysis. The systems analyst is an external observer who tries to improve the scientific understanding of a system. When conceptual and simulation models and the whole model building process are used for learning they mainly serve as tools for intervention. The system analyst is no longer an external observer but becomes part of the social system as facilitator of a model building process and participant in a learning process. However, such social learning processes may also serve as tools for analysis in the spirit of action research.

11.3 Action Research—Building Transformative Capacity

An effective integration of societal concerns into scientific practice may require fundamental changes in the nature of scientific enquiry, and a move towards transdisciplinary research involving stakeholders in the research process. Gibbons et al. (1994) distinguish conventional, "Mode 1" forms of science from a "Mode 2" form in which knowledge production is guided by using values mutually and reflexively constructed by a heterogeneous set of practitioners and scientific experts working together. Providing support for sustainability transformations requires "Mode 2" science. Brandt et al. (2013) pointed out that steering a social-ecological system towards a more sustainable path is an inherently transdisciplinary problem. Identifying and navigating trade-offs and synergies to overcome lock-in situations require the cooperation of different scientific disciplines (interdisciplinarity). It requires the co-production of knowledge between science and actors from outside academia (transdisciplinarity) who engage in joint problem identification and the co-design of solutions. Such scientific practice does not imply that science becomes more societally relevant at the expense of reduced scientific rigour. Argyris et al. (1985) coined the term 'action science' for a mode of scientific inquiry in the social sciences that combines practical problem solving with theory building and testing.

Argyris et al. (1985) highlight the importance of combining the study of practical problems with research that contributes to theory building and testing to the mutual benefit of both. Theories are better grounded in real world foundations, and practice is improved by building on scientific understanding of complex phenomena. However, as Argyris et al. (1985, Chap. 2) point out this an understanding of science may clash with the traditional view of science that makes a clear distinction between theory and practice, and between empirical and normative theory. Action science can be described as a critical social science. "*A critical social science engages human agents in self-reflection in order to change the world*" (ibid. p. 6). Researchers are not distant observers but facilitators of change processes. Even when science does not impose a normative stance on the direction of change scientific inquiry is designed as a form of intervention. Action science may have a normative stance though on the form of learning by explicitly facilitating reflection on beliefs and values underlying action in order to promote double rather than single loop learning (Argyris 1995; Pahl-Wostl 2009). As discussed in the previous section, participatory modelling approaches support these higher levels of learning.

In recent years experimental approaches that can be seen as action science as advocated by Argyris have become quite popular in innovation studies. Living labs refer to platforms or networks that support user driven innovation. In a report developed in the context of innovation in European research programmes living labs were defined as "*...a user-driven open innovation ecosystem based on a* *business – citizens – government partnership which enables users to take an active part in the research, development and innovation process.*" (European Commission 2009, p. 7). Living labs can be seen as a response to the increasing complexity and connectedness of a globalized world. Such labs provide spaces for experiential learning where users create and experiment with innovations that may shape their own future. Scientists can and should be facilitators and honest knowledge brokers supporting fair negotiations and knowledge creations in these innovation settings (Higgins and Klein 2011). Originally the focus of living labs was on technical innovation. They are increasingly applied in a wider setting in the context of societal innovation.

"Real world laboratories" have been promoted as key tools in facilitating sustainability transformations (Schneidewind and Singer-Brodowski 2015). Whereas living labs operate under largely controlled conditions, real world laboratories are situated in a specific real world context. They combine innovation with exploratory implementation. They bring together researchers and stakeholders in an inter- and transdisciplinary research process with the goal of testing and further developing innovative systemic approaches for addressing a sustainability problem. In the spirit of trans-disciplinary research real world laboratories should embrace co-design of the whole process and co-creation of knowledge. They combine analysis and implementation and evaluation of learning as part of the research process. Box 11.5 describes an example of such a real world experiment which aims at a reduction of CO_2 emissions in an urban setting. The example shows that real world experimentation requires many instances of innovation. It needs to be implemented in an adaptive management setting where learning from experience can feed back into the research process.

Box 11.5 Real world laboratory Innovation City Ruhr

This description of the real world laboratory project known as Innovation City Ruhr is based on Schneidewind and Singer-Brodowski (2013, pp. 128–131). The project aims to reduce CO_2 emissions of the city of Bottrop by more than 50 % within a time period of 10 years. Bottrop is a city with 116,000 inhabitants in the Ruhr district of Germany. The Ruhr district is the former coal mining region of Germany. It experienced a serious economic decline with the phasing out of coal production in Germany. The city of Bottrop won a contest to become the model city for testing transformation pathways towards a climate-neutral energy system and for assessing economic and social opportunities associated with such a transformation. The contest was launched by a consortium including the major enterprises of the Ruhr district supported by the government of North-Rhine Westphalia. The transformation project started in 2010. It also embraces a large-scale scientific research process to analyse, monitor and evaluate the progress of the transformation process. Innovation City Ruhr is without doubt a highly ambitious undertaking. Initial experience in the project has demonstrated that the challenges in realizing such an undertaking should not be underestimated. As planned the project would require investments of more than 2 billion Euro until 2020 and thus the coordination of a variety of private and public financing mechanisms. Science is not yet equipped with a tool box that could be mobilized to design and implement a reflexive research process of such dimensions. Hence such a project such as this also implies the development, testing and evaluation of innovative methods in an applied context. This requires good communication and collaboration among scientists and between actors in science and practice.

In terms of using such experimental settings as tools for analysis and tools for intervention both purposes would profit from connecting individual studies in larger networks to support comparative analyses and mutual learning. Such reasoning led Pahl-Wostl et al. (2013) to advocate a learning network of 'Global Water Testbeds'. We suggested the creation of a series of long-term, research-to-action projects at the regional scale with the purpose of demonstrating and analysing how new approaches to research and capacity building can create a tangible move towards sustainability. The realization of such a global network of transformation studies requires also a transformation in how science is evaluated and funded. Funding long-term initiatives would be required since long time frames are necessary to formulate, apply, and then track the progress of transformative change towards sustainable water governance and management. Ideally such a project network would be organized around a core module which serves the function of coordinating the development of shared conceptual and methodological frameworks, establishing shared data and knowledge bases and monitoring the progress of the initiatives in order to allow cross-initiative learning and knowledge integration. An open global sustainability learning platform could connect different transformation regions to foster exchange and transfer experiences. This is also without doubt a highly ambitious undertaking. But ambition needs to match the scope of the transformative change which is tremendous.

11.4 Concluding Comments

Simulation modelling can play an important role both in improving understanding of complex water governance systems and transformation processes and in supporting transformative change. As virtual laboratories they allow the testing of assumptions about the dynamics of governance systems and human behaviour and exploring the potential dynamics in settings that have not yet been encountered in a real world context. Participatory modelling processes support the building of transformative capacity of a collective of actors by promoting double-loop learning and by identifying requirements and options for collective action. Real world laboratories make the step towards triple-loop learning by developing and experimenting with systemic innovations to implement transformative change. To tap the full potential of these methodological approaches requires more support for coordinated long-term research.

References

Argyris, C. (1995). Action science and organizational learning. *Journal of Managerial Psychology, 10*(6), 20–26.

Argyris, C., Putnam, R., & McLain Smith, D. (1985). *Action science—concepts, methods and skills for research and intervention*. San Francisco: Joessey-Bass Publishers.

Balke, T., & Gilbert, N. (2014). How do agents make decisions? A survey. *Journal of Artificial Societies and Social Simulation, 17*(4), 13.

Brandt, P., Ernst, A., Gralla, F., Luederitz, C., Lang, D., Newig, J., et al. (2013). A review of transdisciplinary research in sustainability science. *Ecological Economics, 92*, 1–15.

Brugnach, M., Pahl-Wostl, C., Lindenschmidt, K. E., Janssen, J. A. E. B., Filatova, T., Mouton, A., et al. (2008). Complexity and uncertainty: Rethinking the modelling activity. In A. J. Jakeman, A. A. Voinov, A. E. Rizzoli, & S. H. Chen (Eds.), *Environmental modelling, software and decision support: State of the art and new perspectives* (pp. 49–68). Amsterdam: Elsevier.

Brugnach, M., Dewulf, A., Henriksen, H. J., & van der Keur, P. (2011). More is not always better: Coping with ambiguity in natural resources management. *Journal of Environmental Management, 92*(1), 78–84. doi:10.1016/j.jenvman.2010.08.029.

Chaudhuri, A. (2011). Sustaining cooperation in laboratory public goods experiments: A selective survey of the literature. *Experimental Economics, 14*(1), 47–83. doi:10.1007/s10683-010-9257-1.

Dewulf, A., Craps, M., Bouwen, R., Taillieu, T., & Pahl-Wostl, C. (2005). Integrated management of natural resources: Dealing with ambiguous issues, multiple actors and diverging frames. *Water Science and Technology, 52*(6), 115–124.

Ebenhöh, E. (2006). *Modelling human behaviour in social dilemmas using attributes and heuristics*. Osnabrück: Osnabrück University.

Ebenhöh, E., & Pahl-Wostl, C. (2008). Agent behavior between maximization and cooperation. *Rationality and Society, 20*(2), 227–252. doi:10.1177/1043463108089546.

Engel, C. (2011). Dictator games: A meta study. *Experimental Economics, 14*(4), 583–610. doi:10.1007/s10683-011-9283-7.

Epstein, J. M. (2008). Why Model?. *Journal of Artificial Societies and Social Simulation, 11*(4), 12 http://jasss.soc.surrey.ac.uk/11/14/12.html.

Etienne, M. (Ed.). (2013). *Companion modelling: A participatory approach to support sustainable development*. Netherlands: Springer.

European Commission (2009). Living Labs for user-driven open innovation. Luxembourg: European Commission, DG Information, Society and Media.

Fehr, E., & Gächter, S. (2000). Cooperation and punishment in public good experiments. *The American Economic Review, 90*(4), 980–994.

Fehr, E., & Gächter, S. (2002). Altruistic punishment in humans. *Nature, 415*, 137–140.

Gibbons, M., Limoges, C., Nowotny, H., Schwartzman, S., Peter, S., & Trow, M. (1994). *The new production of knowledge. The dynamics of science and research in contemporary societies*. London: Sage.

Gigerenzer, G., & Selten, R. (Eds.). (2001). *Bounded rationality: The adaptive toolbox*. Cambridge, MA: MIT Press.

Gilbert, N., & Troitzsch, K. (1999). *Simulation for the social scientist*. Berkshire: Open University Press.

Higgins, A., & Klein, S. (2011). Introduction to the living lab approach. In Y. H. Tan, N. Björn-Andersen, S. Klein, & B. Rukanova (Eds.), *Accelerating global supply chains with IT-innovation* (pp. 31–36). Berlin: Springer.
Hulme, M. (2009). *Why we disagree about climate change*. Cambridge, UK: Cambridge University Press.
Jager, W. (2000). *Modelleing consumer behaviour*. Groningen: University of Groningen.
Jager, W., & Janssen, M. (2003). The need for and development of behaviourally realistic agents. In J. Simão Sichman, F. Bousquet, & P. Davidsson (Eds.), *Multi-agent-based simulation II* (Vol. 2581, pp. 36–49). Lecture Notes in Computer Science Berlin: Springer.
Oosterbeek, H., Sloof, R., & van de Kuilen, G. (2004). Cultural differences in ultimatum game experiments: Evidence from a meta-analysis. *Experimental Economics, 7*(2), 171–188. doi:10.1023/B:EXEC.0000026978.14316.74.
Pahl-Wostl, C. (2002a). Participative and stakeholder-based policy design, evaluation and modeling processes. *Integrated Assessment, 3*(1), 3–14.
Pahl-Wostl, C. (2002b). Towards sustainability in the water sector—The importance of human actors and processes of social learning. *Aquatic Sciences, 64*, 394–411.
Pahl-Wostl, C. (2009). A conceptual framework for analysing adaptive capacity and multi-level learning processes in resource governance regimes. *Global Environmental Change, 19*, 354–365.
Pahl-Wostl, C., & Ebenhöh, E. (2004). An adaptive toolbox model: A pluralistic modelling approach for human behaviour based on observation. *Journal of Artificial Societies and Social Simulation, 7*(1), http://jasss.soc.surrey.ac.uk/7/1/3.html.
Pahl-Wostl, C., & Hare, M. (2004). Processes of social learning in integrated resources management. *Journal of Community and Applied Social Psychology, 14*(3), 193–206. doi:10.1002/casp.774.
Pahl-Wostl, C., Craps, M., Dewulf, A., Mostert, E., Tabara, D., & Taillieu, T. (2007). Social learning and water resources management. *Ecology and Society, 12*(2), 5. [online] URL:http://www.ecologyandsociety.org/vol12/iss12/art15/.
Pahl-Wostl, C., Vörösmarty, C., Bhaduri, A., Bogardi, J., Rockström, J., & Alcamo, J. (2013). Towards a sustainable water future: shaping the next decade of global water research. *Current Opinion in Environmental Sustainability, 5*(6), 708–714. doi:10.1016/j.cosust.2013.10.012.
Rittel, H., & Webber, M. (1973). Dilemmas in a general theory of planning. *Policy Sciences, 4*, 155–169.
Schelling, T. C. (1971). Dynamic models of segregation. *Journal of Mathematical Sociology, 1*, 143–186.
Schneidewind, U., & Singer-Brodowski, M. (2013). *Transformative Wissenschaft—Klimawandel im deutschen Wissenschafts- und Hochschulsystem*. Marburg: Metropolis.
Schneidewind, U., & Singer-Brodowski, M. (2015). Vom experimentellen Lernen zum transformativen Experimentieren—Reallabore als Katalysator für eine lernende Gesellschaft auf dem Weg zu einer Nachhaltigen Entwicklung. *Zeitschrift für Wirtschafts- und Unternehmensethik, 16*(1).
Scholz, G. (2014). *How participatory methods facilitate social learning in natural resources management*. Osnabrück: Osnabrück University.
Scholz, G., Dewulf, A., & Pahl-Wostl, C. (2013). An analytical framework of social learning facilitated by participatory methods. *Systemic Practice and Action Research*, 1–17, doi:10.1007/s11213-013-9310-z.
Scholz, G., Pahl-Wostl, C., & Dewulf, A. (2014). An agent-based model of consensus building. *In Social Simulation Conference, Barcelona, 1–5*(09), 2014.
Scholz, G., Austermann, M., Kaldrack, K., & Pahl-Wostl, C. (2015a). A method to evaluate Group Model Building sessions by comparing externalized mental models and group models. *System Dynamics Review*, in press.
Scholz, G., Dewulf, A., & Pahl-Wostl, C. (2015b). Social learning in an agent based model: Using cognitive biases to simulate learning and consensus finding in group discussions. *Journal of Artificial Societies and Social Simulation.*

Simon, H. (1982). *Models of bounded rationality*. Cambridge, MA, USA: MIT Press.

Sol, J., Beers, P. J., & Wals, A. E. J. (2013). Social learning in regional innovation networks: Trust, commitment and reframing as emergent properties of interaction. *Journal of Cleaner Production, 49*, 35–43. doi:10.1016/j.jclepro.2012.07.041.

Sterman, J. D. (2000). *Business dynamics: System thinking and modeling for a complex world.* United States: McGraw-Hill.

van den Belt, M. (2004). *Mediated modeling. A system dynamics approach to environmental consensus building*. Washington: Island Press.

Vennix, J. A. M. (1996). *Group model-building: Facilitating team learning using system dynamics*. Chichester: Wiley.

Vennix, J. A. M. (1999). Group model-building: Tackling messy problems. *System Dynamics Review, 15*(4), 379–401.

Chapter 12
From Understanding to Transforming

12.1 Taking Stock and Thinking Ahead

Many of the reflections on governance in this book do not apply only to the governance and management of water but also to environmental and resource governance in general. Then what makes water so special? Why did I write a book focusing on water and not on resource governance? Water is pervasive—it connects all other resources and other environmental components. By addressing water, the transformation towards sustainability requires and will promote transformative change in resource and environmental governance in general. I argue that it is the most demanding and challenging resource governance problem humankind is facing. Solving the water problem will solve the sustainability problem. Water governance is an agent of change (Pahl-Wostl et al. 2013c).

It is the central assumption of my work and also of this book that the necessary change in paradigm and in the logic guiding water governance and management towards more integrated and adaptive approaches is a, or even *the* key driver for transformative change towards sustainability. However, waves of preferred principles for water governance and management reform have been coming and going. This casts a certain amount of doubt on the power of paradigm shifts as catalysts and drivers of sustainability transformation. Hence, I posed the following core questions in the first chapter: Is the discourse on paradigm shifts in water governance and management, in principle, flawed? Does it mainly fuel symbolic politics and detract from the problems encountered in the implementation of new governance and management approaches? Under which conditions can the discourse on the need for change be translated to the highly needed transformation in water governance and management systems? If the discourse was flawed—either my key assumption on the role of paradigm shifts as drivers of transformative change does not hold or the nature of discourse has to change.

C. Pahl-Wostl, *Water Governance in the Face of Global Change*,
Water Governance—Concepts, Methods, and Practice,
DOI 10.1007/978-3-319-21855-7_12

Global discourses on paradigm shifts in water governance and management have not yet translated into action and transformative change on the ground. Otherwise we would not see all the disturbing trends and developments in ecological systems. General governance and management principles were indiscriminately advocated for the whole planet not taking into account culture or state of economic and institutional development. Even when global discourse as for example on IRWM has been taken up in policy circles and entering policy frameworks in many countries, this has not yet led to transformation in management practice and improvement in management outcomes (Kramer and Pahl-Wostl 2014). The "Plan of Implementation" of the World Summit on Sustainable Development adopted in Johannesburg 2002 stipulated the development and implementation of IWRM and water efficiency plans. Many countries have developed IWRM plans without moving to the stage of implementation. Focusing on IWRM as a highly demanding concept seems to be a requirement of questionable value in countries where even simpler resource problems have not been addressed effectively. In some developing countries water policy seems to be driven externally by donor preferences rather than being driven by their own policy priorities (e.g., Huitema and Meijerink 2010; Schlüter et al. 2010). The global discourse has indeed partly fuelled symbolic politics to satisfy the expectations of donors, activist groups and the public. Lack of continuity in and of internal support for water policy development is not conducive to problem solving. Furthermore, water management has for a long time been dominated by technocratic approaches that have detracted from the political dimension of water governance and management. But the global discourse has started to ask the right questions and to challenge flawed assumptions and address structural problems, such as the state of institutional development and the ensuing lack of respect for good governance principles. These are conditions that transcend the scope of water governance and management. Such developments in the global discourse prompt me to conclude that it is on the threshold of moving from double- to triple-loop learning. Will such developments stall at some point soon or will they pick up momentum? To accelerate progress bottlenecks in the implementation of innovative policy frameworks need to be overcome. These are in particular:

- The legacy of the past—the end-of-pipe, technocratic approach still dominates. Complacency among technocratic elites leading to inertia and aversion to innovation;
- resource and capacity constraints regarding, in particular, financial resources and human expertise;
- disfunctional governance and rent-seeking by powerful elites;
- lack of public awareness of the need for, lack of political will and lack of a sense of urgency to address sustainability problems.

Furthermore, the scientific community does not enter political and public discourse with a strong and united voice. This can largely be attributed to the fragmented scientific landscape in the water field. Engineers, hydrologists, ecologists, economists, and governance scholars are closely attached to their disciplinary communities rather than being part of an interdisciplinary water community collaborating across disciplines and scales. Fragmentation can also be found in practice, in politics and in policy frameworks. Even in the urban realm managing wastewater treatment and drinking water supply have been and still are largely separated activities.

Governance of transformation needs to address these bottlenecks to implementation. I have identified in this book a number of important characteristics of governance systems that support transformative change. Polycentricity combining decentralization of power and effective vertical and horizontal coordination seems to be a promising guiding principle for water governance architectures. Polycentric governance systems should entail a balance between bottom-up and top-down processes, effective coordination between formal and informal approaches and diverse combinations of governance modes (cf. Chaps. 8 and 10).

Governance of transformation shifts the emphasis from idealized outcomes towards effective processes of change. Implementation deficits need to be addressed at the process level rather than at the final stage of reform in water governance. This requires a nuanced diagnostic approach. Here I make direct reference to the medical metaphor. The prevalence of panaceas in water governance reform can be compared to the use of an indiscriminate approach in medical care where the focus is on suppressing symptoms with generic remedies rather than first carefully diagnosing the illness by examining the complex patterns characterising the cause. A diagnostic approach engages the patient in the process. It tries to differentiate between generalizable causes and patient-specific factors. The process of healing requires that the patient be engaged in the healing process rather than simply seeking an external cure with the use of generic remedies only. Obviously there is a significant difference between water governance reform and the recovery process of a patient. Nevertheless the metaphorical comparison can reveal some insights into the nature of different approaches to addressing a complex problem such as the sustainable management of human health or the sustainable governance and management of a resource.

12.2 Promising Global Discourses and Change Agents

There are several global discourses and developments that have the potential to drive transformative change towards sustainable water governance and management. I discuss here, the potential candidates for this including: the “water-energy-food nexus”, “water security”, “bioeconomy and green infrastructure” and “sustainable development goals”.

12.2.1 The Water-Energy-Food Nexus

As pointed out, Integrated Water Resources Management (IWRM) has been the guiding principle for water governance reform for more than two decades

(Pahl-Wostl et al. 2011b; Newig and Challies 2014). IWRM has suffered from implementation problems and has not succeeded in overcoming sectoral boundaries. As a response to these deficits, the Water-Energy-Food Nexus (WEF) nexus concept has entered centre stage in resource management debates within business, policy and practice in recent years (Benson et al. 2015).

Initially, the WEF nexus concept was promoted by business at the World Economic Forum, where water became a central topic of global concern in 2008. Subsequent reports promoted a nexus approach in resources governance and management (WEFWI 2009, 2011). The World Economic Forum emphasizes threats to and opportunities for business and sees market mechanisms and the green economy as effective and efficient solutions for dealing with resource scarcity. The German government took the lead in promoting the WEF concept in policy circles in the run-up to the Rio+20 sustainability summit by organizing a conference on the WEF nexus in Bonn in 2011. This stream within the discourse adopted a broader framing of the concept and emphasized wider policy implications and environmental, social and economic sustainability (BMU 2011; Hoff 2011). Are there reasons to believe that WEF nexus thinking will indeed help overcoming prevailing governance failures (viz. lack of coordination, ineffective implementation)? Allouche et al. (2015) argued against too much optimism in this respect and put forward a more sceptical view. In their opinion, the nexus approach is not really novel, lacks engagement with the respective market logics within sub-nexuses, faces difficulties in integrating sectors and disregards the politics of knowledge in policy framing. Indeed, the framing of the WEF nexus is currently rooted in scientific and technical rationality concerning the requirements for integration. Implementation may fail if the concept is not sensitive to power constellations and political economy issues and their variability at and across different spatial scales. Nevertheless, adopting a WEF nexus perspective may indeed be a game-changer since it implies an entire reframing of the problem perspective. This reframing could support more balanced negotiations of the various interests among sectors and potentially break up entrenched positions and promote innovation and transformative change. A nexus perspective is also essential in the water security debate since it is impossible to achieve water security without full cross-sectoral coordination.

12.2.2 Water Security

The concept of water security has experienced a remarkable increase in popularity over the past decade in both research and practice. Can the global discourse on water security translate into a vision of and integrative process for guiding transformative change? Looking at the current state of the academic debate one may harbour doubts. A wide range of frames and often incompatible approaches can be found in the literature Cook and Bakker (2012). There are some tensions in the debate between support for a broad concept versus a narrow operational framing (Lautze and Manthrithilake 2012), developed versus developing country perspectives (Grey and Connors 2009), engineering/natural science versus social science framing and the corresponding preferred solutions that address water security challenges (Bakker 2012; Pahl-Wostl et al. 2011a). These do not sound like promising conditions for overcoming fragmentation and for supporting integration in science, policy and practice. However, the diversity of interpretations can also be perceived as the strength of the concept since it is an indication that water security has meaning to a wide range of communities. Multiple interpretations such as these can reduce the danger of falling into the simplistic panacea trap. To capitalize on this diversity of approaches to water security it is important to build on pluralistic discourse where exchange and learning may lead to some shared understanding. Is the term water security just another new term to encapsulate the broad and systemic concerns that the scientific and policy community has to deal with in water governance, or is it indeed a significantly stronger term that should replace and/or extend existing notions. While summarizing insights from the contributions to a handbook on water security we concluded that the current state of the debate remained rather undecided in this respect (Pahl-Wostl et al. 2015).

I argue that the concept of water security could support a more risk-based approach to water governance and make the trade-offs between diverse water uses and various dimensions of sustainability explicit. One of the most widely used definitions of water security by Grey and Sadoff (2007) (cf. Sect. 7.1) highlights the economic, social and environmental trade-offs as a matter of concern. The current systems understanding (Grey and Sadoff 2007; Vörösmarty et al. 2010; Pahl-Wostl et al. 2012, 2013a, c) suggests that trade-offs between environmental and human water security cannot be overcome by prevailing governance and management approaches. In their global analysis Vörösmarty et al. (2010) found that only in regions with low human populations and corresponding activities there were no threats to either human or environmental water security. Trade-offs between human and environmental water needs become pronounced in river basins where human uses are high. Table 12.1 presents the relationships between basin characteristics

Table 12.1 Relationship between basin characteristics and water security for humans and for the environment

Basin characteristics		Water security	
Hydro-complexity	Economic-institutional strength	Human water security	Environmental water security
Low	High	High	Medium
Low	Low	Medium	Low
High	Low	Low	Low
High	High	High	Low

and water security in basins experiencing high pressure from human activities (e.g., agriculture, river regulation) and intensive use of ecosystem services.

Hydro-complexity captures the complexity and features of the hydrological regime that are relevant for achieving human water security and maintaining environmental health. Complex hydrological systems that pose greater challenges for achieving human water security (Grey and Sadoff 2007), are characterized by more vulnerable ecological systems and result in increased trade-offs between human and environmental water security. Complexity is represented by water availability, variability and predictability. Governance and economic-institutional strength capture the capacity of the societal system to deal with challenges. This strength is expressed by the state of economic and institutional development of the societal system in general and by the water governance system in place (i.e. integrative legal framework and degree of implementation, respecting good governance principles, polycentric system architecture) in the basin, as a more specific property.

High levels of economic-institutional strength and good governance lead to a moderate trade-off between human water security and environmental health in basins with rather low levels of hydro-complexity. In basins with high levels of hydro-complexity even high economic-institutional strength cannot mediate trade-offs. High levels of human water security are achieved at the cost of the environment. Environmental health is low. In basins with high levels of hydro-complexity low eco-institutional strength and weak governance result in low levels of human water security and low environmental health. These are the most undesirable consequences of watershed modifications threatening both humans and the environment. In basins with low levels of hydro-complexity it may be possible to achieve at least moderate levels of security for humans. Environmental health is still low.

The concept of water security can become instrumental in guiding the diagnosis of syndromes. However, it does not provide a key to governance solutions.

In Chap. 7, I pointed out that water security is a social construct which must be negotiated in a societal discourse. Hence, governance and, in particular, respect of good governance principles are central to defining and implementing a sustainable approach to water security. Sectoral fragmentation and a different system of logic for determining acceptable risks (i.e. for economic production, livelihoods, human health and ecosystems) render a comprehensive analysis of trade-offs difficult. The ecosystem services concept could become instrumental in overcoming fragmentation. Is the increasing awareness for the importance of ecosystem services a development that has the potential to generate synergies with the water security discourse? It could indeed be so if the currently prevailing emphasis on market-based governance is replaced by a more balanced approach (cf. as well Chap. 7).

12.2.3 Bioeconomy and Green Infrastructure

The global discourse on the green or bio-economy has taken off in recent years. International and national policy initiatives abound (UNDESA 2013). In February 2012 the European Commission adopted a strategy which it referred to as "Innovating for Sustainable Growth: A Bioeconomy for Europe" (European Commission 2012) and in 2013 a strategy for "Green Infrastructure—Enhancing Europe's Natural Capital" (European Commission 2013). In April 2012 the US White House published a "National Bioeconomy Blueprint". Subscribing to the principles of a green economy was one of the main messages of the United Nations Conference on Sustainable Development in June 2012 in Rio de Janeiro (Rio+20). On one hand, these are promising developments reflecting the commitment to such principles. On the other hand, the emphasis on economic growth is a reason for concern if the market mode of governance becomes dominant, thus suppressing hierarchical and network governance.

In water policy the push towards market governance in the 80s and 90s (e.g., the wave of privatization) did not lead to the expected positive developments in the delivery of water services and solving environmental problems (cf. Chap. 2). Will the push towards a bioeconomy lead to the sale of nature? In the preparatory phase of Rio+20 warnings have been sounded (e.g., Hall et al. 2012). Contesting the risk of tendencies towards what these voices consider the 'commodification of life', they argue in favour of network and community governance: "*Instead of promoting a socially-blind 'green economy', an alternative world view would recognize the bio-cultural approaches of indigenous peoples and local communities who have long succeeded in developing sustainable livelihoods, a 'buen vivir' in harmony with the ecosystems they live in*" (ibid, p. 3).

This represents a clear clash of paradigms. However, the argument is less about normative goals since proponents of the bioeconomy concept also argue in favour of achieving sustainable development for the whole planet by balancing economic, social and environmental dimensions of sustainability. Proponents of the two streams of discourse hold conflicting views on how such a goal can and should be achieved.

As I have argued repeatedly for the need to combine governance modes (cf. Chaps. 5, 8, and 10) I do not favour either of the discourses dominating the development of water governance. However, the 'threat' that community governance takes over is not high. The power is rather unequally distributed between the discourses on market and network/community governance. However, I see as the main threat the possibility that these discourses and concomitant streams of activities will work in parallel or, even worse, against each other. Instead they should operate in a synergistic mode, supporting but also challenging each other in a constructive way.

12.2.4 Sustainable Development Goals

By the end of 2015 the Millennium Development Goals (MDGs) will expire. The MDGs have guided global development policy for more than a decade. The MDGs will be replaced by the Sustainable Development Goals (SDGs) which are to be adopted in September 2015 by the United Nations.

The MDG process has placed water back on the global political agenda in 2000. MDG 7 has as its target the halving of the number of people without access to safe drinking water and basic sanitation by 2015. In our review of global water governance processes we assessed the MDGs process as being by and large successful (Pahl-Wostl et al. 2013a). It circumvented the lengthy procedures of formal rule-making as a necessary condition for new political attention. By setting clear and measurable targets it helped mobilize resources, commitments, and greater coordination. However, the MDG process also shows clear deficiencies. Policy framing has lacked comprehensiveness which is reflected in the MDG's negligence of universal access. A more comprehensive approach to the water challenge may be more beneficial to long-term sustainability than declaring success on the number of people gaining access to safe drinking water every year. Furthermore, measuring the achievement of targets only with statistics provided by governments casts a degree of doubt on the validity of the assessment of progress. The SDG process has the potential to capitalize on the insights gained during the implementation of the MDGs.

The SDGs adopt a more comprehensive approach by moving away from a development focus towards a broader sustainability framing. Under which conditions could the SDG process become a global process driving transformative change towards sustainability? Hajer et al. (2015) caution against "cockpit-ism". With cockpit-ism they refer to complete reliance on a hierarchical governance mode where national governments and intergovernmental organizations play the key role. Indeed experience from the MDGs and other policy processes suggest a multi-level implementation process. In particular, those societal groups most affected by the implementation process should be empowered and encouraged to actively participate in implementation and monitoring. The SDG implementation process could thus become instrumental in building transformative capacity (cf. conclusions of Chap. 10). It could also unite the as yet antagonistic discourses on market and community governance. The SDG process also poses a significant task for science to develop appropriate indicators and monitoring processes and to become actively engaged in the global governance process of SDG implementation.

One decade of global water research has provided clear evidence of the global dimension of the water challenge and has identified the key problems within it. However, such evidence has not contributed to transformative change in policies and a reversal of global trends. Research in the past has emphasized the identification of problems more than the identification of solutions. Furthermore, current global assessments (e.g., World Water Assessment Programme and their flagship product, the World Water Development Report—WWDR) seem to be insufficient for informing policy leading to effective action. The WWDR is used as source of reference by many scientists and policy advisers but does not have a significant policy impact. I argue that the assessment process is product oriented with insufficient attention given to the political process which it is supposed to inform. In part, this can also be attributed to the absence of a more coherent global governance framework in the domain of water.

Global water governance is fragmented and characterized by the absence of leadership (Pahl-Wostl et al. 2008; Gupta and Pahl-Wostl 2013; Gupta et al. 2013). Some may argue that this is not a problem since water is not a global commodity and should be addressed at national or local levels anyway. However, arguments abound that water governance needs also to be addressed at the global level (cf. Chap. 6). The lack of integration has also been recognized by the United Nations and UN-Water was established 2003 to overcome this coordination gap. UN-Water is an interagency mechanism to strengthen coordination among the 24 UN agencies working on various aspects of freshwater and sanitation. In an assessment of the role of UN-Water in global water governance Baumgartner and Pahl-Wostl (2013) concluded that UN-Water has not yet had any significant impact on global water governance processes. However, it has the potential to act as a bridge between the expert-centered, knowledge-producing background and the political foreground of global water governance. In addition to the formal membership of the UN agencies, UN-Water has established links to a wide network of actors in global water governance. As an interagency coordination mechanism UN-Water lacks the direct control of an intergovernmental governing body and thus lacking formal decision-making power. At the same time, the institutional setup obliges UN-Water to account for concerns related to diplomacy and political correctness. UN-Water cannot, like many other organizations, unilaterally address controversial issues. Instead it has to embrace the broad spectrum of political and scientific complexity of global water challenges and find solutions that are acceptable to all of its member organizations—and ultimately to all member states. The mandate of UN-Water would have to be extended so that is could develop a role as an effective bridging organization linking network and hierarchical governance modes. One may question how realistic this proposal is given the power constellations in the UN-context. However, there is no doubt that such bridging organizations are needed. In our analyses of processes in global water governance we identified some highly important missing links between knowledge generation and policy framing and between knowledge generation and rule-making (Pahl-Wostl et al. 2013a). The scientific community could also play a decisive role in this regard.

Science could and should become more active in the process of SDG implementation and make the transition to developing knowledge for action, and to identifying solutions in a co-production of knowledge process. If the water community would succeed in getting its act together it could establish a think-tank providing global leadership in the identification of knowledge gaps and in promoting recognition of important research findings. To overcome the missing links in global water governance, such a think-tank needs to combine a high level of legitimacy in its role as knowledge generator and assure representativeness.

12.3 The Way Forward

I have identified four global discourses that, if combined in a synergistic way, could be central to providing the impetus for the urgently-needed sustainability transformation in water governance and management. The Water-Energy-Food Nexus supports a reframing of the problem perspective and could support more balanced negotiations of interests between sectors and engage diverse actors. It shifts the emphasis onto relationships and feedbacks between sectors, even when doing so does not yet solve the coordination challenge. Water security can support a push towards operationalizing abstract notions and the development of meaningful indicators at different levels and for diverse social groups. The concept of a 'green economy' is at least an attempt to get the environment on the radar screen of the economic sector. And finally, the process of implementing sustainable development goals offers a great opportunity to develop momentum in the transformation towards sustainability. It requires multi-level governance of transformation and strong agency to make use of this window of opportunity.

The final paragraph of a book is always one of the most difficult to write. In principle everything is said and the writer does not want to repeat recommendations like a mantra. Hence, let me close with a personal statement. In one of my lectures I was challenged in the discussion by the question of whether I really believed that humans/societies would learn—and then fast enough to meet the sustainability challenge. A look at the evidence would suggest that the answer is a definitive no. Nevertheless, I respond with a yes but with a caveat. We need strong leadership, and I expect that the scientific community and funding organizations will create the appropriate enabling conditions to take on a leadership role in this challenge.

References

Allouche, J., Middleton, C., & Gyawali, D. (2015). Technical veil, hidden politics: Interrogating the power linkages behind the nexu. *Water Alternatives, 8*(1), 610–626.

Bakker, K. (2012). Water security: Research challenges and opportunities. *Science, 337*, 914–915.

Baumgartner, T., & Pahl-Wostl, C. (2013). Water and its role in global water governance. *Ecology and Society, 18*(3), doi:10.5751/ES-05564-180303

Benson, D., Gain, A. K., & Rouillard, J. J. (2015). Water governance in a comparative perspective: From IWRM to a 'Nexus' approach? *Water Alternatives, 8*(1), 756–773.

BMU. (2011). *Messages from the Bonn2011 Conference: The Water, Energy and Food Security Nexus—Solutions for a Green Economy*. Bonn: German Federal Ministry for the Environment, Nature Conservation and Nuclear Safety (BMU) and the German Federal Ministry for Economic Cooperation and Development (BMZ).

European Commission. (2012). *Innovating for sustainable growth: A bioeconomy for Europe*. Brussels: European Commission.

European Commission. (2013). *Green infrastructure (GI)—Enhancing Europe's natural capital*. Brussels: European Commission.

Grey, D., & Connors, G. (2009). *The water security imperative: We must and can do more*. Stockholm: Paper presented at the Stockholm World Water Week.

Grey, D., & Sadoff, C. W. (2007). Sink or swim? Water security for growth and development. *Water Policy, 9*, 545–571.

Gupta, J., & Pahl-Wostl, C. (2013). Global water governance in the context of global and multi-level governance: Its need, form, and challenges. *Ecology and Society, 18*(4), 53. doi:10.5751/ES-05952-180453.

Gupta, J., Pahl-Wostl, C., & Zondervan, R. (2013). Glocal water governance—A multi-level challenge in the anthropocene. *Current Opinion in Environmental Sustainability*, This special Issue.

Hajer, M., Nilsson, M., Raworth, K., Bakker, P., Berkhout, F., de Boer, Y., et al. (2015). Beyond Cockpit-ism: Four insights to enhance the transformative potential of the sustainable development goals. *Sustainability Science, 7*, 1651–1660. doi:10.3390/su7021651.

Hall, R., Smolkers, R., Ernsting, A., Lovera, S., & Alvarez, I. (2012). Bio-economy versus Biodiversity. Global Forest Coalition.

Hoff, H. (2011). *Understanding the Nexus, background paper for the Bonn 2011 Conference: The Water, Energy and Food Security Nexus*. Stockholm: Stockholm Environment Institute.

Huitema, D., & Meijerink, S. (2010). Realizing water transitions: the role of policy entrepreneurs in water policy change. *Ecology and Society, 15*(2).

Kramer, A., & Pahl-Wostl, C. (2014). The global policy network behind integrated water resources management: is it an effective norm diffusor? *Ecology and Society, 19*(4).

Lautze, J., & Manthrithilake, H. (2012). Water security: Old concepts, new package, what value? *Natural Resources Forum, 36*, 76–87.

Newig, J., & Challies, E. (2014). Water, rivers and wetlands. In P. Harris (Ed.), *Routledge handbook of global environmental politics* (pp. 439–452). London & New York: Routledge.

Pahl-Wostl, C., Gupta, J., & Petry, D. (2008). Governance and the global water system: a theoretical exploration. *Global Governance, 14*, 419–435.

Pahl-Wostl, C., Jeffrey, P., Isendahl, N., & Brugnach, M. (2011a). Maturing the new water management paradigm: Progressing from aspiration to practice. *Water Resources Management, 25*(3), 837–856. doi:10.1007/s11269-010-9729-2.

Pahl-Wostl, C., Jeffrey, P., & Sendzimir, J. (2011b). Adaptive and integrated management of water resources. In R. Q. Quentin, Hussey, K. (Ed.), *Water Resources Planning and Management* (pp. 292–310). Cambridge, UK: Cambridge University Press.

Pahl-Wostl, C., Lebel, L., Knieper, C., & Nikitina, E. (2012). From applying panaceas to mastering complexity: toward adaptive water governance in river basins. *Environmental Science and Policy, 23*, 24–34.

Pahl-Wostl, C., Conca, K., Kramer, A., Maestu, J., & Schmidt, F. (2013a). Missing links in global water governance: A processes-oriented analysis. *Ecology and Society, 18*(2), 33. doi:10.5751/ES-05554-180233.

Pahl-Wostl, C., Palmer, M. A., & Richards, K. (2013b). Enhancing water security for the benefits of humans and nature—the role of governance. *Current Opinion in Environmental Sustainability, 5*, 676–684.

Pahl-Wostl, C., Vörösmarty, C., Bhaduri, A., Bogardi, J., Rockström, J., & Alcamo, J. (2013c). Towards a sustainable water future: shaping the next decade of global water research. *Current Opinion in Environmental Sustainability, 5*(6), 708–714. doi:10.1016/j.cosust.2013.10.012.

Pahl-Wostl, C., Gupta, J., & Bhaduri, A. (2015). Water Security: A Popular but Contested Concept. In C. Pahl-Wostl, J. Gupta, & A. Bhaduri (Eds.), *Handbook on Water Security*. Cheltenham, UK: Edward Elgar.

Schlüter, M., Hirsch, D., & Pahl-Wostl, C. (2010). Coping with change: responses of the Uzbek water management regime to socio-economic transition and global change. *Environmental Science & Policy, 13*(7), 620–636. doi:10.1016/j.envsci.2010.09.001.

UNDESA (2013). A guidebook to the green economy—Issue 4: A guide to international green economy initiatives. United Nations Division for Sustainable Development

Vörösmarty, C. J., McIntyre, P. B., Gessner, M. O., Dudgeon, D., Prusevich, A., Green, P., et al. (2010). Global threats to human water security and river biodiversity. *Nature, 467*, 555–561.

WEFWI. (2009). Managing our future water needs for agriculture, industry, human health and the environment (World Economic Forum Water Initiative).

WEFWI. (2011). *Water security: The water-food-energy-climate nexus (World Economic Forum Water Initiative)*. Washington: Island Press.

Index

C. Pahl-Wostl, *Water Governance in the Face of Global Change*,
Water Governance—Concepts, Methods, and Practice,
DOI 10.1007/978-3-319-21855-7

FSC
www.fsc.org
MIX
Papier | Fördert
gute Waldnutzung
FSC® C083411

Kolibri 360 GmbH
Ferdinand-Jühlke-Straße 7
99095 Erfurt, Deutschland
produktsicherheit@kolibri360.de